PRAISE FOR ABOUT TIME

"A fascinating and comprehensive survey of ... ing to railways to telegraphy to the internet – has ... every ... concept of time. [Frank] is excellent at showing how our id... of human and cosmic time have evolved hand in hand ... Compelling."
Marcus Chown, *New Scientist*

"Eloquent ... [Frank's] trek through the history of humanity takes a parallel look at how we have gained a deeper grasp of the Universe during our time on Earth."
Nature

"A phenomenal blend of science and cultural history ... Ultimately, Frank argues that recognizing our place in the ongoing narrative of the creation of cultural time and cosmic time – moving beyond the cosmology of the Big Bang (of which 'ours' may be one of many) – is what will allow mankind to enter a new, global era of time and culture."
Kirkus Reviews*, starred review*

"This will fascinate anyone curious about the nexus of astronomy and history and, of course, time."
Library Journal

"Frank ponders fresh ideas in cosmology ... and how the human perception of time will change in the future."
Washington Post

"This one is a must-read! The book does a wonderful job weaving together the story of human history and time in the context of the universe. From the Big Bang to the Renaissance to cell phones to the multiverse, [Frank] takes extremely complex ideas and makes them easily digestible, endlessly fascinating, and fun. About Time *will make you think."*
Culture of Science

"'Time' is the most used noun in the English language, yet we still don't really understand it. Adam Frank tells the fascinating story of how humans have struggled to make sense of time, especially in the context of the universe around us. From prehistory to the Enlightenment, through Einstein and on to the multiverse, this is a rich and inspiring tour through some of the biggest ideas that have ever been thought."
Sean Carroll, author of *From Eternity to Here*

ALSO BY ADAM FRANK

The Constant Fire: Beyond the Science vs. Religion Debate

ABOUT TIME

From Sun Dials to Quantum Clocks,
How the Cosmos Shapes Our Lives – and
We Shape the Cosmos

ADAM FRANK

ONEWORLD

A Oneworld Book

First published in the United Kingdom and Commonwealth by
Oneworld Publications 2012

Originally published in the United States of America by Free Press,
A Division of Simon & Schuster, Inc., in 2011

ISBN 978-1-85168-909-5
ISBN 978-1-78074-060-7 (ebook)

Cover design by Richard Green
Text designed by Paul Dippolito
Printed and bound in Great Britain by Bell & Bain Ltd

Oneworld Publications
185 Banbury Road
Oxford OX2 7AR
England

Learn more about Oneworld. Join our mailing list to find out about our latest
titles and special offers at:

www.oneworld-publications.com

529
ิ₇₀SS6K

For Alana, for all time

CONTENTS

Prologue

BEGINNINGS AND ENDINGS

The girl in the third row raises her hand and I know I'm in trouble.

The lecture hall is packed with students. In front, the perennially scribbling pre-meds have put down put their pens. Usually desperate for any fact that might appear on the exam, this term's crop of grade hunters have stopped blindly transcribing everything I say and are, for the first time all year, simply listening. In the way back, the line of identical frat boys with their matching baseball caps are actually paying attention to the lecture rather than hiding behind their newspapers or whispering to the cute sorority girls clustered around them.

This is the class I cherish. From years of teaching I know this is the subject everyone cares about. I am deep into the cosmology lecture of my Astronomy 101 class. Today it's all Big Bang and cosmic origins, and the students are wide-eyed. For this one hour, the windows of the universe will open up for them. For this one hour, they will climb out of their day-to-day concerns about grades and careers and getting laid and briefly stand in wonder at the deepest questions their species has learned to ask . . . *and answer.*

I don't expect these students to pay such rapt attention to my other lectures: the ones on stellar evolution, the history of astronomy or comparative planetology. But with the Big Bang I know their attention will be fixed for long enough to briefly catch a glimpse of our communal place in the fabric of creation. And, within this hour, I also know that sooner or later someone is going to break the spell and ask that one damn question.

"Professor?" she calls out. Sophie is her name. She is one of the students on fire with the subject this year—earnest, intelligent, alive to the big mysteries an astronomy class naturally washes up against. *OK*, I think, *here it comes.* I tell her to go ahead.

"But Professor," she begins, "what happened *before* the Big Bang?"

The usual vertigo closes in. *Yeah*, I think, *that's a good question. What the hell* did *happen before the Big Bang?* There is a long pause as the class waits expectantly. As if I, or anyone else, really have the answer.

4:08 p.m. I have lost them. Looking across the hall, I can see the mystery has dissipated. The real world has returned. The class is supposed to end at 4:15. Still in the thick of the lecture, I have already strayed too close to that imaginary deadline that marks the end of class. My story of cosmic creation has lost its urgency and become a death march of facts and details. The beginning of time and the nature of time both have become abstractions. Time and the cosmos shrink, congealing into the urgencies of *now*: the next class, the homework review session, the hoped-for hour at the gym, the coffee appointment with friends.

It is still too early for them to gather up their books and begin the shifting and rustling that mark the end of class. Instead, the students sit and feel the minutes collapse slowly—so slowly—into an ooze of boredom. They are caught in a purgatory of waiting, an empty place mediated only by their devices and technology. Some watch the minutes tick off on their open laptops. Others fill the wait by sending instant messages to friends across the quad or across the continent. Others see the abstraction of time become concrete on their mobile phones, each little box connected to a global cadence of milliseconds passing through waves of electromagnetic energy and information. While I continue to lecture about time and the universe, the students feel their own experience of both weighing down on them. If only they knew how closely connected their personal worlds were to the sweep of cosmic history I am recounting. And if only they understood how much it was all about to change.

IT'S ABOUT TIME

This book tells two stories that are braided so tightly they cannot be separated, even if they have never been told together before. Like my cosmology lecture that April day in 2007, the twin narratives I am about to unfold encompass the grandest conception of the universe we human beings have been able to imagine and explore. At the same time they embrace our most intimate and most personal experience of the world—the very frame of human life.

This book is about time, both cosmic and human.

The subject of time can transport us to the deepest levels of reflection. By looking out into the depths of space, we are always looking back in time and so, on its largest scale, our science of the universe is also, always, a story about the depths of time. There are many books—philosophical, technical and popular—on the nature of time as we experience it. There are just as many books telling the story of cosmic time by recounting the grand story of scientific cosmology. But there are few instances where we stop to ask how our stories about the universe's time are intimately wedded to the texture of time in our daily lives. Now there is a compelling reason to recount the braided narratives of cosmic history and human time as a unified whole: *the Big Bang is all but dead, and we do not yet know what will replace it.*

There are those who will tell you that cosmology—the study of the universe entire—has become an exact science. They will tell you that this grand and all-embracing field has, in the last fifty years, moved from the realms of philosophical speculation into the purest domains of science via exacting confrontations between theoretical models and high-resolution data. You should know that they are right. For the first time in the long march of human thinking we are now, finally, able to construct a detailed and verifiable account of cosmic history.

So when I tell you that the Big Bang is dead, I am not referring to the story that begins with a universe far hotter and far denser than what we see today. I do not mean the story of a universe expanding, of matter cooling and congealing over billions of years into stars and galaxies.

That story, the scientific narrative of cosmic evolution over the last 13.7 billion years is, for all intents and purposes, secure.

It is the beginning, the genesis, that stands ready to be replaced. The singular and all-important moment of creation at the beginning of the Big Bang—the beginning of time and existence—is poised to be swept aside. In other words, it's the bang in the Big Bang that we, in our endless quest to understand the world, are ready to abandon. That single moment of creation with no before has been done in by the very precision of the science that gave the idea a measure of reality.

Now it appears that science is ready to go beyond, and before, the Big Bang. Cosmology is waiting at the precipice of its next great revolution. The only question is where—or, better yet, when—do we go from here? We are ending the beginning and beginning down another path.

Cosmology and its impending reformation form the first narrative line of this book. If we are to understand how our grandest theories of the universe are about to change, we must first understand how we got to the Big Bang in the first place. Along the way we will encounter the most potent ideas of modern physics from Einstein's theory of relativity to the powerful but paradoxical realm of quantum mechanics and subatomic physics. In this first story we will explore cosmological foundations so that, when the moment comes, we will be ready to imagine the range and meaning of the Big Bang's bizarre alternatives.

And this is all about time. The roots of cosmology cannot be reworked without a new conception of time, including its origins and its physical nature. In Big Bang cosmology physicists imagined time to simply begin, like God firing up the engine on his cosmic Porsche. Alternative cosmologies, hovering just across the horizon, must replace that vision with something new.

Time, however, is slippery stuff. In both our abstract ideas about time and our attempts to understand its direct experience we are always walking on thin ice. Our scientific theorizing about time must always, at some point, meet our concrete, day-to-day movement through it. But where is that point? If the science of cosmology is about to re-imagine time, then how will that affect the way we experience time from moment to moment?

That question forms the heart of this book's second narrative. If the first story leads us to the precipice of modern cosmology, the second story tells what might be called the social history of time—a history of lived time. And in that second story there awaits a radical truth: as our ideas about cosmology and cosmic time have changed, human time has changed too. The industrial revolution, with its roots in the scientific discoveries of Newton and its radical reformation of human life, is perhaps the most potent and obvious example of the binding of human and cosmic time. Throughout the 1700s, new universal laws of physics pioneered by Newton reworked human conceptions of the heavens. Then, in time, Newton's mechanics became the blueprint for machines unlike anything human culture had built before, laying the ground for the triumph of industrialism. As workers filed into their new punch-clock lives of efficient production their world echoed the new clockwork universe of planets clicking through their orbits governed by economical rules of gravity and motion. Human time and cosmic time had been partnered in mutual transformation. But the two times—cosmological and human—had always been intertwined, and there was never an age when they could be cleanly separated.

The brute facts of time and nature are simple—the day lasts from sunrise to sunset—but from that point onward the simplicity ends. Our sense of social and personal time has been transformed and rebuilt in the many revolutions since we awoke to self-consciousness fifty thousand years ago. From hunter-gatherer tribes to the development of agriculture to the industrial revolution, our encounters with time have been reshaped again and again. Yet unrealized, the resonance between human and cosmic time is the essential instrument in this story of transformation. Cultures need a cosmology to understand their place in the greater framework of creation. But cosmologies—mythological or scientific— are collaborative creations that spring from the collective efforts and resources of entire cultures. When cultural time and cosmic time change, they change together. In an era dominated by scientific advancements, the simple assumption would be that new technologies lead the way, creating new cosmological narratives that also reshape culture. As we will see, the truth is far richer. The imperatives of changing culture or changing cosmology are always pushing back and forth on each other. At some moments in history one side takes the lead in changing time, and at other moments it's the other side that surges forward to initiate change. But always and again, time—both cosmic and human—has changed in ways that we have yet to fully comprehend.

Ask a friend what time it is and he might look at his watch and respond that it is 1:17 p.m. But what is 1:17 p.m.? What is the meaning of such an exact metering of minutes? There is nothing innate, objective or God-given about this kind of time. As we shall see, mechanical clocks did not appear until the fourteenth century, and they did not even have minute hands (an invention that would take approximately another three hundred years to appear). Did 1:17 p.m. even exist one thousand years ago for peasants living in Dark Ages Europe, Song Dynasty China or the central Persian Empire? Was there such a thing as 1:17 p.m. in the long millennia before the vast majority of human beings had access to any form of timekeeping device?

But 1:17 exists for you. As a citizen of a technologically advanced culture replete with omnipresent time-metering technologies, you have felt 1:17 in more ways then you probably want to think about.

How often have you found yourself on time for a train, a bus or an appointment that was scheduled as an exact block on your electronic calendar? Then, somehow, a delay appeared. The bus was late, the train had not yet arrived, the appointment was pushed back. Suddenly you are forced into the purgatory of waiting. Through the mediation of your watch or your mobile phone (with its automatically updating time-zone compliant chronometer), you feel those minutes crawl past just as my students felt the weight of their minutes until the lecture ended: 1:11, 1:15, 1:17. They drag on breeding frustration, boredom and anger. For you, those minutes are real.

Measured against the long arc of human evolution, this experience of time is something new and very radical. You feel time in a way that nobody did a thousand years ago. In 2000 BCE or 850 CE there was no culturally agreed-upon 1:17 p.m. For the vast majority of human existence, there was only "after lunch" or "in the afternoon".

It's a new time that we have created in our hyperdigital, telepresent, instant-messaged society. Connected simultaneously to all points of a GPS-mapped globe, we struggle to get that last batch of e-mails sent out before the 2:30 meeting, only to watch helplessly as a new batch appears. It's a new time we have invented, and it appears to have left us with no time at all.

If the time we live by is something new in human evolution, is it real? If other cultures moved through each moment of their days in entirely different ways, then how concrete is 1:17 p.m., with all the import, urgency and meaning we ascribe to it? As we shall see, the time we imagine for the cosmos and the time we imagined into human experience turn out to be woven so tightly together that we have lost the ability to see each of them for what it is.

Our cosmologies are soaked in time and have shaped the worlds of culture and experience. Our cultures are soaked in time and have shaped our grandest imaginings of cosmology, from myth down to the exacting science and technologies we encounter today. This braiding of science and culture is a story that we are unused to telling. It is easy to think of science as some kind of lumbering giant picking up brute facts and handing them to us in the form of revolutionary technologies (mobile phones, atomic weapons, antibiotics). But the knife-sharp separation of science from other human endeavours such as art, politics and spiritual longing is too abstract to be true or helpful. We want to glimpse the ways our science shapes, and is shaped by, experience and the culture it creates. That task demands we ask the deepest questions of all about the nature of time, the cosmos and their beginnings.

FROM HERE TO THE BEGINNITY: THE STORY OF SCIENTIFIC COSMOLOGY

The story of modern cosmology starts now and moves in reverse. That is how we astronomers and physicists have learned to piece together the story of the Big Bang. We begin with what we see around us—galaxies flying apart, carried along with the flow of space and time—an expanding universe. Then we imagine running the film of that expansion backwards. The galaxies crowd together rather than rush apart. Space becomes dense, the galaxies dissolve, and atoms are slammed into each other, reaching towards infinite density. The heat released drives the temperature of the entire universe towards impossible heights, until we are back at that singular moment—the unimaginable beginning when time itself was born.

The first cosmologies were the myths of our distant ancestors. In their stories of sky gods and mother goddesses, one finds the same explanatory impulse that drives our modern scientific efforts. What is new in our scientific and technological versions of the cosmological narrative is the all-important ability to test our stories against the data. We can ask the universe if we are right and see if it agrees. But Big Bang cosmology is not really one story; it is many. It is an interlocking web of scientific narratives about the nature of reality. Forged in earthbound laboratories, astronomical observatories and the imaginations of theoretical physicists across the last five hundred years, it is a culmination, one of our greatest achievements as a culture. If we are to understand the Big Bang—its triumphs, its failures and the horizon of possibilities that could replace it—we will have to cover a broad landscape of physics and astronomy. We must obtain a complete view of where we are now so that we may be prepared to imagine what comes next.

To understand the Big Bang and its looming alternatives, we must cover a terrain that has a wild topography of remarkable beauty and range, shaped by nature's deepest laws. Passing across that landscape, we will, in the chapters that follow, explore the foundations of modern physics—Einstein's theory of relativity, quantum and particle physics, thermodynamics and astrophysics. We will linger long enough with these fundamental ideas to gain a sense of how the universe has taken the form we see through our eyes and our telescopes.

Crossing this terrain will take us to the precipice we now face. For all their power, our two greatest theories of physics—quantum mechanics and Einstein's theory of gravity (called general relativity)—face a single great failing: they cannot talk to each other. The domains of the very small (quantum physics) and the domains of the very large (gravity) cannot be reconciled. After fifty years of trying, we still lack that holy grail of physics, a theory of quantum gravity—a theory of space and time on scales so small entire universes could be bound in an atom. To understand the bang in the Big Bang, we need quantum gravity. Consequently, our cosmology remains incomplete.

The search for quantum gravity and the ideas it entails will form one part of our story. The problems and paradoxes that have plagued the Big Bang will form the other. To rescue Big Bang cosmology from its own best data, astronomers and physicists were forced to imagine events occurring in the early universe—the barest instants after creation—which have shaken the very concept of a "moment of creation". Together with the attack on quantum gravity, the rescue of the Big Bang has led to a Wild West of new ideas that throw open the frontiers of space, time and creation. The last part of our story will be the exploration of these frontiers.

Could there have been not just one bang but recurring ones? Could our universe be only one in a long line of cycles? Could there be many bangs going off all the time, creating an infinite number of simultaneously existing universes—a multiverse of infinite possibilities? Perhaps, more radically still, our entire conception of time is wrong. Perhaps time is an illusion. Perhaps there is no passage from one moment to the next. Once we have gained a view of where we stand now, in the midst of our cherished but ill-fated Big Bang theory, we can explore these and other possibilities as we look to the future of cosmology and our concept of time itself.

FROM THE BEGINNING TO BEING-HERE-NOW: THE STORY OF HUMAN TIME

Building cosmologies is an old, old business for us. Myths and religion have conceived of Big Bangs before. But that didn't make it any less of a surprise when scientists found that their own pathways of investigation led them towards $t = 0$, with its echoes of a biblical moment of creation. What many of them didn't know was that even alternative cosmological models had antecedents in mythology and religion.

The human engagement, construction and invention of time began with our mental awakening. Archaeologist Steven Mithen calls this, appropriately, the "Big Bang of consciousness", and it remains as mysterious and enigmatic as the origin of the cosmos. Two thousand generations ago, deep in the cold of the last ice age, we humans awoke to the predicament of ourselves in time. In order to cope, we invented new forms of social organization and new ways of thinking that set the species on an unprecedented evolutionary trajectory. We invented culture and in doing so invented ourselves.

It began some seventy thousand to forty thousand years ago with the burying of the dead. Death has always been a portal to time's great mystery. By ending time (at least as we know it) for the self, death acts as an invitation to consider time's reality and its meaning. We felt this even in the early stages of our cultural development. Arranging the departed into huddled postures of repose, we lay our loved ones in graves with precious goods, such as beads and knives, that signified an awareness of death as time. Later, on cave walls and rock cliffs, we began to leave a permanent record of our interior response to the world in art that remains haunting to this day. In these caves covered in paintings of bison and mammoths, archaeologists have also found flutes made of bone, and carvings on bone fragments that seem to trace the phases of the moon. As a species, we awoke not only to symbolic expression through art but also to the explicit experience of time through internal rhythms expressed in music and external rhythms we noticed in the sky.

Personal time and cosmic time have been linked from the earliest origins of culture. When the development of agriculture followed the retreating glaciers, some twelve thousand years ago, a new sense of time emerged with it. Farming led to surplus and wealth, villages turned into towns, towns turned into cities, and cities grew into empires. In each stage, new encounters with time would emerge that were born directly from the material needs of the culture. It was through a direct, embodied engagement with the material world—what we made, how we made it and how that changed the way we organized ourselves—that time itself changed. Each culture shaped its day-to-day life through the technologies it built and through its "institutional facts"—the invented social reality the technologies allowed and supported.

But cultures (with their invented institutions) need justification and support. They need to set themselves against a cosmic background to give individual and collective lives meaning. The central theme of this book will be to explore the enigmatic entanglement that tied human time to the cosmological narratives of sky and stars, origins and final endings.

It is crucial to recognize that each grand change in human history has shifted more than merely ideas about time. Instead, it is the experience of time, its felt contours, that have been transformed. To understand that story, and to see how closely connected our direct encounters with time are to cosmological imaginings about it, we must travel a path parallel to the one we take in our exploration of physics and astronomy. A Palaeolithic farmer moved through his day and experienced time in a radically different way than did a merchant living in the great city of Babylon. The human encounter with time is fluid and malleable. It can and will change again.

Thus, our story of human time will begin fifty thousand years ago with our hunter-gatherer ancestors and move through experiences of the first farmers and city builders. It will take on new themes as the Renaissance begins and clocks are first introduced to town squares. With the industrial revolution, an entirely new form of time comes to dominate culture, and a new politics follows in its footsteps. As the twentieth century begins, the electrified world gives birth to yet another form of encountered time that presages our own wireless world. Then, with the dawn of the space age and the digital revolution, we arrive at our own home in the age of precision, just-in-time, never-enough-time.

By recounting the narrative of our time, in step with our emerging understanding of cosmic time, we will be in a better place to see where we are now and what other times we might create.

We must note that there were (and are) many trajectories of development for human culture. In this book we will focus on the broad sweep of history, science and time but in doing so we focus primarily on the trajectory of cultural development associated with the West.

This makes sense, of course, because science and scientific cosmology emerged from traditions born in what historian Ian Morris calls the "Western core"—Mesopotamia, Egypt, Greece and so on.[1] We should remain mindful, however, that the traditions that emerged in the Eastern core—China, Korea and others—had their own cultural uses for time and their own cosmological visions for it as well. Now that science has become truly global it may be that our future will see metaphors associated with these other traditions finding their way into cosmological theorizing and the cultural constructions of time. It is a possibility we must not lose sight of as we look forward.

THE REDISCOVERY OF MAN

It's only 9:00 a.m. in Lindos and the sun leans hard on my astronomer friend and me as we climb the narrow steps to the top of the cliff. We slept on the beach last night, knowing it would be the only way to reach the temple before the impossible heat and the crowds arrive.

Lindos is a small coastal town on the Greek island of Rhodes. Here, some twenty-five hundred years ago, on a granite promontory that rises dozens of metres above the sea, the Greeks built a temple to the goddess Athena. From the beach below it's a staggering sight—an acropolis hanging in the air. Now, as we reach the top of the stairs and walk out onto the temple grounds, it's overwhelming.

We came to this island not as sightseers but for a business that stretches as far back as this temple. All week we have been attending a conference on astronomy. My companion and I both study star formation—a branch of astronomy that focuses on the assembly of stars and planets from vast clouds of dusty interstellar gas. Along with 150 other astronomers, we gathered at a resort hotel on the other side of the island to share new data, new models and new insights into the early lives of stars very much like the one beating its light and heat down on us now. We must have looked odd to tourists, with our penchant for staying indoors all day, huddled in a dark meeting hall, staring at endless sets of PowerPoint slides.

Locating this conference in Rhodes was not an arbitrary choice for its organizers. Two thousand years ago the city of Rhodes was the home of Hipparchus, the greatest of ancient Greece's observational astronomers. In the days when this temple was home to priests servicing Athena's concerns on Earth, Hipparchus was busy in the city cataloging the starry skies.

I stand under the towering columns of the temple to shield myself from the ferocity of the sun and stare into an impossibly empty sky and an unyieldingly blue Aegean. Here, at a temple where each day would be metered by prayers to the gods, on an island where true astronomical investigation gained an early foothold, the glue binding human and cosmic time seems as concrete as the giant stone columns standing guard over the ocean.

What began here has been continuously reshaped in a long march leading directly to the cosmos my colleagues and I explored at last week's meeting. Now all of us—scientist and nonscientist alike—are about to start this march anew even if we do not recognize it. We are ready to end one kind of time and one kind of universe. We are ready to end the beginning and to begin again.

ABOUT
TIME

Chapter 1

TALKING SKY, WORKING STONE AND LIVING FIELD

From Prehistory to the Agricultural Revolution

ABRI BLANCHARD, THE DORDOGNE, FRANCE · 20,000 BCE

The shaman stands before the opening of the cave and waits. Night is falling now and the piercing cold of autumn easily penetrates her animal-hide cloak. Outside, beyond the sheltering U-shaped bay of low cliff walls, the wind has picked up. These winds originated hundreds of miles to the north at the blue-white wall of glaciers that cover much of northern Europe.[1]

Winter is coming and the shaman's people will have to move soon. The hides and supplies will be gathered and they will begin the trek towards the low-hanging sun and the warmer camps. But tonight the shaman's mind is fixed on the present and she waits. It is her job to read the signs the living world provides. It is her job to know the turnings of Earth, animal and sky. The shaman's people depend on this wisdom and so she waits to complete the task her mother suggested before she died. She waits, massaging the reindeer bone fragment in one hand and the pointed shard of flint in the other.

Now she sees it, the glow over the eastern horizon. The great mother rises. The shaman waits to see the moon's face—pale, full of power. There, see! She is complete again.

From the crescent horns of many days ago, she has now returned, completed, to life. The full circle of the moon's face, promising rebirth and renewal, has returned. The shaman holds the bone fragment before her in the

1

grey light. With her index finger she traces out the long serpentine trail of her previous engravings on the bone. Then at the end of the trail she makes tonight's mark with the flint-knife, carving the shape of tonight's full moon into the hard bone.

Two rounds of the moon's dying and rebirth have now been traced out. The shaman's work is complete. The round of life and death in the sky, like the rounds of women's bleeding, have been given form, remembered, noted and honoured. She returns the bone to its place beneath the rocks where she keeps the other shamanic tools her mother gave her so long ago. Now she has added to the store, a marking of passages, that she will use and pass on to her own children.[2]

WANDERING TIME:
THE PALAEOLITHIC WORLD

The origins of human culture are saturated with time but we have only recently learned to see this truth. The evidence of the age when we grew into an awareness of ourselves, the cosmos and time itself had always been just out of reach. For most of our remembered history, the clues to the birth of culture and cultural time lay forgotten, buried a metre or so below the ground. Then, in the late 1870s, the discoveries began and we started to remember.

We first encountered the great awakening of our consciousness in 1879 in Spain when the nine-year-old daughter of amateur archaeologist Marcelino Sanz de Sautuola led him to a cave at Altamira. Venturing inside, he found its walls covered in vivid paintings now known to be twenty thousand years old. As archaeologists began systematic explorations, new caves were discovered, many of them also covered in paintings. The caves are a bestiary: bear, bison and mastodon appear on the walls along with other species. Sometimes these animals appear alone. Other times they are in herds. Sometimes we appear too—human figures set against the herd, spear at the ready.

These paintings gave voice to a heretofore silent past, showing that early humans were anything but "savages" incapable of representation,

FIGURE 1.1. The Recognition of Time. The Abri Blanchard bone fragment (dating back between 12,000 to 20,000 BCE). Archaeologist Alexander Marshack proposed that the distinct pattern of engravings was an early record of lunar cycles.

abstraction or response. This vivid prehistory, the time before written records, was set into a temporal sequence through painstaking work across the twentieth century. Radiocarbon dating allowed scientists to see when each cave was inhabited and when the paintings were laid down. As the sequence became complete, scientists came to understand how rapidly our ancestors had re-imagined what it meant to be human. With that recognition, conceptions of evolution and the origins of the human mind were thrown open. Scientists began searching for some purchase on a new theory that could explain how the mind rapidly woke to itself and to time.

Buried next to the painted walls, archaeologists discovered artefacts worked by human hands. There were statues of women with wide hips, pendulous breasts and finely articulated vulvas, seeming to focus on the mysteries of fertility. Spear points, needles, flint daggers and hammers spoke of cultures skilled in the creation of a diverse set of tools for diverse needs. And on the floor of a cave at Abri Blanchard, a rock shelter in the Dordogne region of central France,[3] archaeologists found a "small flat ovoid piece of bone pockmarked with stokes and notches . . . just the right size to be held in the palm of the hand".[4]

The Abri Blanchard artefact would sit in a French museum for decades until amateur archaeologist Alexander Marshack chanced upon it. Marshack was a journalist hired by NASA in 1962 to recount the long human journey leading up to the Apollo moon landing programme. In his efforts for NASA, Marshack dug so deeply into the history of astronomy, he found himself at the very dawn of human time. He first became obsessed with the Abri Blanchard bone when he encountered it, as part of his NASA project, in grainy illustrations of forgotten archaeology tracts. Eventually travelling to the imposing Musée des Antiquités Nationales near Paris, he found the small artefact all but ignored in a "musty . . . stone chamber" among "accumulations of Upper Palaeolithic materials, crowded under glass with their aged yellowing labels".[5]

Marshack dedicated years of intense study to the Abri Blanchard bone fragment, rejecting the accepted wisdom that the notched carvings were nothing more than early artistic doodling. He traced over the curved trail of finely etched crescents and circles again and again. Finally he discovered a pattern echoing his own interests and his ongoing studies of the moon's impact on human life, time and culture. In 1970 he published a closely reasoned and convincing argument that the etchings on the bone fragment were one of the earliest tallies of time's passage, a two-month record of passing lunar phases.[6] Rather than random doodles, the creator of the Abri Blanchard fragment had been marking time in a systematic way, for perhaps the first time ever.

In caves at other sites across the world, fragments of humanity's first encounters with time were also found. Sticks with sequential notches, flat stones engraved with periodic engravings—each artefact gave testimony to an increasingly sophisticated temporal awareness.[7] It was in those caves, tens of thousands of years ago, that the dawn of time broke in the human mind.[8]

Scientists call this era the Palaeolithic—meaning "old stone age". The origins of human culture (and human time) occurred during the so-called Upper Palaeolithic, from roughly 43,000 to 10,000 BCE.[9] All human beings during this period lived as hunter-gatherers and left us no written records of their thoughts (writing would have to wait until

FIGURE 1.2. Artist's depiction of Palaeolithic dwelling at cave formations in the Dordogne region where the Abri Blanchard bone was discovered.

around 4000 BCE). Through painstaking analysis and considerable imagination, however, scientists have pieced together the outlines of the Palaeolithic human universe. Using the artefacts found at sites around the world and the narratives of contemporary hunter-gatherers, scientists have gained some understanding of these first modern humans.

The Abri Blanchard bone fragment dates back from twelve thousand to twenty thousand years ago, a time when the planet looked very different from the relatively warm, wet globe we now inhabit.[10] It was an era when ice covered much of the northern hemisphere. France, as experienced by the unnamed palaeoastronomer marking the moon at Abri Blanchard, was a realm of polar deserts, tundra and towering glaciers.[11] Humans were scarce on the ground, and they struggled to endure the hostile climate. This was not the first time members of the genus *Homo* had endured climate change of this kind. Ice sheets had come and gone in the past, but something remarkable happened to *Homo sapiens* during the planet's last deep freeze that would, in time, alter the face of the planet itself.

Somehow the depths and deprivations of the last climatic winter drove our ancestors to make revolutionary leaps in behaviour. It was early in this period that human beings began burying their dead. Later they invented art and music. Clothes were soon fashioned by sewing animal pelts together.[12] People began to invent systems for counting and, most important, started to codify the passage of time.[13] It is in this period that the modern mind—with its penchant for analysis and allegory, concretizing and abstraction—emerged. Compared with earlier eras, the speed with which these radical changes swept through populations is startling and has led scholars to call the Upper Palaeolithic the Big Bang of consciousness.[14] Any understanding of the inseparable narratives of cosmic and human time must begin here, with the rapid expansion of consciousness that would one day embrace the very cosmos in which it was born.

TIME OUT OF MIND: THE ORIGINS OF MODERN CONSCIOUSNESS

The explicit recognition and representation of nature's repeating patterns were radical developments in hominid evolution. Thus, to understand the emergence of time in the mind, we must first understand the emergence of mind.

The hominid species that came before us faced challenges similar to those facing our ancestors. From the fossil record, it appears that the earliest versions of *Homo sapiens* may have overlapped for a brief time with both *Homo erectus* and *Homo heidelbergensis*.[15] It is also clear that until a mere fifteen thousand years ago we were in direct competition with a group of hominids who were likely also the descendents of *H. heidelbergensis, Homo neanderthalensis*—the Neanderthals.[16] This overlap with other species, particularly the Neanderthals whose brains were essentially the same relative size as ours, raises a pressing question: what spurred the Big Bang of consciousness in the Palaeolithic era?[17]

It wasn't the environment that woke us up. In the long stretch of time before the Palaeolithic's cognitive revolution, the same challenges presented themselves over and over. Ice sheets grew and ice sheets retreated. Early hominid species adapted by changing behaviours as the situation demanded. What was lacking, however, was rapid innovation and learning. The toolkit of the human species was impressive compared to that of nonhominid species: the domestication of fire, deliberately shaped stones for cutting and scraping.[18] But the pace of innovation in the development of these tools and the mental processes likely used in their development seems positively glacial.

The basic stone tools our ancestors used 1.5 million years ago look fairly similar to those used by their descendants a million years later. Almost no technological innovation appears over the course of thousands of generations.[19] On the other hand, comparing the brute stone scrapers used in 100,000 BCE to the needles, harpoons, arrow points and axes developed in 20,000 BCE is akin to comparing a raft to a nuclear submarine.

There are many competing theories for the rapid origin of the modern human mind. One of the most fruitful lines of research comes from evolutionary psychologists who, in the 1980s and 1990s, began thinking of the prehistoric mind as a kind of Swiss Army knife. From this perspective the mind is remarkably complex and does many highly specialized tasks that it can only have developed in response to very specific evolutionary pressures. Researchers Leda Cosmides and John Tooby have each argued that, like a Swiss Army knife, the mind has different blades—that is, different cognitive modules.[20] Each module adapted in response to specific challenges our hunter-gatherer ancestors faced 100,000 years ago. These modules come with their own data and instruction sets—they are content-rich. A *Homo sapien* child born fifty thousand years ago (or fifty years ago, for that matter) already had modules for hunting, social interactions, tool building and so on. Thus, each module comes hardwired with a considerable storehouse of data.

There are compelling arguments supporting this claim. When a tiger leaps towards you from behind a rock, a mind built like a generalized computer sorting through all possible responses may not be the optimal evolutionary setup. A brain built like a Swiss Army knife, however, might very well provide the evolutionary adaptation to keep you alive. Given the notorious speed of hungry tigers, a Swiss Army knife mind with preloaded modules governing tiger recognition and the run-like-hell response would appear, on the face of things, to be a more successful evolutionary strategy.[21]

There is also compelling evidence for this type of hardwiring. According to some lines of research, we appear to be born with at least four distinct domains of intuitive knowledge. These content-rich modules govern language, human psychology, biology and physics. In each of these domains there is evidence that humans evolved with an internal guidebook of understanding, a degree of hardwiring imprinted by evolution, to help us deal quickly with communication, social interactions, the living environment and the behaviour of the material world.

Experiences of secondary school science lessons may convince many people that they have no intuitive understanding of physics. Psychologist Elizabeth Spelke would disagree. Along with researchers such as Renée Baillargeon and others, Spelke has explored the notion of a preexisting "folk physics" innate to us all. Spelke has shown that even very young children have a clear understanding of the behaviour of physical objects.[22] Though children's lives are built around (and depend on) interactions with other people, they can clearly distinguish the properties of people, other living things and inanimate objects. Most important, children seem to be born with concepts of solidity, gravity and inertia. From an evolutionary perspective, intuitive physics makes sense. From the flight of a projectile to the impact of two stones against each other, an intuitive grasp of physics serves as cognitive bedrock for learned skills such as tool making and weapon use.

While the Swiss Army knife story of the mind's evolution is powerfully suggestive, it is likely not the whole story. Archaeologist Steven Mithen has argued that the "architecture" of the mind—its specific cognitive structure—had to evolve in specific ways to drive the Big Bang of consciousness. In particular, Mithen sees the ability to move information between modules as the innovation that led to the modern mind and the all-important capacity for culturally invented time. Moving information between modules allows for metaphor and analogy, which are the essence of human creativity. When information is shared between modules, a lump of clay that is roughly in the shape of a human form becomes the spirit of a departed ancestor, some coloured fluid splashed on a wall becomes a symbol for a bull felled in a fierce hunt, and markings engraved on a bone become a record of the moon's phases.

Mithen likens the developing mind to a building with different rooms, each housing a different cognitive module. The story of the mind's evolution is a narrative of reworking its architecture. Only when the architectural plan of the human mind changed by removing walls between isolated modules could the rapid evolution of consciousness and culture begin. As Mithen puts it, "In the [modern] mental architecture thoughts and knowledge generated by specialized intelligence can now flow freely around the mind . . . when thoughts originating in different domains can engage together, the result is an almost limitless capacity for imagination."[23]

This new architecture of mind did not, of course, emerge on its own. The fact that evolution instilled in us an intuitive physics means that one cannot ignore the physical aspect of the developing mind. If, as Mithen imagines, channels are opened between modules to create culture, then there must also exist feedback loops where encounters with the world's physical reality through culture amplify the opening of those channels. What we did with the "stuff" of the world that we found and shaped for our own uses—bone, wood and reed—changed what we could do and what we could imagine. The mind that would eventually come to imagine and organize time was a product of the looped interaction between the physical world it encountered, the physical culture it created and the symbolic culture it imagined linking the two.

It begins with the reindeer bone in one hand and the flint in the other—the bone's solidity in the palm, the rough feel of its edges on the fingers, the sharp bite of the flint's point on the thumb. Then it rises in the mind to the realm of the symbolic—carved notches on the bone fragment become a representation for shared experience that can be shown to others in the tribe. The leap between physical object and cultural creation—the idea of the bone and its markings as a symbolic notation of time's passage—closes the loop.

What began millennia ago continues today: the circulation between a physical encounter with the world, the cultural forms engendered by that encounter and the shape of consciousness determining how we think and what we experience. The evolutionary modules with their hardwired understanding of physics were a starting point. But what happened in the Neolithic was a braided process flowing between the outside world and the interior response. We can call this poorly understood but essential connection between the physical world and cultural invention *enigmatic entanglement*. Through this enigmatic entanglement, a remarkable dialogue between mind and matter was begun— forever linking cosmic and human time together.

CYCLES IN THE SKY: THE RAW MATERIAL OF TIME

Human life is set against the natural rhythms of the sky. These celestial changes are the raw material of time. In the braided history of cosmological and cultural time there are four cycles most important to our development: the day, the month, the year and the periodic motion of planets.

– The Day –

The most fundamental astronomical period we experience is day's journey into night and back again. Deeply imprinted in our biology as our circadian rhythms, the day/night, light/dark cycle sets the ebb and flow of our sleep and wakefulness.[24]

Concerning the day, we must recognize one vital point in order for our social history of time to make sense: the day's length varies. The sun spends more time above the horizon during the summer than it does during the winter. There is more daylight—more time to work— in summer than in winter. Thus, the natural experience of the day, and the unnatural attempts to slice it up into precise divisions, runs into a conflict: what is one to do with the difference between the length of a summer day and that of a winter day? This conflict would eventually pit the facts of astronomy against the needs of culture. Should the day's divisions be of fixed length or should they expand and contract with the season? This question is of no small importance if, for example, you are a guildsman paid a wage that depends on accountings of the day's length. As we shall see, the astronomical/cosmological recognition of the day as a by-product of a spinning Earth emerged during a long transition. It is no accident that this transition accompanied the increasing economic need for an accurately metered day.

– The Month –

The next cycle imposed upon us by nature is the motion of the moon. The lunar cycle has two distinct aspects: a variable position in the sky (where you see the moon relative to the sun each day) and changes in its appearance (its phase). The synodic month, as the astronomers call it, begins with the phase called the new moon, which occurs when the moon crosses the sun in the sky. The synodic month lasts approximately 29.5 days.[25] Since the moon shines only with reflected sunlight, a new moon is a virtually invisible moon. But as the days pass, the moon moves eastward in its orbit (relative to the sun's position at noon) and we see the moon's transformation from the sickle-shaped waxing crescent phase to the familiar D of the first quarter moon and then on to the full moon, the third-quarter moon and finally the waning crescent.

Next to the day, the cycle of lunar phases defining the month is our most obvious and visceral experience of celestial time. "Many moons" was the preferred means of time reckoning for hunter-gathering cultures. Most early cultures kept lunar rather than solar calendars. Thus it was the round of the moon's phases repeating themselves after approximately twelve or so cycles that defined a year. The moon provides an easily recognized measure of time cycles short enough to count but long enough to measure durations stretching across many days. It is a great loss to us moderns, with our skies lit up by electric light, that we rarely notice the moon and its passage through a cycle of phases.

– The Year –

The yearly cycle of the sun—or seasons—is the longest repeating period imposed by the heavens that most people experience in a single life.

Few of us are as intimate with the yearly changes in the sun's position as we are with the daily solar turn of night into day. We may not pay close attention to those annual shifts, but we all feel the effect of the sun's motion in the sky through the change of seasons. Everyone living in temperate climates knows the feeling of the sun beating down at noon on a summer's day or of the feeble heat of the winter sun hanging low in the sky. Both experiences are unconscious measurements of the sun's movement through the sky over the course of a year. On its daily journey, the sun climbs higher into the sky during the summer than in winter. Here, "higher" means closer to directly overhead, a point astronomers call the zenith.

The sun's daily arc takes it from rising in the east to setting in the west. The changing height of the sun at noontime over the course of a year actually reflects a change in where it appears to rise and set on the horizon (though of course it is the Earth's turning that creates this effect). Most of us are far too removed from the experience of the natural world to notice where on the horizon the sun rises and sets. Our ancestors, however, couldn't help but notice. Beginning in the depths of winter they watched morning after morning—the sun rising progressively further north on the horizon as the days passed. Then in the middle of summer the steady northward march of solar risings (and settings) would stop and reverse itself. The sun's rising begins moving further towards the south on the horizon line until it reached a southern extreme and repeated the cycle once again.

The seasons, the length of day and the strength of the sun's warmth are all tied to this yearly cycle. The year's shortest day (December 21 in the northern hemisphere) occurs when the sun rises at its southernmost point on the horizon; this is the winter solstice. The longest day comes when the sun is at its northernmost rising, called the summer solstice (June 21 on our modern calendars). The spring and autumn equinoxes (March and September 21) mark days of equal length.

The yearly cycle of long days turning to short days, warm months turning to cold months and growing seasons turning to seasons of decay has been the pulse of human life since we were hunter-gatherers. The direct connection between this lived, embodied experience of repeating celestial patterns made time and the sky intimate partners. But it was not only the sun that mattered; the stars themselves acted as the original cosmic metronome.

Like a child being turned outward on a merry-go-round, we humans get a different view of the night sky in winter, spring, summer and autumn as the Earth wheels around the sun in its orbit. The constellations we can see each night change with the seasons. By the era of the Greeks at least the ancients had mapped out the positions of the fixed stars on the starry sphere. They were clever enough to imagine the noontime sun set against that background of stars even if those stars were blotted out by sunlight during the day. In this way they knew the sun was tracking a path against those stars in a line called the ecliptic. They saw that the sun, the moon and even the wandering planets stayed close to the ecliptic as they moved across the sky. The twelve distinct constellations the sun passed through in its motion along the ecliptic were called the zodiac. The movement of sun, moon and planets against the fixed pattern of stars, along the ecliptic, was a cosmic dance that anyone could see before artificial light stole the night from us.

– The Periodic Motion of the Planets –

The final celestial period imposed on us is one that few other than astronomers would notice today. Each of the five visible planets (Mercury, Venus, Mars, Jupiter, Saturn) makes a slow march along to the ecliptic and against the background of fixed stars. Night after night each planet moves slowly through the constellations, speeding up and slowing down at different times in the year and for different durations. Each planet also steadily brightens and then steadily dims as it completes its motion against the stars. Strangest of all, each planet executes a loop in the sky—called retrograde motion—halting its usual eastward march in a short pirouette that takes on the order of a few months to complete.

FALLING INTO TIME:
THE PALAEOLITHIC COSMOS

The daily turn of night to day, the monthly cycle of lunar phases, the yearly journey of the sun through the zodiac and, finally, the strange wanderings of the planets—each of these celestial dance steps formed a raw physical encounter with time. It was seen on the sky and felt in the seasonal changes of warmth and cold. It is from this most basic experience that humans built their stories of the cosmos, its origin in time and its meaning for their lives.

The world we inhabit today makes a clean separation between time in our daily experience and the scientifically defined time of our cosmology. No one today connects her 12:15 appointment at the dentist to the fractions of a second in which Big Bang cosmology plays out. Daily time is lived through the digital time on mobile phones and our electronic calendars. Cosmological time is the domain of scientists, observatories and university graduate study. This separation, however, is an illusion.

The arc of cultural evolution has hidden the binding of human time and cosmic time from us, but in the Palaeolithic world the separation never existed. The cosmos of our Palaeolithic ancestors was of a whole, and that included time. The hunter-gatherer peoples of the Neolithic knew their place in the world, for they had yet to fully distinguish themselves from that world or its movement through time.

Our understanding of the Palaeolithic cosmos (and cosmology) relies on an understanding of myth. Myth, in this essential context, does not mean a false story (as in the urban myth of an old lady drying her poodle in a microwave oven). Myth, in the domains of cosmology, is far more essential and powerful. Every culture has its mythology, its potent narratives of origins and endings. To follow the roots of modern cosmological theorizing back in time to the imaginative territories of prehistory we must turn to the universe of myth, for within it we will find the first responses to the mystery linking time and being.

Myth came before all our forms of religion and before the practice of science.[26] Its function was to recount "sacred" stories that set human beings into their proper context within the universe. In myth the experience of the world as sacred was codified into stories so old that they embrace the origins of both religion *and* science. As the great scholar of religion Mircea Eliade put it: "Through myth, the World can be apprehended as a perfectly articulated, intelligible, and significant Cosmos."[27] Thus it is through myth that the cosmos of prehistory becomes apparent.

Since our Palaeolithic ancestors left no written records of their cosmological myths, archaeologists must piece together their worldview from other materials. One important source is the repository of mythological narratives transcribed many millennia later, once writing was invented around 3000 BCE.[28] These stories contain traces of the cosmological narratives of prehistory. The mythologies of existing hunter-gatherers, such as the Aboriginal people of Australia or the Inuit peoples of the Arctic, provide their own insights.[29] From these sources a clear picture emerges of a cosmology that is dominated by ideas of a time without time, and the separation of elements once joined.

The world of the hunter-gatherer does not split cleanly between man and animal, culture and nature. Instead those paired worlds remain integrated. The Mbuti people of Zaire, for example, identify the forest they inhabit as a person. It is not "the environment" but another sentient being, a giving parent or trusted kin.[30] The Inuit of Greenland take a similar position with respect to the animals they hunt. The polar bear is not simply a lower animal but is, instead, a member of the tribe. Once it has been successfully hunted and killed it must be treated with the same respect as any other deceased member of the society.[31]

Just as the separation of man and nature does not exist for modern hunter-gatherers, it likely did not exist in the Palaeolithic. As anthropologist Tim Ingold has written, "For modern hunter-gatherers there are not two worlds of persons (society) and things (nature) but just one—one environment—saturated with personal powers and embracing both human beings and the plants and animals on which they depend, and the landscape in which they live and move."[32]

This seamless continuity between humans and their environment is also reflected in their conceptions of time and cosmos. The link, however, comes with a twist. Palaeolithic narratives of creation reflect the recognition that an immutable separation was imposed on humanity when we awakened in the Big Bang of consciousness. Thus hunter-gatherer myths will often speak of a lost paradise. In the "before" humans were immortal and lived in an ongoing balance with the world and the divine forces that shaped it. Human and animal could speak to each other and transformations from animal to human form were common.

The most important aspect of Palaeolithic cosmological myths was the Fall, which was the loss of a perfect, pre-existing harmony between humans and the nonhuman world. Somehow, the myths tell us, that harmony was shattered. As Karen Armstrong put it, "At the centre of the world there was a tree, a mountain or a pole linking Earth with heaven, which the people could easily climb to reach the realm of the gods. Then there was a catastrophe; the mountain collapsed, or the tree was cut down and it became more difficult to reach heaven."[33]

In this myth, the pre-existing Eden was a timeless place. It was complete and existed without change. People in the golden age lived either exceedingly long lives or forever. Thus the Fall was also the fall into time. In many myths, when the harmony was shattered, time and death entered the world together. But in these myths the golden age was not truly gone. Instead it remained as an archetypal, timeless state that could be—and needed to be—recovered. The Aborigines of Australia, for example, conceive of returning to the primordial cosmos in the Dreamtime. As Armstrong wrote, "Dreamtime . . . is timeless and 'everywhen'. It forms a stable backdrop to everyday life, which is dominated by death, flux, the endless succession of events, and the cycles of seasons."[34]

The myth of the golden age and the Fall are all but universal in early human cultures.[35] Like the Australian aboriginal emphasis on Dreamtime, the point of these cosmological myths was not to recount history; it was to recover original time. As Armstrong puts it, "Today we separate the religious from the secular. This would have been incomprehensible to the Palaeolithic hunters." These myths had purpose. They "show people how they could return to this archetypal world, not only in moments of visionary rapture but in the regular duties of their daily lives."[36] Thus the time of origins existed not in the past, as we imagine it now in cosmology, but in this ever-present "everywhen".[37] For the hunter crouching low in the brush, waiting for the herd to pause, the time of creation and his experience now were never far apart. For the women filling baskets with wild grain, their actions and the primordial divine acts, which set the world in motion, were always closely paired.

The daily time of Palaeolithic people was set against a cosmos that was not "out there" but rather was close by. It was a cosmos without birth or final death or linear time. It was a universe alive with animate and divine powers that were always present, re-energizing the world each day with the return of the sun, each month with the return of the moon to fullness and each season with shifts in light and warmth.

But culture and its needs would change and the powers animating the universe would grow distant and more difficult to engage. Divine entities that were once proximate and personal slowly became remote and wilful. It is in this understanding that the sky was the first retreat of the divine. The development of more complex cultures led to myths of the sky as the first and distant domain of the divine. Armstrong notes that the sky god—a distant but powerful first cause—begins to appear in this era. She observes, "The sky towered above them [the Palaeolithic people], inconceivably immense, inaccessible and eternal. It was the very essence of transcendence and otherness." The sky with its wheeling, repeating patterns was the first house of the father god, the original seat of divine power.[38]

The Palaeolithic cosmos was a direct reflection of our first awakened experience of the daily experience of time. As hunter-gatherers, people followed the herds and watched as the seasons ripened the best edible plants. In the Big Bang of consciousness they had begun to watch the cycles of the world around them and from it they lifted out the idea of time. In this way time was a creation of culture, just as culture was a creation of the embodied mind. The enigmatic entanglement between experienced physicality and the symbolic, cultured imagination created a time that existed in both daily life and myth. But as the world warmed, humanity changed, and our universe would change with it. A new kind of time and a new relation to the world would emerge with the next great human revolution. Our focus and our consciousness would be re-created as the Earth was put to the plough.

STAYING HOME: THE NEOLITHIC REVOLUTION

The world would not remain locked in ice forever.

Around twelve thousand years ago, the climate shifted, as it had countless times and countless millennia before. The planet warmed, the glaciers retreated and the Palaeolithic era ended.[39] But unlike previous interglacial periods, something new was born with the return of climatic spring. Our species learned a fundamentally new way of being human.

As warm seasons lengthened and the ice disappeared, some bands of hunter-gatherers invented a new way of life. They stopped following the herds into warmer pastures. They settled down and began domesticating themselves—switching from hunting and foraging to the deliberate cultivation of grains. They built houses that would last from one year to the next.[40] They grouped these houses together to create villages that would endure for generations. All these changes fed upon one another and were possible because people had learned to till the land and reap its harvests. Cosmos and time were each, separately and together, essential facets of this revolution. In this Neolithic era (new stone age), a flowering of human creativity began. It was a time of profound change that would not be matched again until the age of the machine.

The archaeologist Colin Renfrew identifies at least seven key features of the Neolithic revolution:[41]

1. The development of food production through domesticated plants such as wheat, lentils, barley and flax
2. The use of tools such as grindstones for processing these plants
3. The domestication of animals such as sheep, goats, cattle and pigs
4. The emergence of settled village life with permanent dwellings
5. The appearance of ritual practices involving shrines and human representations

6. The interment of the dead in cemeteries, sometimes featuring monumental tombs
7. The development of long-distance procurement systems for raw materials such as obsidian

Each of these cultural innovations required a fundamentally new way of organizing human activities, as well as a new way of imagining culture and its place in the cosmos. And, just as important, each one required a daily engagement with time unlike anything that had come before.

According to some researchers the Neolithic saw the completion of a transition that had its beginnings in the Palaeolithic. In the eyes of archaeologists such as Renfrew, the artistic revolutions of forty thousand, thirty thousand or twenty thousand years ago were local and uneven. The astonishing cave art found in Spain and France were not universal phenomena spanning the entire human population. In comparison, after the great global warming, our self-domestication and adoption of agriculture rapidly swept across every continent. It was a revolution that transformed almost everyone, almost everywhere. Thus, the Palaeolithic might be seen as a long series of skirmishes in a cognitive revolution that found completion in the Neolithic.

What made this final step possible? The answer, which continues to shape culture down to the modern era, is our physical embodiment. We live in the world through our bodies and their materiality.[42] What altered the human mind was not simply the introduction of new ideas in our heads but new encounters with the world through what we built with our hands.

Many pivotal inventions were developed in the Neolithic era: planting and harvesting technologies, the construction and deployment of grinding wheels, the mastery of metallurgy. All of these changes represented fundamental shifts in the way people encountered the material world. It was, literally, the act of shaping the raw stuff of the world into these inventions that enabled new ways of thinking and new ways of organizing human activity. This process of material engagement completed the Big Bang of consciousness and is the root cause of all the innovations and revolutions that followed.

Brute facts are where material engagement begins; new ways of being human is where it ends. The early farmers of northern Europe circa 5500 BCE had surpassed the culture of their hunter-gatherer ancestors with new ways of handling material that superseded the natural world. Housing for hunter-gatherers, for example, had "required no more than promoting and combining the existing suppleness of hazel, the stringiness of willow and the sheets of birch bark that grew ready-made".[43] The timber-frame homes the new farmers built required nature "to be torn apart and the world built anew".

Changes in material engagement redefine culture by altering what are called its institutional facts. Institutional facts define the human world into which we are each born. From punch-clock jobs to jury duty appointments, it's the institutional facts that define how culture organizes itself and then imposes that system onto our individual lives. But cultural organization derives its power in the mental realm of symbols. Thus the changes material engagement wrought meant more than just a farmer figuring out a new way to fashion a sharpened axe. These shifts in material engagement implied shared understandings within a community that were at once social and cognitive—in that way they drove the creation of new institutional facts. For the farming villages, the materials that made this new life possible were reflected in the day-to-day organization of the community. A Neolithic farmer, looking back across her tilled fields and the chores that defined her day, was moving through an entirely new world.

With the advance of material engagement came new ways of experiencing time. By kneading clay with the hands, pushing iron ore around in a fire and stretching wool across a wooden frame, people engaged with the material world in fresh ways, and time was an integral part of this process. How long did it take for clay to be baked into pottery? How many differently timed steps were involved in forging an iron plough? Just as each invention made new forms of culture possible, cultural imagination also developed alongside the technology. Because time always exists at the interface between the physical and the imagination, it would be closely tied to material engagement and the changes it drove in culture.

Nowhere is the effect of material engagement on institutional facts more apparent than in Neolithic megaliths such as Stonehenge. Megaliths are massive, highly structured stone works. These imposing stone monuments, as well as massive earthworks, are associated with prehistoric cultures across the globe. The construction of megaliths is one hallmark of the change from a hunter-gatherer culture to an agrarian culture. The eighty-metre-wide circular mound at Newgrange, about thirty-five miles north of Dublin, for example, appeared sometime after the advent of farming in Ireland. The earliest structures at Stonehenge were constructed later but were still part of England's Neolithic agrarian past.

The construction of these monuments required the intense and co-ordinated effort of many individuals to transport materials across hundreds of miles. The megalith at Newgrange, for example, is fronted by a quartz façade composed of stones found on beaches near the coast of Dublin, many miles away.[44] At the centre of the enormous mound is a vaulted central chamber only accessible through a narrow twenty-five-metre-long passageway. The stone supports for this chamber and passageway also had to be dragged across many, many miles. In a similar manner, the imposing central blocks of Stonehenge, each weighing forty-five metric tonnes, had to be transported as much as fifteen miles, possibly from quarries at Marlborough Downs.[45] Thus, the decision to build these megaliths demonstrates entire communities willing to dedicate time and treasure in the effort to make symbols out of stone. And in that effort time and symbol were reborn.

The construction of Stonehenge likely required more than thirty million hours of labour. So much work that the effort must have spanned generations.[46] With this new form of material engagement passing from father to son, mother to daughter, the construction of megaliths was itself an agent for imagining new forms of culture and time. As a direct result of building the megaliths, the monuments became the axis around which a new type of living community would be born.

The megaliths rebuilt culture and time through the very process of their labour and time-consuming construction. And time was always an essential aspect of the megaliths on both a physical and symbolic level.

FIGURE 1.3. The Neolithic megalith at Stonehenge. *(Top)* Each upright stone weighs seventy-two tonnes. Stonehenge's construction required millions of hours of labour and, in the process, created new cultural encounters with time. *(Bottom)* Astronomical orientations of Stonehenge. Each arrow shows an alignment between stone placements and a repeating celestial pattern. The long arrow (A) shows alignment with the northernmost extension of the sun's position on the horizon.

On the one hand, men would be called away from the farms to scramble over tilting stone slabs or bend low in sodden holes dug into the earth. On the other hand, concerns with time in a mythic, cosmological context was undoubtedly a core motivation for building many of these megaliths in the first place.

There is ample evidence that Newgrange and Stonehenge were built to correspond to astronomical events such as the winter and summer solstices. Without a torch, the central chamber of Newgrange is darker than death. But for a few days each year near the winter solstice, the rising sun is perfectly aligned with the ancient passageway, allowing a shaft of sunlight to pierce the darkness. For these few minutes, the central chamber glows in warm ochre—a promise of the light and life to come with the approaching spring. Newgrange's builders constructed their monument with its solstice-illuminated chamber for cosmic rituals whose meanings may now be lost to us but whose astronomical orientation cannot be missed. Thus this solar alignment of Newgrange's entrance is one potent example of a concern with the raw facts of the sky and its movements.

The megaliths were, however, more than prehistoric calculators. In addition to its many astronomical orientations, Stonehenge may also have been a burial site for chiefs and tribal elders.[47] In this way ritual and religion must have been central to the symbolic demands of megalith construction. Their builders were clearly aware of cosmic time and willing to dedicate effort and riches in creating monuments that called to the vaulted sky and its repeating patterns.

TIME AND AGAIN: ETERNAL RETURN AND THE NEOLITHIC COSMOLOGY

The cosmology of a farmer did not look like that of a hunter-gatherer. The material engagement of hunter-gatherers was so fundamentally different from what people in an agrarian culture experienced that the very concept of a cosmos, and the symbols used to represent it, had to change in a fundamental way. The mobile hunter-gatherer lived out in the open amid the forest and plain, populated by free-roaming animals considered equals and cousins. The farmer accepted a tamed, sedentary life with the closed stability of a homestead roof replacing the infinite, dynamic arch of the night sky; for the farmer, beasts of burden were possessions and worked as slaves to the human master. Where the hunter-gatherer lived through time as an unbroken whole, the farmer lived within a time marked by the daily rounds of animal husbandry, home maintenance and village life. Thus time and the cosmos had to change because the ways in which people encountered the material world had changed.

Looking at the change from Palaeolithic to Neolithic societies, Karen Armstrong notes that myths that cease to be useful will have to be abandoned. Thus, with the development of farming, new cosmologies and new conceptions of cosmic time would have to be invented.

Agriculture was the product of analytic thought, an early kind of science. But unlike our technological revolutions, the agriculture of the Neolithic farmers was never "a purely secular enterprise".[48] The mythic universe of the Neolithic was a mix of new activities and a changed understanding of the wider universe that surrounded them. Farming was as much a sacred act in the Neolithic as hunting had been in the Palaeolithic. Crops, emerging from careful seeding and cultivation, were a new form of symbol and a new representation of time. Cultivated plants grew through material engagement with the Earth and its unfolding powers in time. Thus, crops were both food and divine power. Time acted as an intermediary between the concrete world of farming and the symbolic world of the gods. Unseen energies, manipulated through time, were the focus of new mythic narratives, as the tilled Earth became a fertile womb for the community.

The rites, rituals and myths of the Neolithic peoples responded to the requirements of this new agricultural cosmos. The whole community came to the fields, standing together as farmers ritually discarded the first seeds of the year's sowing as an offering to the divine powers that would animate the new crop's growth. Cosmic time was manifested in the fields for these men and women months later when the first fruits were left to drop, recharging the hidden forces animating the agrarian cycle. Ritual sexual union sometimes preceded the planting in order to signify the sacred marriage of soil, seed and rain.[49] In all these rituals people were tapping into vast cosmic creative powers that, like the seasons, were clearly periodic in time. The close observation of the world that made agriculture possible had its internal complement in the new sacred narratives of mythology.

Awe and wonder before the sky had led humans to cosmic myths of sky gods in the previous Palaeolithic era. In the Neolithic, the Earth became central in the form of a mother goddess. Myths throughout the Fertile Crescent tell stories of goddesses and their connection to farming. The heroic hunter's journey, which is so widespread in the Palaeolithic, gave way to the dangerous travel of the mother goddess descending into the underworld of death and returning to bring new life. These myths expressed both the recurring terrestrial bounty an agricultural people experienced and its devastating alternatives such as drought, famine and flood. Mythology was never a form of escape. It was an honest expression of people's personal knowledge, and it forced them to face up to the reality of life in the midst of death and transition, a life given meaning in new contexts of agricultural time.

Most important, the new myths were sensitive to the cycles of the agrarian year and its implications for conceptions of time. The yearly cycles of life and death appear, for example, in the myth of Demeter and Persephone, which dates back to the Neolithic era. Demeter controls the fertility of the Earth. When Hades, god of the underworld, abducts her daughter, Persephone, in her grief Demeter starves humankind by shutting down the growing season. Zeus is forced to rescue Persephone. But once the God-King learns that Persephone has tasted the fruit of the underworld, the girl cannot be allowed to return fully and must spend several months of each year with Hades. It is Demeter's grieving during these months that is meant to explain the bareness of winter. Time, bound up in the seasons, is essential to this sacred narrative's plot.[50]

The great scholar of religion Mircea Eliade articulated this development in his work *The Myth of Eternal Return*. As social arrangements were built around agriculture, time itself was regenerated with each round of the year. "On the occasion of the division of time into independent units, 'years', we witness not only the effectual cessation of a certain temporal interval and the beginning of another, but also the abolition of the past year and of past time."[51]

The extended festivals of new harvests and first plantings were more than just an enactment of creation; they were, literally, the re-creation of time. Costumed actors would embody the myths as a kind of sacred theatre in which the entire community participated. "Every new year is a resumption of time from the beginning", said Eliade. Thus at every new year, the community repeated, in ritual form, the origins of the cosmos. In many agrarian cultures these rituals of time regeneration required active human participation via the death of the king. The rituals would require the "king" to be felled either symbolically or in actuality. Spilling the king's blood and returning his creative energies to the Earth enabled time and the cosmos to be born again.

The revolution that put the earth to the plough reshaped human consciousness and pulled us out of a time defined by now, by whatever was happening in the present. Cyclical time and cyclical universes, reborn anew each year, met the needs of our new ways of life and held a powerful grip on the human imagination across these long millennia. Thus in the Neolithic we see, for the first time, the switching of one cosmology for another. With the advent of agriculture a different cosmological time emerged, one mediated by our new material engagement with the world. This story of cosmos and culture enigmatically entangled through material engagement will be repeated again and again to the current day. It is still at work now as we face the end of the Big Bang and our own cosmological revolution.

Chapter 2

THE CITY, THE CYCLE
AND THE EPICYCLE

From the Urban Revolution to a Rational Universe

BABYLON · 1560 BCE[1]

He would have to hurry or all would be lost. The afternoon shadows had grown long and he was still unprepared to make observations or record them on the tablets. The sight of a high priest in flowing robes running through crowded city streets was bound to draw attention but he could not worry about that now.

He was young for his advanced position, rising quickly in the order because of his skills sighting the wandering planets and his startling capacity for calculation. Many of the older priests had made their jealousy apparent. The king, however, favoured him. His readings of the planets had given King Ammisaduqa firm guidance in the suppression of the valley revolts. So far his place in the high court was secure.

But now all was at risk. It was the king alone who had tasked him with completing the Venus tables. Records of the fair planet's risings, relative to the sun, were the key. Those records, held in stone, had begun almost twenty-one years ago. But when he had reviewed the sightings of previous high priests he found ambiguities and inconsistencies that threw his astrological calculations into chaos, making it difficult to interpret the long-term path of Venus and what it meant for the king's future. Without accurate observations, he could not predict the king's future or the fate of his plans—such as the proposed marriage of the king's daughter to that oaf of a Sumerian prince,

27

which was risky in spite of the grand alliance it would bring.[2] *But if he obtained a precise sighting of the planet as it set today, he was sure it would be the key to a proper reconstruction. Then he could correct all mistakes in the records and prepare an accurate reading.*

He hurried past stalls and shop fronts. The market was packed with throngs of tradesmen, merchants and farmers from the valley. He struggled to push through the bustling dusty streets. Up ahead, he could see the imposing steps of the ziggurat rising high above the one- and two-storey mud brick buildings of the city. Quickly, quickly now. He must be in position when the white planet emerged above the horizon.

FALLING INTO HISTORICAL TIME: INVENTING THE URBAN EMPIRE

It happened at different times in different places. In the Fertile Crescent the transition to city-based empires began at least five thousand years ago.[3] China made the shift some time later, as did the cultures of the Indus River valley on the Indian subcontinent.[4] Though the timing was different, the results were always the same.

Moving from the small farming settlements that characterized the Neolithic era, human beings began creating great cities and the extensive empires that supported them. This was the urban revolution and it marked a second great reorganization of human culture and human time. As power shifted to densely inhabited cities, the time that ruled human concerns and cosmic architecture would, once again, be re-created and re-imagined.

For the western world—Europe, the Mediterranean and eventually the Americas—all roads in the urban revolution lead back to the Fertile Crescent and the civilizations that would grow in the broad plains between the Tigris and Euphrates rivers. It was in Mesopotamia that the first great cities, such as Uruk, would emerge.[5] At about the same time, to the west, the vibrant Egyptian kingdoms sprang to life.[6] With these complex urban societies, daily life would become structured in ways impossible to imagine in the smaller, more widely distributed villages of

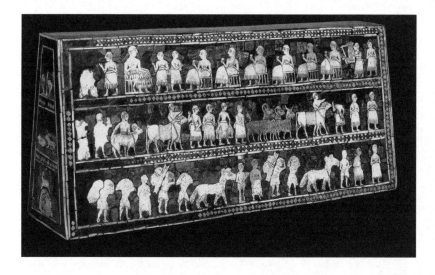

FIGURE 2.1. Life and time in the urban revolution: the Standard of Ur's "Peace Side." While the purpose of the Standard of Ur is unknown, it offers a beautifully stylized depiction of ancient Babylonian life.

the Neolithic. Babylonian cities, for example, maintained populations ranging from ten thousand to fifty thousand inhabitants. To support so many people not directly involved in growing their own food, the pattern of human society would have to be rebuilt. Agriculture became subservient to the demands of the city, and work within the city become subservient to the demands of specialization. Trades for everything from leatherwork to fabric production emerged, each with its own specialized training, economic needs and political interests.

These changes in consciousness created by the urban revolution moved material engagement to its next level. People began to think in more sophisticated symbols. The growth of cultural complexity (and the cosmological complexity it would allow) depended on the ability of people to store information externally as well as internally in individual memory. Thus began the development of written language and written records. In the evolution of human culture and its approach to time, writing was the most critical result of the urban revolution.

The history of written language is closely woven into the history of time and economic necessity. Around 4000 BCE, in the ancient cities of Sumer, people began to use tokens in economic transactions.[7] These tokens were small, flattened clay forms (around the size of coins) and at first were used to represent something like a day's work or a basket of wheat.[8] Sumerians soon realized, however, that the tokens could represent any kind of time or material. Archaeologists have found hoards of the tokens in Sumerian grain houses. As Sumerian civilization became more complex and sophisticated, the number and type of tokens grew.

Keeping track of these tokens—all of different forms—became a burden for Sumerian merchants, and they were eventually bundled and stored in clay wrappers. Markings pressed into the wrappers before they were fired and hardened allowed the Sumerians to keep track of the number and type of tokens enclosed. As generations progressed, however, the system evolved, and the imprinted markings came to replace the tokens as symbols of time and quantity. Why use cumbersome tokens to remember how many bushels of wheat or days of labour you were owed when you could just look at the markings on the clay wrapper? In this way, the wrappers became the prototype for the clay tablets on which cuneiform—the first true writing system—would emerge.[9]

From these origins it appears that writing and the use of time for economic gain were intimately connected in the new city-states of Mesopotamia. In Egypt (and thousands of years later in Mesoamerica), a similar story would play out as the origins of writing were linked directly to time via the political necessities of recording dynastic lineages or historical events.[10]

While the invention of true systems of writing was a gradual process, their emergence represented the increasingly sophisticated relationship between the realm of the material and the realm of the symbolic. This rising complexity would have direct consequences for the forms of both human and cosmic time. Once again material engagement with the world would serve as the source of cultural and cognitive innovation. One critical domain illustrates this innovation during the urban revolution: the creation of measurement standards.

Developing a system of measurement requires a highly focused form of material engagement. Physical objects must be abstracted. They must rise above their individual identities and serve as an objective standard of their own properties. If that sounds slightly confusing, it should. To turn a particular stone into a universal unit of measure requires an abstraction that is remarkably subtle. So nuanced is the idea of a "unit of measure" that its development alone signified a radical shift in human thinking.

A particularly vivid example of this higher form of material engagement comes from archaeological studies of Indus Valley Harappan cultures.[11] From the Harappan city of Mohenjo-Daro in the Sindh region of Pakistan, scientists have found polished stone cubes that vary in size in a uniform progression. The cubes are thought to be an early system of weight measurement, and their importance cannot be overstated.[12] Each cube symbolizes an inherent property that had not been quantified before in and of itself. By developing the stone cubes as a measure of weight, weight itself became "isolated for study and measured for the first time".[13]

The cognitive transformation that turns the weight we experience into a weight we can conceptualize occurs through the need to engage with weight in a new way. The same process must have occurred with time. Just as weight was re-imagined in the development of stone cube standards for economic needs, time would be re-imagined during this same period for much the same reason. Time emerged as a separate property of the world during the urban revolution, forever altering the topography of culture and cosmos.

Physical embodiment is the pivot point on which this kind of cognitive transition turns. The symbolic thinking required to develop a system of weights begins with the actual, embodied handling of heavy objects. The next step goes beyond the individual to the shared needs of the emerging culture. The labelled stone cubes that serve as symbols of weight were the result of more than a process of "playing with words". Its vital force was the ability of the human imagination to isolate specific properties in material. While standardized weights are symbolic of themselves—a weight as a symbol of weight—this symbolism would be meaningless without the cognitive and cultural changes on which the urban

FIGURE 2.2. Set of Harappan weights with scale. These polished stones mark an early step in human attempts to abstract material engagement by standardizing the concept of "weight" to make commercial transactions easier.

revolution depended. It was a new kind of conceptual thinking that grew out of real world concerns when, for example, merchants needed some standard against which the cost of goods could be agreed upon.

What worked with weight would prove equally effective for time. The urban revolution enriched our interaction with the world by transforming how time was encountered and used. The tokens of ancient Sumer were used to represent time in the same way that stone cubes had come to represent weight. While time is not physical in the immediate sense, as is a polished stone, it is nonetheless part of our encounter with the physical world. In our early evolutionary development, time was folded into the hardwired physics modules that Steven Mithen described in his account of the Palaeolithic's cognitive Big Bang. By the era of the urban revolution, however, culture expanded far beyond evolution. Genetic hardwiring was augmented by the symbolic thought city dwellers had inherited.

It is during the urban revolution that the first written calendars and explicit divisions of the day emerged. The Sumerians, for example, used a lunisolar calendar, which tracked dates based on lunar phase and time of the solar year. Abundant cuneiform tablets show that explicit and accurate calendars were required to meet Mesopotamia's agricultural, economic and political needs. Calendars were needed to time the tax collections that supported the sprawling civil engineering projects and hungry armies.[14] They were also needed to set the timing of religious festivals with their explicit political function of supporting the king and his state.[15] For all these reasons and more, the new urban cultures needed accurate calendars that matched the way they had begun to use time. To meet that demand, the properties of time would be abstracted and the human encounter with time re-imagined.

The material engagement with time meant dealing with its raw material—the patterns of season and sky. The Mesopotamians, for example, were aware of the basic astronomical fact that twelve months (measured by twelve cycles of the moon through the sky) and one year (determined by one cycle of the sun's motion through the sky) are not equal.[16] This discrepancy demanded accurate calendars lest midsummer months drift through the seasons until they appeared in midwinter.

To keep time in joint, a process of periodic intercalations was invented—inserting a thirteenth month into the calendars to set lunar and solar cycle in step.[17] As we shall see, this balance of months and years through intercalation would haunt calendar makers and calendar reformers for millennia to come. Intercalation demonstrates the origin of humanity's material engagement with time as a shared civic resource. As each culture evolved into greater complexity, its institutional time had to be hammered into a shape that could fit its needs. In the process urban dwellers created new experiences of time via tax collection schedules, religious festivals and periodic demonstrations of civic power. On a more intimate level, the shape of the lived day for the blacksmith or the potter flowed from new forms of material engagement that were possible only through new forms of social organization. Lived time as experienced from day to day also changed in ways as simple as the regulations for the opening and closing of a market, and as complex as the increased demands on the productivity of tradesmen as city-states grew rich from trade.

It is important to understand how these cultural innovations— created from new forms of material engagement with time—were flowing in two directions at once. On one hand, they moved downwards, shaping time in everyday life. But they also advanced upwards, towards new conceptions of a time developed in new cosmologies through myth. This process is fundamental to human cultural evolution; in the enigmatic entanglement between material engagement, culture and cosmos remains an open circle. Cultural change allows for the development of new kinds of technology (material engagement), which then allows for new forms of individual experience, which in turn allows new forms of cultural change.

A CREATED UNIVERSE: COSMOLOGY AND TIME IN THE URBAN REVOLUTION

Cosmological mythologies always change with experience. As the urban revolution progressed, humanity grew more self-aware. We were no longer part of nature but had stepped out of its frame. In response, the city builders invented a universe that, like their cities, came into being from chaos. Their cosmos were ordered, created places formed in the often contentious interactions of the gods.

There are at least three distinct creation myths from ancient Egypt, and each begins with dark, infinite, primeval waters. In these stories, the Egyptians imagined a cosmos that mirrored their own city-building experience, in which order was hewn from the cosmic wilderness.[18]

It should be no surprise that highly political city dwellers imagined creation stories of personified, deeply political forces shaping the universe. In Egypt it was the god Nun who embodied the primal, chaotic waters.[19] From Nun emerged Atum, the true creator god, who would then spawn Shu of the air and Tefenet of the rain and moisture. Then came Geb, the Earth, and Nut, the sky. An A-list and B-list of gods then appeared, each one the personification of a different animate power in the cosmos. As historian of science Helge Kragh said, "These [myths] depict the universe as a dynamic entity, something which was created and is full of life, change and activity."[20] Most important, this activity

FIGURE 2.3. In this depiction of the Egyptian creation myth, Shu, the god of air, separates Earth and heaven. Ships passing over the firmament symbolize the day's passage.

was interpersonal (or between gods and humans). Cosmological change mirrored human change.

The cosmological myths of the Mesopotamians also reflect the intense, conflict-prone negotiations needed to maintain a complex urban empire. The Mesopotamian universe was ruled by three gods. The earth and waters were the domain of Ea, the god Enlil mastered the heavens and Anu ruled the air in between. Below these three triumphant high gods dwelt an army of quarrelsome lower deities, each responsible for his or her own domain of the universe. As in the Egyptian cosmos, the Mesopotamian gods also emerged from a primeval chaos. Humans, when they arrived, found themselves inhabiting a universe in which formlessness had already been beaten into form. Their universe was also a place of political tension as the gods' endless squabbling shaped the day-to-day world humans experienced.

For the Mesopotamians even the creation of humanity flowed from political infighting. The Babylonian myth of human origins recounts the great god Enlil's enslavement of the lower deities. For eons, the story goes, Enlil had press-ganged the lower gods into service for his incessant canal building and irrigation projects. When the lower gods finally rebelled, Enlil was forced to free them from their bondage. In need of a new workforce, he slew a single rebel deity. Mixing the ill-fated god's blood with clay, Enlil created the human race, who then took on the burden of further shaping the world through canal building.[21]

Thus the cosmos of the Mesopotamians resembles the cities they created. No longer were humans enmeshed in the natural world, as they had been in the Palaeolithic. The era when humans and animals could speak to each other was long gone, left over only in tales of a distant Eden. The purely agrarian world of the Neolithic had also faded. Though many myths of this era continued to imagine a cyclic time and each year still required renewal through rituals, a critical change had reshaped those rituals. In the urban revolution the rites ensuring time's continuity were now performed in the service of a king and his state. It was the institution with its presiding god-king that now required renewal and continuity. The state had become joined with the cosmos.

A critical aspect of this shift was the development of a priestly class.

As religion became the province of the state, a trained cadre of priests emerged. It was their job to mediate between the temporal realm of humanity and the divine eternal realm of gods. In the great cities of Babylon these priests would become the first true astronomers, maintaining the first truly long-term records of celestial events. Such records were necessary because, for example, tracking Saturn's orbit requires at least thirty years of continuous observation. Cosmic time and human time were explicitly joined as astronomer-priests began keeping track of these long-term celestial motions across many human lifetimes. Though the observations were carried out in the service of what we now call astrology—predicting the future of the king and his state—they marked a turning point in our attitude towards the sky, and more important, our attitude towards the cosmos.[22] The dedicated priestly class served a cultural purpose that was deliberately cosmic. In their work, material engagement with time would grow to new levels as the roots of astronomy as a science took hold.

MECHANISMS IN MATTER AND MIND: GREECE AWAKENED

The Antikythera mechanism, as it would come to be called, waited two thousand years at the bottom of the Mediterranean before its genius was revealed. Discovered in 1900, it was part of a first-century BCE shipwreck off the coast of a small Greek island named Antikythera. The mechanism—a mass of seawater-fused gears, levers and pins—was promptly packed off to a museum in Athens, where the discovery of its true astronomical purpose would have to wait for another century. Then, in 2007, a team of researchers brought sophisticated X-ray tomography machines to the Athens museum to create a detailed 3-D map of the device's inner workings.

Within their digital re-creation of the Antikythera mechanism, scientists found Greek text engraved on the gears that read like a how-to manual from antiquity and made clear the mechanism's purpose. It was an astronomical clock of extraordinary precision designed to predict

FIGURE 2.4. The Antikythera mechanism, a Greek astronomical calculator lost in a shipwreck circa 100 BCE. The mechanism's gears have been fused by millennia under water. Reconstructed replicas show the mechanism's precision in tracking celestial events such as eclipses.

the motion of the planets, the phases of the moon and even the timing of solar eclipses. It was a device of such exacting manufacture and design that it put even modern clockmakers to shame. As one researcher stated, "The design is beautiful, the astronomy is exactly right. The way the mechanics are designed just makes your jaw drop."[23] It was as if all the genius of the ancient Greeks had been poured into a single object.

There are simply not enough words in all the books written to fully embrace what happened in Greece from 700 to 100 BCE. The Palaeolithic, Neolithic and urban revolutions of the previous fifty millennia were distributed affairs; they occurred in separate places on the globe and along distinct trajectories for different cultures. The revolution in mind and culture initiated by the Hellenistic Greeks was a unique, localized flowering of human genius. No doubt the history of human culture has seen other explosive moments of creativity, like the Tang Dynasty in China in the eighth and ninth centuries, or the Renaissance of Western Europe in the sixteenth century.[24] The Greek experience is still singular. The Greeks are our pivot point in the interlocking narratives of human and cosmic time. Across the loose association of city-states we call Hellenistic Greece, an entirely new conception of nature, order and time would be invented. The Greek genius for seeing pattern and order in the world, for seeing mechanisms in nature and fashioning mechanisms as the material basis for culture, was unprecedented. The vision they established would become a foundation supporting intellectual inquiry for the next twenty centuries.

THE PERSONAL COSMOS: INTIMATE TIME IN HELLENISTIC GREECE

The Antikythera mechanism, with its finely machined gears and dials, demonstrates how the Greeks imagined a well-ordered world in both matter and mind. The many histories, commentaries and theatre that survive from classical Greece also give testimony to a culture with a sophisticated sense of ordering time for its own uses. Nowhere is this order more apparent than in the seventh-century BCE farmer-poet Hesiod's *Works and Days*.[25]

Hesiod's *Works and Days* is an equal mix of poem and farmer's instruction manual. According to Anthony Aveni, *Works and Days* reads like an "archaic self-help text in which the poet tells how to lead an orderly, structured life" in the mountainous farming environment of the Peloponnese.[26] Time appears as subtext everywhere in the poem as Hesiod guides the listener (the poem was meant to be sung) through the difficulties of the agrarian year.

Hesiod begins the poem with the Greek version of creation mythology. The narrative tells of a golden age when the world was in harmony under the god Kronos—the embodiment of time itself. Echoing the earlier myths of the Palaeolithic, this golden age is lost through a steady decline into the silver and brass ages. In each successive era, humans fall further from their noble origins, becoming more decadent and more disrespectful to one another and the gods. Hesiod's own era is named for the basest of metals—the iron age. Only through hard but honest work, the poet tells us, might the Greeks of the iron age redeem themselves and forestall the complete dissolution of the world.[27]

After setting the cosmological stage, Hesiod advises the reader on how to order their own lives, working from a kind of "morally based farmer's clock".[28] Nature, embodied in celestial cycles, provides guideposts for ordering the work of the year:

> *At the time when the Pleiades, the daughters*
> > *of Atlas are rising,*
> *Begin your harvest, and plough again when*
> > *they are setting.*
> *The Pleiades are hidden forty nights and*
> > *forty days,*
> *And then, as the turn of the year reaches that*
> > *point*
> *They show again, at the time to first sharpen*
> > *your iron. (383–387)*[29]

Here, Hesiod is using the stars as a calendar. The Pleiades are a clearly visible cluster of seven stars that appear in the constellation Taurus. As Taurus is one of the constellations of the zodiac, it will only be visible in the night sky for a fraction of the year. Thus, the first appearance of the Pleiades in Taurus in the evening sky makes an excellent and explicit time marker for the farmer. In using the Pleiades this way, Hesiod gives us an explicit example of material engagement with time. The seasonal appearance of stars is just as surely a material to be engaged with in the ordered world of Greek agrarian life as is the hard iron blade that needs sharpening.

Throughout "Works", the first book of *Works and Days*, Hesiod provides numerous time-based formulas for sowing, reaping, pressing grapes to wine, taking to sea, gelding horses, even having sex. For bringing in the harvest, Hesiod uses the familiar constellation of Orion:

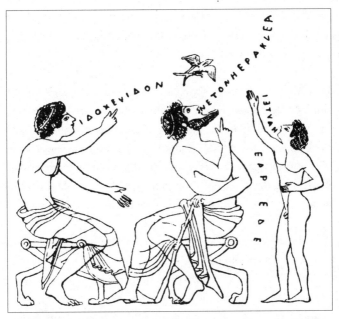

FIGURE 2.5. Early Greek representation of natural signs used as time markers. The text reads "Look at the swallow," "So it is by Herakles," "There it goes," "Spring is here!"

> *Rouse up your slaves to winnow the sacred*
>   *yield of Demeter*
> *When powerful Orion first shows himself.*
>   *(597–598)*

Speaking of sex, but by no means love, Hesiod once again uses the sky to set rhythms. It is the appearance of the star Sirius (Seirios) in the summertime sky that sets both women's desires and men's failings:

> *Then is when the goats are fattest and when*
>   *the wine tastes best*
> *when women are at their most lascivious but*
>   *men's strength fails them*
> *most, for the star Seirios shrivels them, knees*
>   *and heads alike. (585–587)*

As Anthony Aveni has shown in his book *Empires of Time*, the structure of Hesiod's "Works" gives us a powerful insight into the Greek experience of time. As Aveni puts it:

> Hesiod's word imagery shows us that the early Greeks did not think of time as some abstract phenomenon rated on a clock, the way we think of it today. For them, time was the ordered cycle of sensible natural events to which human beings were meant to relate the events in everyday life from tilling the soil to worshiping the gods. For Hesiod the true essence of time lay in a dialogue continually going on between nature and culture.[30]

Thus throughout the poem Hesiod is explicit in setting the order of life in the context of both celestial and human time. The ability to relate the rhythms of the sky to the rhythms of experienced life in a reasoned pattern mark the initial conditions of the Greek cultural trajectory. By relating to the patterns of nature in and of themselves, without recourse to divine explanations, the Greeks soon would create an entirely new framework for cosmos building.

While Hesiod watched the natural world of stars, weather and animals with a keen and discerning eye, it would be a mistake to see him as either taxonomist or a scientist.[31] In his imagery, Hesiod presents the daily life of a farmer managing his land, mindful of the way time manifested itself through nature. Across the centuries that would follow, in the great cities of Greece—Athens, Corinth, Rhodes—a wider imagining of time would emerge. This new cosmic vision would lift Hesiod's order of natural and celestial signs into new and more abstract realms. While the agrarian Hesiod was unconcerned with "*compartmentalizing* nature's happenings into astronomical, ornithological, horticultural or meteorological categories", the emerging philosophical schools of the Greek cities would do just that.[32] They would categorize with a vengeance—imagining an entirely new vision of the cosmos and its place in time.

A GEOMETRIC UNIVERSE:
THE ORDERED *KOSMOS* OF
GREEK COSMOLOGY

For six centuries, Greek cosmological thinking established an ever more radical vision of the world. The flourishing order the Greeks injected into culture, arts, politics and daily life would be reflected in a confidence that the cosmos (a concept embodied in their word *kosmos*) was also rationally ordered.[33] The city-empires of the urban revolution had created universes of myth where warring gods ruled the skies *and* the affairs of men. The new breed of Greek thinkers vaulted past this ego-centred worldview and invented a universe of mind, in which the mysteries of matter, space and time itself would yield to humanity's ability to reason.[34]

This idea took root in Ionia. The Ionians were one of three populations that made up the classical Greek world. Living on islands and coastal communities near modern-day Turkey, the Ionians were joined by a common dialect and a unique turn of mind concerning the natural world.[35] The sixth-century BCE Ionian philosopher Thales began the revolution with a determination to explain the world without recourse to the gods. Using geometry learned from Egypt and astronomy learned from Babylonia, Thales revealed a new frontier. In 585 BCE he predicted a solar eclipse using only mathematical models for the motion of the sun and moon. Though modern scholars believe this story may be more myth than reality, it was recounted by later Greek writers to demonstrate Thales' genius. Even if the story is not true, the mythologizing of Thales' mathematical feat by enthusiastic Greek philosophers tells us a great deal about the swift but seismic cultural shift that had occurred in the Hellenistic world.

It was Thales' students who carried the revolution forward, articulating foundational ideas of natural philosophy that, in some cases, would wait two millennia to be further articulated by modern science. Thales' pupil Anaximander proposed the world was built of "intermingled opposites: hot and cold, dry and wet, light and dark".[36] The tension between these opposites produced a dynamic, evolving world. In Anaximander's account, all animals and humans evolved from lesser ocean creatures—a prototype of Charles Darwin's vision.

What made Thales and those who followed him so utterly different than what came before was their assertion that the universe was comprehensible to the human mind. Nature was rationally constructed; there was no need to turn back to the supernatural for explanation, counsel or divination. The Greek philosophers picked up on Thales' approach and added new tools by describing nature with an expanding language of mathematics instead of the language of priestly divination.

Pythagoras, another native Ionian, forged many of the mathematical tools Greek philosophers would use to build their new universe. It was Pythagoras who began formulating theorems with economy and rigour, developing geometry to a level that later thinkers such as Euclid would inherit. Born on the Ionian island of Samos sometime in the early 500s BCE, Pythagoras is reputed to have travelled widely across the civilizations of the Mediterranean, metabolizing what he learned into a new mode of thought.[37]

The philosophical school Pythagoras established around the mid-400s BCE exerted enormous influence over the rest of classical history. No strangers to deeply felt mysticism, the Pythagoreans were a secret fraternity. Unlike the cult religions that had preceded them, however, the basis of Pythagorean religious rapture was the contemplation and exploration of mathematical beauty. "All is number" was their creed. For the Pythagoreans, reality *was* mathematics. The concept of "Kosmos"— a universe that could be studied and contemplated mathematically—is believed to be a Pythagorean invention. Those who followed Pythagoras' tradition found enlightenment in mathematical explorations of this cosmos. For Pythagoreans, both ancient and modern, the universe is "suffused with arithmetic divinity".[38]

Central to the Pythagorean vision were the five geometric constructs that would later be called the Platonic solids. The Platonic solids constitute five highly symmetrical three-dimensional forms constructed using the rules of geometry. The simplest is the perfectly symmetrical sphere. Next is the four-sided pyramid called the tetrahedron. Above it on the ladder of complexity is the cube with its six sides, followed by the octahedron (eight sides), the dodecahedron (twelve sides) and the ico-sahedrons (twenty faces).

Each Platonic solid embodies a wealth of elegant geometrical re-lationships. For example, with a proper choice of sizing, all five solids can be nested into each other with the vertices (tips) of each inner fig-ure gently touching the interior surface of the next, outer figure. Better yet, each solid can be nested into that most perfect of all forms—the sphere—with the vertexes just glancing the sphere's inner surface. For the Pythagoreans, the Platonic solids were the embodiment of beauty and harmony. This was an essential point, for in their view of the cos-mos, beauty and truth were one and the same. The elegant symmetries of Platonic solids led these Greek mathematician-philosophers to asso-ciate each figure with an elemental property of the physical world. The cube was the element earth, fire was bound into the tetrahedron, air was embodied within the octahedron and water's essence lay in the icosahe-drons; the dodecahedron was associated with the whole of the cosmos, which Pythagoreans believed was bound into the perfect sphere.[39]

The association of elemental mathematical form with elementary physical essence was a profoundly new development for cosmological thinking. With it came an association of underlying mathematical reali-ties with the forms and behaviour of the observable world. The logic this association set in motion still drives scientific attempts to under-stand the world, as seen in the emphasis on finding ever more abstract mathematical models to describe subatomic particles or space-time it-self. With the cosmology of today now standing on the edge of its own transformation, beauty and elegance in competing mathematical mod-els plays an important, and perhaps exaggerated, role in capturing sci-entists' attention. Such is the legacy of Pythagoras and his followers.

"Let no one enter here who is ignorant of geometry." This admo-nition crowned the gates of the Academy, Plato's famous school of phi-losophy. Plato, an Athenian who lived during the city's golden age of the fourth and third centuries BCE, continued the Pythagorean enchant-ment with numbers and mathematics. While he never developed detailed models of the cosmos like other Greek thinkers, Plato contributed an all-encompassing idea that would be critical in shaping Western science.

Building on theories of the Pythagoreans, Plato argued that behind the appearances of reality lay a purely perfect, fully mathematical world. This flawless realm of mathematical forms acted as a blueprint for all that we see. For Plato, the time-bound world we experience so vividly is a corrupted version of the ideal and timeless world of mathematical forms. This concept was sometimes called the Doctrine of Forms (or Doctrine of Ideals), and it would lodge in the minds of philosophers like a burr, beginning a millennia-spanning search for the world's pure, underlying architecture. After Plato, mathematics was seen as a skeleton on which the flesh of the world was to be hung. Even today a strong, if tacit, background of Platonic idealism characterizes the efforts of modern physicists and cosmologists.

One of the most enduring consequences of Plato's Doctrine of Ideals was his exhortation of Greek astronomers to "save the appearances", or find mathematical models of the heavens that fit the Greek ideal of order and beauty. Rather than seek the nature of the cosmos through observation alone, these models must begin with the assumption that nature was ordered along mathematical, or geometrical, form. Using geometry, the models must account for, and predict, all celestial motions—including the looping retrograde motions of planets against the stars.

The Greek geometric image of perfection was the circle. Thus the true motion of a planet should be a constant, stately march along a circular orbit. The platonic demand that any true model of the solar system would invoke all planets on circular orbits, moving with constant velocities. The apparent planetary disorder seen on the sky—the speeding up, slowing down and puzzling loop-the-loops—must be shown as an Earth-bound illusion. It would be Plato's greatest student who fully answered his challenge and built the foundations for an Earth-centred, or geocentric, universe that would last more than a millennia.

This student was Aristotle. After time spent tutoring Alexander the Great, Aristotle returned to Athens to found his own school, the Lyceum. There he developed a philosophy that differed strongly from his teacher's, and those differences would lay their own brand on Western civilization. Aristotle considered all aspects of the natural world—

biology, physics, astronomy—but he was not a scientist in the modern sense of the word. He did not validate his theories through rigorous experimentation but instead combined a set of core beliefs he held to be self-evident with select observations about the world's behaviour.[40] In this way Aristotle deduced an elaborate account for the cosmos built on an unequal balance of reason and evidence.

Ruling minds for more than fifty generations, Aristotle's cosmos was a divided kingdom. The spherical Earth was the centre of creation, but it was a polluted realm—according to Aristotle, the sublunar domain was home to decay and imperfection. Aristotle's vision for physics (as important to history as his astronomy) depended on this division of sub- and superlunar cosmic domains. Five basic elements existed in Aristotle's cosmos: earth, water, fire, air and aether. Each element had a "natural" motion associated with it. Earth and water "naturally" sought movement towards the Earth's centre. Air and fire naturally rose towards the celestial domain. The aether was a divine substance constituting the heavenly spheres. These "natural" inclinations seemed self-evident to Aristotle and did not require separate tests. Only many centuries later would a new breed of scientists such as Galileo (in the late sixteenth and early seventeenth centuries) demand that a hypothesis such as natural motion be validated through experiments.

To reclaim order from the chaos of apparent celestial motions (retrograde loops most of all), Aristotle built his cosmos on a model proposed by Eudoxus, another of Plato's students.[41] It was a fully geometrical, geocentric vision of the cosmos with the Earth surrounded by a concentric set of rotating spherical shells. The sun and each of the planets revolved around the Earth attached to a different crystalline shell, each of which guided its heavenly body through mathematically perfect circular motions at a constant speed. By tuning each shell's rotation to observations, Eudoxus could use his "universe mechanism" to recover many of the appearances Plato had demanded be saved.

Aristotle's model of the solar system captured the Greek imagination with its mix of geocentric egotism and geometric harmony. It was, however, ill-suited for detailed comparisons with the ever more accurate observations of astronomers such as Hipparchus of Rhodes. The final crowning achievement in the creation of the Greek cosmos would come not from Greece but from the great library of Alexandria in Egypt—the intellectual glory of the classical world and its version of a fully funded research institute. It was in Alexandria that astronomer Claudius Ptolemy solved Plato's challenge once and for all.

Ptolemy created a truly accurate geometric, geocentric model of celestial motion that could stand up to Hipparchus' data. The Ptolemaic universe was built using only uniform circular motions (though Ptolemy was forced to stretch the meaning of "uniform" somewhat). Any astronomer of average competency could use Ptolemy's work to predict the motions of the sun, moon and planets with an accuracy that matched the best naked-eye observations (telescopes were still fifteen hundred years in the future).

So powerful was Ptolemy's work that his book *Syntaxis Mathematica*, written in 140 CE, became the standard astronomy textbook for more than a millennium. The Arabic astronomers, who carried science forward while western Europe huddled in the Dark Ages, called Ptolemy's book *Almagest*, or simply "The Greatest". For the next 1,300 years, the study of astronomy was the study of Ptolemy.

Claudius Ptolemy achieved his accuracy at a price, however, and that price was simplicity. The appearances could be saved in his geocentric model only by adding an impressive array of geometric bells and whistles. Capturing retrograde motion, for example, demanded that each planet not move directly on its circular orbit about the Earth. Instead, planets tracked along on a smaller circle called an epicycle, and it was the centre of the epicycle that moved with uniform speed around the Earth. When the planet's motion on the epicycle was in the same direction as the epicycles' orbital motion, it would appear to an Earthbound observer to move fairly steadily across the sky. Recall that this motion is always against the fixed stars. As the planet steadily marched against the stars towards the eastern horizon, its direction on the epicycle would match the epicycle's direction along the orbit. But things changed when the planet looped around to the other side of its epicycle; its motion now moved in opposition to the epicycle as a whole. Thus the planet would appear to change direction in the sky and move backwards towards the western horizon. It was the combination of motions—the planet on its epicycle and the epicycle orbiting the Earth—that allowed Ptolemy's model to get Plato's homework problem right.[42]

Only in hindsight does Ptolemy's vision of the solar system look like a convoluted Rube Goldberg contraption. For astronomers of the classical world (and for the next fifteen centuries), the sway of Greek philosophical preferences for circles and circular motion made Ptolemy's work seem a triumph of reason. Using geometry alone, Ptolemy had mapped out the perfect mechanics of the perfect heavens.

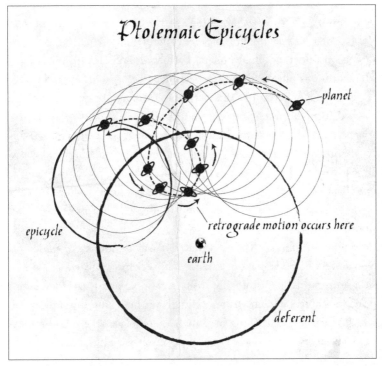

FIGURE 2.6. Retrograde motion and epicycles in Ptolemaic astronomy. Ptolemy explained the periodic loops in planetary motion by imagining each planet riding an "epicycle" whose centre orbited the Earth on its deferent. For an observer on the Earth the combination of circular motion on the epicycle and circular motion on the deferent created the appearance of retrograde motion on the sky.

TIME, CHANGE AND THE FIVE COSMOLOGICAL QUESTIONS

While the later Greek thinkers, with their emphasis on maths and geometry, created sophisticated models of the cosmos—which for them meant the solar system—there were deeper questions that needed to be addressed. Lurking behind the philosophical and astronomical achievements of Plato, Aristotle and their descendants was an essential question touching the very nature of time itself: is change, and therefore time, real, or is it an illusion?

The polarities arising in response to this over-arching question still echo down to our day as cosmology searches for its next step beyond the Big Bang. The conflict is cleanly embodied in the teachings of two famous pre-Socratic philosophers: Parmenides of Elea (in Italy)[43] and Heraclitus of Ephesus (now the Turkish coast).[44]

Parmenides looked at the world of change and saw it as nothing more than an illusion of the senses. At the deepest levels of reality there was no change, no transformation and hence no time. He reached this conclusion not by considering what is but by contemplating what is not. In *The Way of Truth*, a section of his sole surviving work, Parmenides considers the relationship between existence and nonexistence. He concluded: "It is necessary to speak and to think what is; for being is, but nothing is not."

With the phrase "Being is, but nothing is not" Parmenides rejected the very idea of nonexistence and was forced to conclude that things cannot come into being from nothing; neither can they disappear. "Nothing can come from nothing" was his claim. But movement, which must involve moving into a void—the place where nothing existed a moment before—cannot be possible either. Thus, behind the appearance of change and the appearance of movement must be a single unified timeless reality. Change and time are illusions.

Heraclitus drew the opposite conclusion. "All is flux" was his doctrine, and the famous dictum "One cannot step in the same river twice" is attributed to his works. In Heraclitan philosophy the world was composed of ceaseless movement and transformation. Every object was considered new from moment to moment. The basis of this endless flux was a constant conflict between different elements of the world. Fire was primary among the four elements and its transformations were the basis for the apprehended world: "All things are an equal exchange for fire, and fire for all things, as goods are for gold and gold for goods."

Thus the dichotomy was set early. Either an eternal, timeless world existed behind appearance, or time and change were the only essential reality. Plato, with his Doctrine of Ideals, was naturally an admirer of Parmenides. Heraclitus had his devotees as well; the Stoics, a school of philosophy that subsequently gained influence in both Greece and the Roman Empire, would claim Heraclitus' rule of change as their own. In their cosmology, the world emerged through *ekpyrosis* (literally, "out of fire") and would return to fire again.[45]

One far-reaching attempt to reconcile the different views of reality and change emerged in the work of the atomists.[46] In a display of remarkable prescience, Greek philosophers such as Democritus argued in 400 BCE that everything visible was caused by the movement of tiny, indivisible specks of matter called atoms. The eternal nature of these atoms satisfied Parmenides' stand against change. But while Democritus accepted most of Parmenides' conclusions, he and the other atomists rejected the idea that change is an illusion. Atoms did move, and they moved through an infinite void.[47]

The themes of Heraclitus and Parmenides would reverberate across millennia to our own time, as would the ideas of the atomists. One view would gain favour over the other only to be discarded again later. Thus, echoing the debate between Heraclitus and Parmenides, cosmic time in its human context shows both constancy and change. There is constancy in the questions humans ask about the universe and there is change in the answers that dominate one culture or another as the centuries pass. Understanding this balance is critical. Just as important is seeing how the questions and answers cosmology offers predate even the Greeks.

Humanity's first thematic encounter with time and the cosmos began with the mythological imagination of the urban revolution. And, in a sense, it ended there as well. Looking broadly across creation myths from the Egyptians to the Hittites to the early Chinese civilizations and on to the cultures of Mesoamerica (Aztec, Inca and so on), we find almost all potential explanations for time and cosmos clearly mapped out. Thus, there are only so many ways to think about the origin of time.

This is a point cosmologist Marcelo Gleiser eloquently explored in his book *The Dancing Universe*. Myth forms a storehouse of possible solutions to questions of time and the universe's origins. As Gleiser wrote, "In their variety these myths encompass all the logical answers we can give to the question of the origin of the Universe, including those found in modern theories of cosmology."[48]

Humanity's first rational encounter with time and the cosmos began with the Greeks, producing results that have stayed with us until the present day. The dichotomies of Parmenides and Heraclitus, the prescience of the atomists, the arguments about infinite space and time, the nature of the vacuum and the mathematical models of the solar system as cosmos—these all set the stage for what would follow. While it would be foolish to argue that no novelty emerged in the next two thousand years of cosmological thinking, it would be equally detrimental to ignore the way the Greeks established the palette of colours that later thinkers would use to paint their universes.

Thus we have myth on one side and the Greeks on the other, each defining a set of responses to the cosmos and time that would endure even while they responded to cycles of cultural change. For Gleiser, the impulse behind the earliest creation myths is the same one driving the modern science of cosmology. In one key respect the result of both efforts has been the same. Gleiser distils the range of our responses to the cosmos and time into two distinct forms, two distinct geometries of time: creation myths and no-creation myths.

A universe born with time constitutes one mythic response to origin questions. These creation-based myths were quite common in civilizations spawned by the Western urban revolution. The universe is created and time begins with that creation. This may occur through the action of a creator or emerge from a timeless background of chaos. In either case, time and the cosmos are joined. "Before" creation there is no time, no duration and no ordered sequence of events.

By contrast, no-creation myths imagine a universe without a beginning. Some proposed a cosmos with an infinite past and an infinite future that would need no beginning. There was always a cosmos and there was always time. Another possibility explored in myth is a universe of eternal cycles. Time simply forms a loop in these myths, with cosmic history being played out over and again throughout eternity. There is a beginning but there is also a before.

In the Bible, the origin narrative of Genesis clearly falls into the first category. The creator, Jahweh, wills the universe into being from nonbeing. An undifferentiated whole soon separates into opposite components. It is noteworthy that Genesis differs from other creation myths by its absence of an initial state. The darkness and the waters of Genesis do not exist before God's decision to initiate the act of creation. In order to emphasize God's absolute dominion over creation, the later Church would adopt the doctrine of creation ex nihilo (creation from nothing), though it is not mentioned explicitly in the Bible. This has often been taken to mean that time was created with the universe, since God himself must be timeless.

The Hindu mythologies of the dance of Shiva fall into the second category. In this myth, endless cycles of creation and destruction are manifest through the great god Shiva's cosmogenic choreography. A remarkable aspect of the Hindu mythic imagination was its attempt to grasp the impossibly large numbers associated with the cycles of cosmic history. One cosmic turn of creation and destruction was called a *mahayuga* and was reckoned to last 12,000 gods' years, with each divine year being the equivalent of 360 human years. Thus, each cycle lasted more than four million solar years. A *kalpa* was a single day in the life of the divine godhead Brahma (a deity below Shiva) and it consisted of two thousand of these *mahayugas*. There would be many Brahmas, each destined to die and be reborn. The lifetime of a single Brahma lasted one hundred Brahma years, or more than three hundred trillion human years. In these vast stretches of time we can see an early attempt, written in myth, to embrace the intuitive sense that creationless cosmic histories must stretch across vast horizons of time.

Gleiser's schema allows us to see how the legacy of our mythologies maps out two basic geometries of cosmic time. Creation myths imagine time as a straight line with a beginning point. No-creation myths imagine time either as a straight line extending infinitely in both directions or as a circle. As we will see in the chapters that follow, the history of cosmology describes our varied explorations of these models. And, as we shall see, the dominance of one model over the other from one era to the next is not random. Rather, it is critically dependent on the way each culture engages with the material world and how the material world pushes back.

The Greeks first expressed these possibilities in myth, then abstracted them so that their logical possibilities in a fully natural, fully rational universe might be explored. The basic possibilities for the dimensions of time were fleshed out in Greek debates about change versus timelessness and an eternal cosmos versus a created cosmos. Later, the Greeks saw conceptual dilemmas implied by the existence of a true vacuum and the possibilities of other universes. By the end of their era the bedrock of future cosmological debate had been established.

Thus from the era of myth to the era of the Greeks we can see a finite set of questions emerge that would come to guide nearly all future debate and future cosmological imaginings.

Question 1: Is there a universe or a multiverse?

Is what we see, all the stars and galaxies, all there is to the universe? Are there other collections of matter we can't see? Are there other universes with laws that differ from our own?

Question 2: Is space infinite or bounded?

Does space extend forever? Does it have a boundary? If so, what lies beyond the boundary?

Question 3: Does space exist by itself?

Is a true vacuum possible, or does space exist only relative to the matter that fills it? Would space exist in a universe without matter?

Question 4: Does time exist by itself?

Is time a real property of the universe? Or does it exist only relative to changes in matter? Would time exist in a universe without matter?

Question 5: Does the universe have a beginning and/or an ending in time?

Did the universe come into being? Will it come to an end? If so, what came before and what will come after? If the universe is eternal, how is such a thing possible?

These five questions will appear again and again in the braided narrative of cosmic and human time. In each culture's response—from the age of myth to the digital age—we will see the way lived experience and the seemingly abstract realms of cosmic time are entangled.

Chapter 3

THE CLOCK, THE BELL TOWER
AND THE SPHERES OF GOD

From the Medieval Monastery to the Renaissance Cosmos

MONTE CASSINO, ITALY • 1100 CE

The Frenchman was in trouble again.

The abbot looked at the sleeping form slumped forward in the pew and asked God for patience. Outside it was dark. The monastery was still deep in its night-time stillness. The abbot was not usually awake so late but his ongoing tussle with the Vatican, eighty miles to the north, had robbed him of sleep once again.[1] Every year was the same but he fretted over it nonetheless. With sleep impossible, the abbot had decided to join the late-night prayers. So it was, with considerable dismay, that the abbot had entered the sacristy to find Brother Jacques—the monk assigned timekeeping—asleep again and missing the call to prayer.

The abbot bit his lip. He was not one to indulge in anger. Still, the offices of the Church must be maintained. It was St Benedict himself who had established the Rule, breaking the day into horae *(hours), the regular rounds of prayer that formed the meter of monastic life. Each day, every day, year in and year out, the Benedictine monks at Monte Cassino moved through their ordered days of prayer. It was the heartbeat of their worship. "Truly are they monks, if they live by the work of their hands," the abbot would tell his charges, quoting St Benedict.[2] The division of the day into periods of work and periods of prayer was a mirror of God's own divine order. Some of his*

56

monks listened and applied themselves in earnest exaltation. But some were burdened by the sin of sloth, none more so than his dear Frenchman.

Father Jacques had come to him as a refugee from the Cistercians and had begged to join the order at Monte Cassino. His love for the Lord and for monastic life was so apparent that the abbot could not turn him away. But Father Jacques had failed at almost every job the abbot turned him to. In desperation he assigned the Frenchman the role of sacrist—keeper of the horae, ringing the bells that alerted brothers to the change of time and called them to prayer. The position of sacrist was an important one.[3] There was, of course, no exact measure of the change. He had heard that some monasteries used sandglasses to mark the hour but he found such practices distasteful. God's time could not be marked off like an account book. Instead he taught them to watch for the signs, the rising morning stars in the east and the waking of birdsong. "God's signs from Nature for God's prayers," he called it. That was enough. But, of course, one had to be awake to read those signs.

The abbot drew a long breath, let compassion wash over him and said, "Jacques, Jacques. Dormez-vous? Wake up. Wake up!"

FIGURE 3.1. Monks keeping time in a monastery. The monasteries provided the model for a new form of "time consciousness" as Europe emerged from the Dark Ages. From Heinrich Seuse's *Horologium Saptieniae*, c. 1406.

TURNING OF THE WHEEL:
FROM SKY TO CLOCK

The day we live now, made of hours rigidly fixed like moths pinned in a collector's display case, began at the intersection of man's world and God's cosmos. The first intimation of our modern metered day was born in the insular world of the Dark Ages. Medieval monastics provided the archetype of our time in their ritual of the canonical hours—the divisions of the day into separate rounds of prayer. From sunrise to sunrise, the monks marched through the *horae canonicae*, the rounds of worship beginning from sunrise (matins) through midday (sext) and sunset (compline) and around through the night to matins again.[4] The *horae* were announced by watchful brothers. Each *hora*, with its prayers and duties, had its place within the ordered day of work and worship of God.

Europe's Dark Ages began sometime around the fifth century CE and did not retreat until after 1100 CE.[5] In the long years after the fall of Rome, the baton of learning and inquiry—and a trove of Greek and Latin texts—had passed to the Islamic empires. Ptolemaic astronomy continued its dominance as Muslim scholars charted the skies with increasing accuracy and invention. Then Europe slowly reawakened to its own ingenuity, and time and cosmos were shaped again.

CALENDARS RULE: POLITICS, RELIGION
AND TIME MATERIALIZED

The European Middle Ages were, for the most part, lived on Roman time (as we are still today).[6] Our word *calendar* derives from the Latin *calends*, denoting the Roman designation of the months beginning at new moon.[7] The earliest Roman calendars were based on counting monthly cycles of the moon rather than marking the sun's cycle through the sky. According to archaeoastronomer Anthony Aveni, the first Roman calendars were conveniences for farmers and may have been little more than listings of festivals and "rights days" when business could be conducted. While counting the astronomically imposed monthly cycles that formed the basis of the early Roman time reckoning, a purely economic time unit also emerged in the form of an eight-day "market week". Altogether, the early Roman year was based on a ten-month cycle, and each month contained thirty days.[8]

While the Romans were well aware that a solar cycle contained approximately 365¼ days, their shorter 300-day year was not considered a problem. The extra two months occurred during the seasons when fields were not productive. In a sense it was a time of no time; at least, time did not need to be tallied.

By the last century BCE, the Romans were well on their way to controlling much of the Western world. One of the principal demands of the nascent empire would be a more rationalized time. Our word *rationality* takes its original meaning from the Latin *ratio*, and indeed, calendars came to focus on the ratio of two critical numbers.

Attempts to fit the 29.5306-day lunar cycle into the 365.2422-day solar cycle animate much of the calendar's story.[9] Divide the length of the year by the length of the month (365.2422 ÷ 29.5306) and you end up with the number 12.3683.[10] As Roman society became more complex it demanded continuous time reckoning. It was relatively easy to build a socially agreed-upon year with twelve months. But the remaining 0.3683 month could not be ignored. It left approximately eleven days of each solar year consigned to limbo. If unattended, those eleven days could set the year adrift, sending months sliding through the seasons until the winter months begin to appear in midsummer. To overcome this dilemma, the Romans gave most of their months twenty-nine or thirty-one days, leaving only February with twenty-eight days. But this manipulation was still not enough to bring the months into perfect temporal alignment with the year. A fraction of a day still remained unaccounted for. Wait long enough and even that sliver of a day would push the months through the seasons until summer festivals were celebrated in snow.

By the end of the Roman republic, a special class of calendar "priests", or pontifices, was created. A pontifex's job was to periodically manipulate the calendar in order to keep the months and seasons in line by inserting an extra twenty-seven-day month known as Mercedonius.[11] In a clear example of material engagement yielding new institutional facts, time and politics were swept together as pontifices purposely manipulated the intercalation, delaying or speeding up the insertion of the bonus month to benefit themselves or their friends. Then, in 63 BCE, an ambitious young politician with dreams of military glory was elected pontifex maximus. His name was Julius Caesar. In 46 BCE, after conquering Gaul and winning Rome's bloody civil war, Caesar undertook a dramatic temporal reform. The Julian calendar, with its six 31-day months, five 30-day months, and 28-day February, became the standard.[12] Every fourth solar cycle became a leap year with an extra day added to February to keep the months and seasons in line.[13]

The Julian calendar became the de facto map of time for the next sixteen hundred years, undergoing only minor modifications and adjustments during the long centuries. The most nagging problem facing Caesar's construction emerged with the advent of Christianity.

Ever since the calendar's invention in ancient Mesopotamia and Egypt, its mythoreligious role in marking festival days had been paramount. The emergence of Christianity as the dominant power in Europe would be no exception to this rule. The great failing of the Julian calendar for the Christian cosmos was its inability to fix the all-important holy day of Easter. With a cycle of 365 days and six hours, the Julian calendar's year was still eleven minutes too long.[14] That may not seem like much, but over centuries those extra minutes add up. After 138 years, the Julian calendar ends up a day ahead of itself. Easter, intended to occur on the spring equinox (March 25 in the Julian calendar), slowly drifted to earlier dates. In the third century CE, the Church in Alexandria reset the date to March 21 using new calendar calculations that were intended to solve the problem. But as the centuries clicked by, the slide of Easter away from the equinox became ever more apparent. It was not until the famous papal bull of Pope Gregory XIII in 1578 that a new calendar reform process was set in motion.[15]

The maintenance of the Julian calendar throughout the long centuries after the fall of the western Roman Empire was one of the few examples of astronomy-based thinking left in Europe during this period.

This narrow focus is indicative of the European state of mind during the Dark Ages. Astronomical learning and the cosmological imagination stalled. In the life of the mind, time had regressed.

ATHENS AND JERUSALEM:
THE MEDIEVAL COSMOS

The Middle Ages are rightly called "dark" when compared with the triumph of Greek culture. Many European Christian thinkers during the first millennium CE had abandoned the Greek tradition and no longer considered reason to be a sufficient form of investigation. Nowhere is this "closing of the Western mind" more apparent than in cosmology and astronomy.[16] Many in the newly ascendant Christian Church were hostile towards natural philosophy and the Greeks who championed it. As the early Christian writer Tertullian put it, "We want no further curious disputation after possessing Christ Jesus."[17] The old era of inquiry was gone and the achievements of the Greeks were dismissed as irrelevant. "What indeed has Athens to do with Jerusalem?" Tertullian asked mockingly.[18]

So intellectually apathetic had the Western world become that the Earth was made flat again. Lactantius, a fourth-century bishop, dismissed ideas of a spherical Earth as heretical and absurd: "Is there anyone stupid enough to believe that there are men whose footprints are higher than their heads?"[19] While many of the early Church fathers were content to simply dismiss Greek cosmology without offering anything in response, the sixth-century Byzantine merchant Cosmas Indicopleustes did propose such a map of the cosmos in his work entitled *Christian Topography*. In this early work Cosmas claimed the universe was built like a vaulted tent or a tabernacle and heavenly bodies were moved by the will of angels, with the sun and moon disappearing each night behind a huge mountain located at the centre of the cosmic box.[20] After the intricacies of Ptolemy's astronomy, with its finely tuned epicycles, a vaulted tent with a mountain in the middle was certainly a step into the dark. But while the Greek emphasis on reason was abandoned, mate-

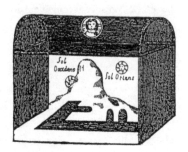

FIGURE 3.2. As Europe fell into its Dark Ages, cosmological models stepped backwards. Early Christian cosmology held that that the universe was a vaulted tent with a mountain at the centre.

rial engagement was still present and still shaping the human worlds of mind and matter.

When the Roman Empire collapsed in the western domains of Europe, the capacity to engineer culture on a massive scale was lost as well. At the same time, Christianity became the new cosmological material of everyday life. The structure of the previous society had been dismantled, and order and stability disappeared with it. The vaulted roads crumbled, enduring institutions failed and power devolved locally to whoever was strongest.[21] The world had become dangerous in entirely new ways. Thus the Church, with its promise of eternal salvation beyond the chaos of life in the here and now, did not just augment life, but came to utterly define it.

Just as agrarian myths were the very stuff of life in the Neolithic, the cosmic religious vision offered by the Church was not simply a shackle that had to be thrown off before an intellectual life could be lived again. The time after the collapse of Rome may have been a retreat from the Greeks' "first scientific revolution", but people did not stop thinking or learning or picturing the universe. They may not have approached the world in the same way as the natural philosophers, but they were not stupid and they certainly were not all as ham-fisted in their cosmological thinking as Cosmas and Lactantius.

The English monk known as the Venerable Bede was, for example, a man of considerable learning and imagination. Having worked hard to provide reliable adjustments to the Julian calendar, he turned his attention to cosmology and the nature of the world as a whole in a text called *De natura rerum* (*On the Nature of Things*). Bede had the intellectual sophistication to understand the five consistently appearing cosmological questions detailed in the last chapter.[22] He was also wise enough to acknowledge the validity of arguments for a spherical Earth. Still, the lack of access to the majority of Greek thinkers left Bede, and most other European natural philosophers, without the tools to venture far from established Christian doctrine.

Light first broke into the Dark Ages in Spain. After centuries of relatively tolerant Moorish rule, Spain in the twelfth century had become a crossroads of Islamic, Christian and Jewish cultures. The Greek texts that had been kept and copied in Islamic libraries, as well as the many Arabic translations of Aristotle, Ptolemy, Hipparchus and other Greek thinkers, had found their way to Spain. In the city of Toledo, Gerard of Cremona began to translate Arabic versions of Greek texts into Latin. Through his and others' efforts, the so-called translation movement was launched.[23] Soon Latin versions of everything from Aristotle's cosmological text *De Caelo* to Euclid's mathematical masterpiece *Elements* began to spread through Europe.

By the 1200s, European astronomical and cosmological thinking was again illuminated by the progress of Greek thinkers.[24] Once more in Europe, the fine scalpel of human reason would be used to separate subtle issues on the nature of time, space and existence. The shadow of Christian dogma, however, lay heavy on this revival. Aristotle's natural philosophy was fine as long as it could be squeezed into a Church-sanctioned conception of God and his creation. As scholars began to revisit issues such as the Earth's location in the cosmos, the spatial extent of the universe and the ever-present question of time and its geometry, their imaginations remained constrained by the fear of heresy charges.

As was true of the Greek era, cosmology and astronomy were known to be related but often were pursued without reference to each other. Astronomers made mathematical models for accurate predictions of celestial motions. The reality of these models was considered debatable. Many scholars saw the models, with their circles within circles, as nothing more than computational tools. They were good for determining the positions of planets in the sky but they were not descriptions of reality. Cosmology, therefore, could be pursued without deference to astronomy. From the perspective of these scholars, reason alone could infer the true structure of the universe in both space *and* time. But reason would have to defer to scripture. During this epoch, cosmological thinking was allowed only if it passed biblical muster.

Thus Ptolemaic astronomy became the poster child for the synthesis of Greek-based learning with the longstanding biblically based authority of the Church. A geocentric universe with the corruptible Earth placed at the centre of a perfect celestial realm dovetailed neatly with the Church's theological vision. In this way, both Ptolemy and Aristotle became Christianized. By the 1300s most reasonable scholars accepted a spherical Earth but few were willing to imagine the globe twisting in daily rotation. After all, the Bible told of Joshua commanding the sun to stand still, not the Earth.

The Church and the new scholarship found agreement on the geometry and physical extent of the universe—the second and third of the five cosmological questions. Most scholars believed the physical world must be finite. A nonmaterial, spiritual domain beyond the last of Aristotle's crystalline shells was accepted by some, but it was almost universally agreed that the created material world must be of limited extent. Thus most scholars believed a true vacuum, an emptiness beyond the last celestial sphere, was impossible.[25] Arguments appearing in this era against the vacuum and for the finite universe demonstrated how much sophistication in physical reasoning had returned. For example, the fourteenth-century Parisian scholar Jean Buridan argued that an infinite body in circular motion must have infinite rotational speed far from the centre, and since this was clearly impossible, he reasoned that the physical universe must be finite.[26]

Time was, however, a problem. Aristotle's importance had surged as translations of his works spread across Europe in the thirteenth and fourteenth centuries, and he was soon referred to as "*the* philosopher".

Scholars were loath to disagree with his writings. But in *De Caelo*, Aristotle argued that the universe was infinite in time—it always had existed and always would exist. How could scholars square this view with the Church's official doctrine of creation ex nihilo? The Church had, after all, decided that both time and the universe began with God's act of creation.

Then, at the University of Paris, a group of radical scholars declared that they were willing to "carry Aristotle's rationalism and naturalism as far as possible".[27] Faculty members Siger of Brabant and Boethius of Dacia challenged Church authority on issues such as eternity and the age of the cosmos. The Church responded by issuing a list of propositions that were "declared false and heretical". Of the 219 positions the Church warned against, at least twenty were directly related to cosmology. All five of the great cosmological questions came into play. Scholars were warned against arguing that God could not create more than one universe. The Church reiterated that something could be created from nothing (if God was the creator) and scolded scholars for thinking that a vacuum must have existed before God created the universe. Chief among the heretical views, however, was number 87: "That the world is eternal as to all species in it; and that time is eternal."[28]

The question of "before" had returned. Even in this deeply Christian era there were Christian scholars who could not abide a cosmic beginning. In the thirteenth century what came before creation was an issue that remained vibrantly alive.

The old adage that there is no bad publicity was just as true in the 1200s as it is today. The Church's 219 articles of heresy merely brought more attention to the new cosmological debates. To avoid trouble with the Church, many scholars adopted literary devices when discussing forbidden topics such as infinite time, the existence of other universes or even the reality of Earth's rotation. After analysing new cosmological or astronomical ideas in great detail, writers quickly backtracked, bowing to scriptural authority. Nicole Oresme used this trick in his impressive analysis of arguments for a rotating Earth. After working his way to the very edge of convincing his reader that our planet rotates, he stopped and doubled back:

However, everyone maintains, and I think myself, that the heavens do move and not the Earth: For God hath established the world which shall not be moved . . . What I have said by way of diversion of intellectual exercise can in this manner serve as a valuable means of refuting and checking those who would like to impugn our faith by argument.[29]

Within two centuries, these kinds of half steps would no longer be needed. The wheels of creative investigation were turning, and with new intellectual daring, Europe was preparing for a tectonic shift in astronomical perspective. But this turn of mind could not occur in a vacuum. Once again material engagement—the specific details of the human encounter with the world as embodied in the stuff we built—would lay the groundwork. The wheels of the coming revolution turned not on chariots or trebuchets but within the clock towers of cities across the continent. Clocks and clockwork were remaking the European experience of time, and the universe would have to follow.

MONS, FRANCE · 1188

"The coward does not come," said the knight. "I have prevailed and I demand justice now!"

Gerardus of St Obert was cousin to local royalty, the count of Hainault, and not a man to be denied. He was arrogant and prone to spasms of violence. Still, it was not Gerardus who had demanded the duel he and the other nobles had assembled for that morning. Robert of Beaurain, a knight known for his valorous acts and his pride, had insisted on the duel. Robert claimed that Gerardus had insulted his honour, ignoring his freeman status and boasting to others that Robert had been a serf. The count of Hainault had no choice but to allow a duel so that right might be made of the situation.[30]

The day had broken warm and clear. As the monks assembled for prayers at prime, the first hour of the day, the count and his nobles arrived at the central square of Mons. There, before the monastery, Gerardus and his retinue were already assembled and waiting. Robert of Beaurain, however, was nowhere in sight. The sun rose higher into the sky and still he did not show.

As the wait continued, Gerardus grew impatient and angry. Finally the bells of the monastery rang announcing none, the third division of the day and the end of the period when duelling was allowed.

"The coward fails to show himself," cried Gerardus. "I am victorious."

"Not yet," said the count's councillor. "The time of none has not yet passed."

"But the bells," insisted Gerardus. "They have rung for none."

In a low, patient voice the councillor explained to Gerardus that the time for prayer called none, which was recognized by the monastics, and the time of day called none recognized by the court were two different things. "At court we have always used different understanding of none. The monks have their hours, we have ours and the two are not the same. In legal matters I am afraid our time is what matters."[31] With Gerardus standing to the side, the councillor and clerics from the monastery huddled together and pointed to the sun, now beginning to sink in the west. They pointed to the bell in the bell tower. They argued and drew figures in the dirt. Finally Gerard could stand it no longer.

"I have waited long enough. Can none of you fools tell the hour?"

VICTORY OF THE CLOCKS

While a consensus on the hour would be hard to achieve in a square at Mons in 1193, within a few hundred years it would be as easy to find as a glance towards the centre of town. In 1193 the majority of humans had no need for, or access to, accurate time reckoning. But by 1393, many cities in Europe would be equipped with that most modern of devices: a mechanical clock. With its prominent face showing the hour and bells demarcating the time in loud peals, the clock towers left no facet of European life untouched. The day was divided, never to become whole again.

But for hunter-gatherers and farmers living outside the immediate influence of a city, there was little need for accurate divisions of the day. It was good enough to indicate the time by the event that defined it. The Konso people of southwest Ethiopia, for example, still break the day into six unequal divisions. Each of the six sections is given a name describing what is scheduled to happen in that time. Thus, for example, the interval from 5:00 to 6:00 p.m. is called *kakalseema,* or "when the cattle return home".[32] Such a rough parcelling of the day is all that is required. As Anthony Aveni observes,

> Though simple morning, noon, and afternoon can easily be sorted out simply by observing which side of the sky the sun is in, even 8- or 10-, perhaps even 12-, fold divisions of the day can be reckoned unambiguously simply by gesturing with the arms. Were an hour's accuracy all we ever needed, the entire world might be completely devoid of clocks today.[33]

The need for greater accuracy occurs only in more complex societies. These needs, however, do not arise on their own. Instead, material engagement allows the culture as a whole to imagine such accuracy and what to do with it. Culture evolves with these imaginative innovations by creating new institutional facts that then demand temporal accuracy. As always, the enigmatic entanglement between cosmic (or scientific) time and personal (or experienced) time meant both would be implicated in the evolution of a new time consciousness. This braiding is exactly what occurred in the history of the clock and the hour.

The division of the day into twenty-four equal hours was a Babylonian invention and may be linked to the division of the zodiac into twelve constellations.[34] It was likely a response to the well-known fact that simple sun dials (such as a stick placed upright in the ground) show divisions of the day of unequal length—the shadow moves slower at midday then it does in the late afternoon or early morning. But the equally spaced twenty-four-hour Babylonian day was used only by astronomers—a scientific elite—and did not gain widespread acceptance until the appearance of mechanical clocks.[35]

In ancient Rome, a timekeeper announced noon based on the sun's location between the city's most prominent buildings.[36] Crude sun dials were also placed in public spaces, allowing some measure of public time. Still, what these measures offered was a far cry from the accurate, abstracted and omnipresent time we live with today. That time would have to wait for the creation of clocks.

FIGURE 3.3. The wheel of time of common folk. Each month is associated with activity set against the natural world. The zodiac surrounds the cycle of the year, creating its cosmic frame.

While it is important to note that attempts to create accurate divisions of the day—as well as mechanical timepieces that could monitor them—predated the European invention of the clock, they were limited to small segments of the population. Slow-burning candles with colour-coded wicks were employed both in monasteries and in the courts of kings such as Alfred of England.[37] Water clocks driven by the flow from a large tank were cumbersome and inaccurate and they belonged only to the wealthy and powerful. Thus most of the European population, and indeed most of the world's population, "floated" on a time that was still directly enmeshed with the physicality of daily life. The shift from temporal flotation to rigid regimentation began in the monasteries, where a holy cosmic order would find reflection in the hours of the day.

"One is not straining the facts when one suggests that the monasteries . . . helped to give human enterprise the regular collective beat and rhythm of the machine." So wrote Lewis Mumford in his famous essay "The Monastery and the Clock".[38] The Church initially adopted the unequal Roman divisions of the day, and with the development of monasteries and the so-called Rule for ordering the day, prayer times were fit into seven periods: matins, prime, terce, sext, none, vespers and compline.[39] Matins began with sunrise, sext was approximately midday, none fell around midafternoon, vespers marked the workday's end and compline set the evening prayers. It's worth noting that the twelve-hour division of the day was rarely used, either inside or outside the monastery.[40]

In the monastery, the hours of prayer were the metronome that ordered monastic life. As Gerhard Dohrn-van Rossum puts it in his *History of the Hour*, "The monastic day was divided into an almost unbroken sequence of divine offices, meditation, reading, work, meals, and periods of sleep."[41] While authors such as Lewis Mumford may have gone too far in seeing the "iron discipline" of the monasteries as a metaphor for the machine age to come, the monks' vigour and vigilance in ordering the day likely provided an example others hoped to follow as western Europe awoke from its Dark Ages slumbers. The greatest influence of the monasteries may have been the simple act of keeping their regular offices of prayer in a world that had little use for regulated, complex time. In other words, it was the example of order rather than a demand for accuracy that may be their most important legacy when it comes to time.

For the centuries that make up the Middle Ages there was no rigid rule or technical capacity to say when, exactly, terce might end or none begin. In general, monasteries used natural (and naturally vague) time signals such as the crowing of the cock to determine the hours. Thus considerable latitude existed in timekeeping within monasteries. "If the monks rose late", Dohrn-van Rossum explains, "the liturgy of Vigils would be shortened."[42]

What mattered was not the act of keeping exact units of measured time; instead, monasteries were developing a fundamentally new role for time as a background to an ordered life. A new sense of time was being established. "The monastic time system was meant to set itself apart", writes Dohrn-von Rossum. "Regularity and repetitiveness in regard to year and day and the collectively lived life produced a special rationality and required a special discipline. . . ."[43] Regularity and repetitiveness also demanded a submission to a temporal order that was imposed by the collective. Note that the idea of an "order imposed by the collective" is an excellent definition for the institutional facts that had changed millennia before in both the Neolithic and urban revolutions' response to material engagement's evolution. What emerged from those changes were new forms of human time (and new visions of cosmic architecture). The same process was beginning once more in the twelfth, thirteenth and fourteenth centuries. For many scholars looking at the change in time consciousness destined to emerge in the Renaissance and then permeate the industrial revolution, the medieval age serves as a bridge linking one temporal order to the other.

While the monastics were developing a new form of time-ordered life within their walls, the outside world was mostly tuned to the activity of their bell towers. As European cities became more complex and wealthier in the thirteenth and fourteenth centuries, bells took on a great role in regulating civil time and its uses.

Bells were used to signal the beginning of the workday for different guilds.[44] A bell pealed for the sheep shearers to start work. A different set of bells, or a differently timed bell signal, rang for the carpenters, the joiners and the armourers. There were bells used to signal the beginning of market and bells used to announce the end of market. In some cases there were separate bells to tell different groups, such as the nobility or the Jews, when to enter the market.[45] Bells rang to regulate the political life of the city, from the start of court to the public announcements of its verdicts. The bell and the bell tower became the pulse of the city.

The braiding of material engagement and institutional facts became explicit as the bell towers turned into a direct symbol of economic and political vitality. Whoever "can ring the bells at will can easily rule over the city", wrote Galvano Fiammo of Milan in the fourteenth century.[46] When the French town of Hesdin used its bells to call peasants to revolt against local authority in 1179, the count of Flanders toppled the bell tower in retribution.[47] The bells ruled time and time ruled the new urban enterprise.

A bell signal for the courts, a bell signal for the markets, a bell signal for the guilds—the use of bells to order urban life grew more confusing as cities became more complex. New trade and economic activity forced the institutional facts of European life to shift. Cities needed a simpler and more universal means of ordering temporal flow. In the midst of these shifting demands a radical new form of material engagement made its appearance. It was, without doubt, the most important invention of the last thousand years: the mechanical clock.

No one knows who invented the clock, though it was likely someone associated with a monastery.[48] In particular, no one knows who invented its key component—the escapement, the notched metal rings that allow gravitational energy stored in a hanging weight to be regulated and regularly released. As an eighteenth-century writer commented, "It is certain that if we knew who first invented the means of measuring time by the movement of toothed wheels . . . this person would deserve all our praise."[49] No matter who came up with it, the invention quickly surged through the culture like a wave of possibility. Within a single century mechanical clocks would transform life in Europe.

The list of cities with public clocks begins with Orvieto, Italy, in 1307, followed by Modena in 1309, Parma in 1317, Ragusa in 1322, Milan in 1336 and Padua in 1344. By mid-century, mechanical clocks began to appear outside of Italy: Windsor Castle gained a clock in 1353, Avignon in 1353, Prague in 1354, Regensburg in 1358.[50] By the beginning of the fifteenth century, public clocks had evolved into the standard even in smaller settlements. The town clock became a matter of both civic need and civic pride.

The mechanical regulation of time by machines freed cities from the cumbersome menagerie of bells and their various tolling patterns. Knowing the hour number was all that was needed for the new system. The establishment of twelve equally spaced hours for day and twelve equally

FIGURE 3.4. Erfurt with clock. Europe builds a new experience of time as clocks set the pace of civic life. This manuscript illumination from Hartmann Schedel's *Weltchronik* (1497) shows the city of Erfurt with a clock tower in the background. Such towers were a new addition to cityscapes and artists were keen to include them in their works.

spaced hours for night began in Italy and was applied unevenly throughout Europe. With the mechanical clock setting its rhythm, the regulation of time became abstracted, as did people's relationship to time. Instead of a special set of bell signals to tell the armourers to start work, it was the bell tolling a numerical hour that initiated work (and pay).

The first clock towers were still bell towers; they had no clock faces. In these towers a clockworks drove the timing of the bells. By the beginning of the fifteenth century, however, the acoustic signal announcing the abstraction of evenly spaced hours had become visual with the introduction of clock dials. According to city records, in 1410 the keeper of the Horloge du Palais in Paris complained to the authorities about the extra burden of working with the complex machinery of the dial.[51] And yet in these first clock faces only an hour hand appears. Smaller divisions of the day were not yet required.[52]

The hour itself, eventually imposed uniformly across Europe, was the medium of momentous change. "Abstract time became the new medium of existence", wrote Lewis Mumford.[53] By the end of the fifteenth century, the affairs of the urban world were fully regulated by clocks striking the hours. Communities became enslaved to the peal of clock bells and the march of the hour hand. "One ate, not upon feeling hungry, but when prompted by the clock; one slept, not when one was tired, but when the clock sanctioned it", said Mumford. What began as the ordering of a prayerful life transformed into an ordering of life itself.

The human experience of time had been entirely redesigned by the invention and diffusion of the clock. The reawakening of learning, the growth of a wealthy new merchant class, the opening of new trade routes by ever more sophisticated seagoing vessels—each of these factors contributed to this newly imagined, clock-inspired form of lived time. As humans had done thousands of years before in the Neolithic and urban revolutions, time was reconstructed for our own imagined needs. And, as in those previous revolutions, the reformation of the universe in new cosmologies would have to follow.

THE REVOLUTION REVOLUTION: COPERNICUS MOVES THE HEAVENS

In 1497 Luca Pacioli, an Italian monk and mathematician, published the first description of double-entry accounting. A means of tracking both debits and credits, it would soon become the standard for Venice's rapidly rising merchant class.[54] In 1517 the firebrand theologian Martin Luther nailed his attack on the Vatican's corruption—his Ninety-five Theses—to the door of Wittenberg's church, effectively firing the first salvo of the Protestant Reformation.[55] Two years later, five ships under Ferdinand Magellan's direction set sail in what would become the first voyage to circumnavigate the planet.[56] In the midst of these events, a young Polish lawyer and astronomer, Nicolaus Copernicus, was travelling through Italy and beginning work on a new vision of the heavens. There were many revolutions occurring at the beginning of the sixteenth century, and it would be Copernicus who gave them all a name.

Just as clock-driven time spread across Europe, a new power rose in the cities to make use of it. As the fifteenth and sixteenth centuries unfolded, the Catholic Church and its political allies faced a challenge from the rising merchant class. Wealth from seagoing trade (among other sources) allowed this newly empowered population to push against the century-old dictates of Europe's social, political and intellectual structure. Renaissance ideals that individuals carried their own ability to parse the world rose alongside the Protestant Reformation's rejection of papal hegemony.

As the Renaissance set its sights on a new order for man on Earth,

FIGURE 3.5. Medieval representation of Ptolemy's geocentric universe. Geocentrism, Aristotle's physics and Catholic dogma were blended together during the late medieval period.

the Protestant Reformation demanded a new order for man's relation
to God. Both movements were driven by material wealth and the new
forms of material engagement they brought to a society rapidly gaining
confidence in the ability to shape its own fate. Soon Earth itself would
be moved to make way for the new era, and Copernicus would give the
word *revolution* its revolutionary meaning.

Born in Poland in 1473, Copernicus studied law, medicine and astronomy. By the time he reached his late thirties, he was already exploring alternatives to Ptolemy's planetary model, which still prevailed after
thirteen hundred years. He was sure that Ptolemy had put the sun in
the wrong place, and that the sun—not the Earth—deserved the central
position in the solar system. Though he still held firm to the Greek bias
for uniform circular motion, he was offended by the sun's subservient
position as well as by the mathematical bells and whistles of Ptolemy's
geocentric model. Copernicus was sure that God, the author of the universe, had been more economical in his designs.

While simplicity may have motivated Copernicus to begin his work,
the shift from a geocentric model of the cosmos to a heliocentric one
required a heroic mathematical effort. He had set himself the task of reordering the solar system and, most important, developing a working
mathematical model that could make accurate predictions about planetary motions.

Copernicus' model had three critical features. The first and most
obvious is that the sun was located at the centre of the planetary system. All planets, including Earth (he left the moon in orbit about the
Earth), moved around the sun. The second critical feature of Copernicus' model was the daily rotation of the Earth. Even by Copernicus' era,
the rotation of Earth remained a difficult idea for many scholars to swallow. The third and most important aspect of Copernicus' model was its
ability to cleanly recover the retrograde motion of planets, which had
vexed astronomers for so long.

Copernicus' explanation for the retrograde loops was the essence
of simplicity compared with Ptolemy's vision of epicycles. By assuming
that the periods of planetary revolution increased with distance from
the sun (Earth orbits more slowly than Venus, which orbits more slowly

than Mercury, etc.), Copernicus turned all retrograde motion into a simple catch-and-pass effect. Faster-moving inner planets lap the outer ones. As an inner planet, such as the Earth, passes an outer one, such as Mars, the apparent motion of Mars—what an observer sees from the Earth's surface—becomes a retrograde loop. The same effect occurs for observers looking from an outer planet towards an inner one. Thus for observers living in a heliocentric solar system, the mystery of retrograde loops reduces to nothing more than simple relative motion of planets against the fixed background of stars.

The geometrical manipulations Copernicus needed to illustrate this catch-and-pass effect were complex, but the underlying idea was much simpler than Ptolemy's. And there was an added benefit to Copernicus' system. In a geocentric model, the size and speed of each planet's orbit was arbitrary. Bigger orbits with faster-moving planets worked just as well as smaller orbits with slower planets. Both configurations could fit the observations. The model had no built-in yardstick to gauge the dimensions of the heavens.

In the final versions of Copernicus' heliocentric vision of the solar system (brought to fruition by Johannes Kepler), the timing of retrograde motions forced a single set of orbits. The observed retrograde periods allowed these new heliocentric models to translate time into space and set sizes and speeds for all the planetary orbits. Thus the model yielded an approximate size of the solar system based solely on its motions. When the calculations were completed it was clear that the new Copernican universe was much roomier than the Ptolemaic one.

The difference in size between the Ptolemaic and Copernican models was startling. The heliocentric cosmos was at least four hundred thousand times bigger (in terms of volume) than Ptolemy's.[57] This vast enlargement of the universe would be the first of many times that scientific astronomy would inflate the cosmos. With each step outward, humanity appeared to shrink in significance.

Copernicus' book *De revolutionibus orbium coelestium* (*On the Revolutions of the Celestial Spheres*) was not published until just before his death in 1543.[58] The work spread quickly throughout Europe thanks to a vibrant trade in manufactured (printed) books. Copernicus' ideas

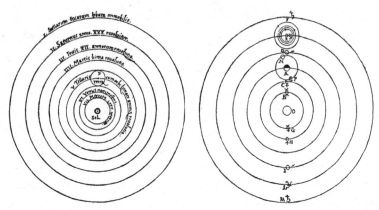

FIGURE 3.6. Copernicus places the sun at the centre of the universe. The diagram on the left is from Copernicus' *De revolutionibus* (1543). On the right is a figure from Galileo's *Dialogo* (1632) after his telescopic studies showed moons orbiting Jupiter.

were the essence of controversy. For some the new heliocentric model was heresy. For others, it did honour to the greater glory and conception of God. But in a changing world new cosmologies were dangerous. There was a background of political, theological and economic tumult that made debates over the Copernican universe flicker between metaphor and cosmic reality. Europe had been pushed off the centre of the map with the discovery of the New World. The Vatican was being pushed aside as the sole arbiter of both earthly and divine power. And the Earth had been pushed aside to make room for a new cosmic architecture.

Ironically, even though they told completely different stories about the solar system's architecture, Copernicus' new model did not predict the motion of planets much better than Ptolemy's. It is remarkable that the greatest scientific revolution in the history of humanity was driven, at first, more by aesthetics than by questions of data and predictions. Pythagoras and Plato would have perfectly understood the spirit of Copernicus' effort. This emphasis on elegance and simplicity, dating back to the Greeks and revived in the Renaissance, would have a profound effect on the development of modern cosmology and physics. Today, scientists working the boundaries of both fields routinely invoke beauty, elegance and simplicity when considering rationales for choosing between competing theories.

The geometrical completion of Copernicus' vision required one step beyond perfect circles, and that task fell to the quiet, mathematically oriented astronomer Johannes Kepler. Kepler was a kind of sixteenth-century Pythagorean who felt a rational mysticism in the workings of mathematics. Through his patient studies Kepler became convinced that a core of sublime and deeper mathematics must run through the foundations of the world's structure. This conviction led the young Kepler to attempt modelling the Copernican solar system by nesting planetary orbits on the five Platonic solids. It was a glorious geometrical creation, and while it brought Kepler some fame, he soon realized its impossibility. Resolving to abandon the planetary prejudice for circular motion, Kepler spent the next ten years searching for the true geometry of orbits.[59] Working with the exquisitely sensitive naked-eye observations provided by the Danish observer Tycho Brahe, Kepler found a different form—the ellipse—that could describe the motion of the heavens.

Ellipses, which appear as squashed circles, were a geometrical form known since the time of Pythagoras. But it was not until the sixteenth century that Kepler discovered that all planets travel through space on elliptical orbits as if guided by the hand of God the mathematician. It was cosmic order at its simplest. With the ellipse, Kepler was able to embrace all aspects of planetary motions—the variable speeds, the retrograde loops, the brightening and dimming of the planets—in an elegant heliocentric system expressed in the most compact mathematical expression. All of Kepler's observations could be conveyed in just three simple laws, each based on elliptical orbits. The descriptive economy of these relations, which came to be called Keplers Laws, became a model for scientific descriptions of nature. The cosmos was sparse, elegant and built on an invisible superstructure of higher mathematics.

Material engagement would play a particularly explicit role in the final step of the Copernican revolution. Sometime around 1608, in a dusty workspace in the Dutch town of Middleburg, Hans Lippershey was hard at work grinding lenses for glasses. Trolling for treasures at his workbench, two of his children picked up different lenses and held them before their eyes. "Look," one of them called out as he held two lenses apart and peered through them both. "The church steeple, it looks bigger!" The telescope had been discovered.[60]

News of the new optical device—the spyglass—spread rapidly among the scientific cognoscenti of Europe. A thousand miles to the south, an ambitious young mathematician and astronomer quickly taught himself how to shape glass lenses for his own use. Placing these lenses at the opposite ends of a ninety-centimetre tube, the young Galileo Galilei understood that he had found the key to his own future.

Galileo was born in 1564 into a highly cultured family in Pisa. His father was a well-known music theorist whose books on harmonies in musical scales had deeply affected Kepler's thinking about the structure of the solar system. Galileo's talent was obvious to his teachers and as a young man he climbed the academic ranks becoming one of the most famous astronomers in Europe.

Galileo's genius was equal both in his scientific and his social ambitions. From an early age Galileo had set out to win fame for himself, and the telescope gave him the tool he needed. After building his own telescope in 1609, Galileo demonstrated its uses to the lords of Venice, who were so impressed with its capabilities they commissioned the young Galileo as a university professor and gave him a profitable trade in constructing telescopes for local merchants. These merchants and nobles understood immediately that bringing the far distance into view meant economic and military advantage. A thousand images of admirals, whalers and pirates standing tall on the foredeck, spyglasses raised, had been launched.

But Galileo was interested in astronomy, not conquest, and he began systematically observing the night sky with a series of ever more powerful instruments. In 1610 he published a short report of his work called *Sidereus Nuncius* or *The Starry Messenger*. The book caused a sensation, turning Galileo into a kind of Renaissance rock star. Proud and ambitious, Galileo was confident he could single-handedly win official Church acceptance for Copernicus' theory.

He soon established a substantial body of evidence supporting the heliocentric universe and rejecting scholars' slavish adherence to Aristotle. Galileo saw mountains on the moon and spots on the sun. Both observations contradicted Aristotle's physics and cosmology, implying that celestial objects were just as imperfect and rutted as the terrestrial realm. Galileo watched Venus pass through a full round of phases, just like the moon, contradicting explicit predictions of the Ptolemaic model. Most shocking of all, Galileo discovered a family of satellites orbiting Jupiter. These "Galilean" moons of Jupiter offered proof that not everything orbited the Earth, shaking people's confidence in the idea of a geocentric cosmos.

Galileo's astronomical discoveries and his relentless arguments for the validity of the Copernican model made him a hero to many. With his newfound fame, he was certain the Catholic Church could be swayed from its adherence to Ptolemaic astronomy. But while he was a scientific genius, his political sense was clearly lacking. With a few infamous false steps Galileo found himself dragged before the Church's most feared judicial body—the Inquisition—on charges of heresy.

Galileo's trial was a turning point in the history of culture and science. The Church convicted him of failing to heed its orders by promulgating heliocentrism. Though the Church had won its battle by sentencing Galileo to house arrest, it was fated to eventually lose the cosmological war. Across Europe a new tide was rising. Astronomers, natural philosophers and scholars of all stripes were willing to put aside doctrine in order to use observation and experiment to guide their thinking.

Galileo, however, was stunned by his inability to influence the Church and stung by the sentence of house arrest. "This universe that I have extended a thousand times has now shrunk to the confines of my body", he wrote in a letter to his beloved daughter.[61] Within the confines of his villa, an unbowed Galileo began a remarkable period of experimentation and discovery that would provide the outlines of a radical new physics. Just as he had challenged the worldview of Ptolemaic astronomy early in his life, in his last years Galileo finally overturned the physics of Aristotle.

Galileo's greatest achievement during these last years was the development of an experimental method for physics. For centuries scholars had relied on the authority of classical authors, and issues such as the relation between force and motion were still pinned to Aristotle's philosophical works. Galileo began to investigate nature directly, developing ingenious experiments using wooden wedges, balls and water clocks that allowed ideas about the nature of motion to be tested rather than simply debated. In this way he established a scientific practice that would be emulated across disciplines. Fifty years later, Isaac Newton would use Galileo's discoveries to develop his own groundbreaking theories.

THE RENAISSANCE REAWAKENING
TO COSMOLOGY

The universe as a whole and the origins of time were not the explicit concern of Copernicus' grand project, which only addressed a subset of the five questions. "The Earth is to the Heavens . . . as a finite is to an infinite magnitude", he wrote. "It's not at all clear how far this immensity stretches out."[62] While Copernicus recognized the implications of his expansion of celestial dimensions, he did not care to conjecture how far the universe ultimately extended. Others, however, were ready to take on questions of both cosmic architecture and the infinity of space.

The immense size of the Copernican universe revived discussions about infinity. Throughout the medieval period, scholars debated what lay beyond the last, starry sphere in Ptolemy's model. Many were loath to admit the possibility of a true void or vacuum, a physical space devoid of matter. By shifting the Earth away from the centre of creation and replacing it with the sun, Copernicus raised the possibility that there might not be a true centre to creation. And with the Earth thrown from its perch, scholars began to reconsider the full dimensions of space and its relation to matter.

Part of the revival of infinity lay in the revival of atomism—the original Greek theory that matter was composed of tiny particles separated by pure void. Natural philosophers such as Francis Bacon and Edward Sherburne derived their interest in atomism from their growing interest in chemistry.[63] Reading classical Greek philosophers in the context of contemporary debates about the heavens, they conceived of an infinite universe of atoms, and of the infinite voids that separated these particles.

Others found inspiration for infinity from different sources. Giordano Bruno, the radical philosopher and former Dominican friar, also advocated for an infinite universe of stars, planets and even life. But while famous for his heretical ideas—for which the Catholic Church burned him at the stake—Bruno was never a systematic natural philosopher. William Gilbert, the English doctor who carried out early studies of magnetism, was more coherent in his thinking about infinities. Though it was not clear if he fully accepted the Copernican world system, his contemplation of the idea drove him to question the location of the stars on a finite outer sphere: "What, then, is the inconceivably great space between us and these remotest fixed stars? And what is the vast immeasurable amplitude and height of the imaginary sphere in which they are supposed to be set?"[64]

Infinite space was by no means universally accepted. Kepler, unlike Copernicus, was more than willing to address questions of the spatial dimensions of the universe. Believing an infinite number of stars could be ruled out by logic alone, Kepler rejected a universe of infinite extent; "all number of things is actually finite for the very reason that it is a number".[65]

While a debate raged about infinite space, cosmic questions of time found more agreement. Like most of his contemporaries, Kepler stuck to biblical orthodoxy in considering time and infinity. The cosmic clock had begun with the creation. Extrapolating the patriarchs' ages from the stories in the Bible, he calculated the exact date of Genesis, arriving at the date 3992 BCE for the creation of the world.[66]

It is both remarkable and telling that a man of Kepler's intellectual capacity, a man so willing to challenge orthodoxies on the spatial structure and dimensions of the universe, would be so unwilling to push against those same orthodoxies as they related to the origins of time and the cosmos. And he was not alone in this regard. Throughout the sixteenth and early seventeenth centuries few scholars were willing to challenge biblical authority on these issues.

In 1650, Bishop James Ussher of Ireland published what would become the definitive biblical chronology of the world. Even here, the shifting institutional facts driven by the world's material changes played their role. Ussher's work was inseparably bound to the Protestant Reformation. An English Protestant, Ussher was driven to his study of time's origin by Catholicism's claims that it alone embraced the true account of God, man and the world.[67] In a remarkable feat of literary scholarship, Ussher's chronology drew on Persian, Greek and Roman sources in order to create a history of the ancient world that remains in remarkable agreement with modern accounts. It was the truly ancient history, of course, that Ussher got wrong, pinning the date of God's creative act to October 23, 4004 BCE. While scholars such as Kepler, and even Isaac Newton, created their own biblical chronologies, Ussher's work was seen as the authority. His version of cosmogony—a timeline for cosmic history—would be difficult to shake even as the scientific revolution began in earnest.

BEGINNINGS AND ENDINGS: CLOCK, COSMOS AND THE CADENCE OF INFLUENCE

The failure of Renaissance thinkers to break free of biblical time might at first seem surprising. How could men and women who were so willing to challenge orthodoxy on the nature of the universe's architecture be so unable to move beyond biblically based claims of its five-thousand-year history? But as we explore the braided history of cosmic and human time, we should take note of the surges and swells of influence driving changes in both. Like waves rolling back and forth on a stormy channel, at some historical moments cultural influences surge forward, pushing cosmological science and its vision of time onward. And at other moments changes in science surge forward, forcing culture and its use of time to respond.

Changes in the experience and conception of time always originate in our encounters with the material reality of the world. They begin as we find new ways to process matter—wood, metal, fibre, glass—and in turn these new forms of material engagement allow us to change how we organize culture through time and in time. "Through time" is the way human institutions evolve out of new forms of material engagement, taking on a life of their own and propagating across generations. "In time" is the day-to-day organization of our lives that these institutions force on us: the structured school day, workday and so on. Together, they represent the flow downward from material engagement to new institutional facts and new human experiences of time. The invention of the clock and its rapid diffusion across Europe is a key example of this downward flow. But the changes in European culture opened up, even demanded, new accounts of the cosmic order. The beginnings of scientific practice, so evident in the Copernican revolution, represent an upward flow from new forms of material engagement into the realm of ideas, theories and cosmological conception.

Nowhere is this braiding of cosmic and human time more evident than in the metaphor of the clockwork universe. It is in this era that the first description appears of the cosmos as an intricate machine like a clock. As early as the thirteenth century, John of Sacrobosco would refer to the universe as the *machina mundi*, the machine of the world.[68] But this was before the diffusion of clocks. By 1377 mechanical time had made sufficient inroads into Europe that Nicole Oresme, in his *Book of the Heavens and the World*, could make the link between clock and cosmos concrete. Oresme described the world "as a regular clockwork that was neither fast nor slow, never stopped, and worked in summer and winter". The movement of the planets was akin to the well-balanced escarpment driving a precision clockwork. As Oresme wrote, "This is similar to when a person has made an *horloge* [a clock] and sets it in motion, and it then moves by itself."[69]

People had refashioned their daily, intimate worlds to the beat of the clock, so it was only natural that their conception of the surrounding universe should follow. In the centuries to come, however, Newtonian science would make the clock metaphor concrete by rebuilding the laws of physics so that they marched to a steady beat of cosmic time. As Newton's mechanics drove the age of the machine forward, the braiding of time, culture and cosmology would tighten.

Chapter 4

COSMIC MACHINES, ILLUMINATED NIGHT AND THE FACTORY CLOCK

From Newton's Universe to Thermodynamics
and the Industrial Revolution

CROWLEY IRONWORKS, WINLATON, ENGLAND · 1701

Damn the clock and damn the clock warden.

The bellowsman was tired as he looked up at one of the always present clocks mounted on the high wall. It had been a long shift at the fire—four hours straight, keeping the flames high and the iron ore flowing like blood in the devil's veins. The man looked at his swollen hands, blackened with the fire's smoke. It had been more than six years since he came north to County Durham and Crowley Ironworks.[1] When his Sarah died he had been forced to leave farming for the sake of his children. "Crowley has something different going on," they said at the tavern. "There's regular pay from him." Pay was what he needed to keep his girls from going hungry, so he packed what they had and travelled north.

Crowley was good to his families, that was for certain, and the bellowsman was thankful for that. There was schooling for the children and no family was allowed to starve even if one of the men was injured. But the price they paid was steep.

He stood at the bellows cursing under his breath as the clock warden passed. His pay was going to be reduced again. He had broken another of the damn codes. There were so many, who could keep track? Almost all of their codes and laws were envious of his time, as if the old bastard watched over his shoulder for every breath he took.

"Makin' sure you don't waste Crowley's time," the clock warden would say, and drop his pay because he took too long to relieve himself. For God's sake, was it his fault it took time to let his water down? The clock warden didn't care. Order 103 of the code! Order 40 of the code! He leaned down hard on the bellows and looked across to the other end of the ironworks. Everywhere he looked he saw men busy with their duties and everywhere they looked was the clock, the damn clock warden and old man Crowley's code.[2]

TWO MEN, ONE NEW TIME

Sometimes revolutions begin in the mind of a single genius. Sometimes they come from the flesh of the encountered world—wood and iron and the shop floor. Sometimes it can be hard to tell the difference.

Isaac Newton and Ambrose Crowley were contemporaries. Each was born in the middle of the seventeenth century and each would live into the early years of the eighteenth century.[3] A genius physicist, Isaac

FIGURE 4.1. English razor factory, circa 1783. The industrial revolution made time an explicit part of material engagement as working life became ever more bound to the clock.

Newton would re-create time and space, establishing the theoretical mechanics that became the skeleton of the industrial revolution's machine-dominated world. Ambrose Crowley's name did not echo down history as Newton's did but in his way he too saw the future of time and set it into living form.

BOUND TO TIME:
CROWLEY'S IRON LAWS

A devout Quaker born into a family of ironwrights, Ambrose Crowley was both respected and feared, ridiculed and lauded.[4] Sensing a growing spirit of innovation and expansion, Crowley as a young man was "not prepared to accept the restrictions of the trade" as it existed in the West Midlands, the so-called Black Country.[5] In the seventeenth century, cumbersome geography and poor transport had kept English iron making a small-scale affair. Rising from obscurity, Crowley overcame these restrictions by setting up his ironworks in the village of Winlaton, near Newcastle. His operation was close to local supplies of raw "bar iron" and near seaports allowing access to Swedish iron supplies. Newcastle also boasted proximity to coal (in the north) and markets (in London, to the south).[6]

Crowley's new factory model of ironworking succeeded due to his genius for organizing human activity across space and time. Originally created to produce nails, the Winlaton ironworks diversified in 1690, adding a long list of products including stoves and cookware for domestic markets. Significant naval contracts in 1707 allowed Crowley to buy up a rival operation at Swalwell.[7] His iron empire soon became the largest industrial operation of its time outside of the sprawling London shipyards. In 1702 the Crowley Ironworks employed 197 workers. In the second half of the eighteenth century it would employ more than a thousand. In scale at least, it was the first vision of the modern factory.

Crowley was a "hard, demanding man, too forceful to win friends", and he brought an uncompromising sense of purpose to his project of re-creating iron manufacturing.[8] Organization and labour were the major limitations Crowley encountered in developing his new model for iron production. Agents were sent far and wide (even across the Channel) to find able workers. To manage so many men working on a single site, Crowley enforced a set of rules to organize and manage the operation.[9] In these rules, *The Law Book of the Crowley Ironworks*, a first vision of the new industrial time, would be glimpsed.

In order to maintain the flow of the operation and ensure adherence to his rules, Crowley created a new position, that of the warden.[10] Also called the monitor or timekeeper, the warden slept on the premises and kept exacting records of each employee's comings and goings. Under the warden's watchful eye, the factory was to be precisely timed. As the *Law Book* states,

> Every morning at 5 a clock the Warden is to ring the bell for beginning to work, at eight a clock for breakfast, at half an hour after for work again, at twelve a clock for dinner, at one to work, and at eight to ring for leaving work and all to be lock'd up.[11]

Labour was long and hard at the Crowley Ironworks, with employees expected to put in six-day weeks of eighty hours.[12] In spite of these long hours, workers had to be punctual and avoid wasting time—that newly recognized precious commodity. The warden was to deduct from a worker's pay time that was "wasted" in drinking, smoking or conversation. A greater offence was to alter time. Severe penalties were laid down for tampering with the master clock. Informing against others who sinned against the clock (as well as informing against loiterers in general) was encouraged.

Crowley's organizational genius made him a wealthy man, but he was more than the caricature of the uncaring industrialist. With his wife, Mary Owen, Crowley had eleven children, only six of whom would survive past infancy.[13] The trauma of such sustained loss, combined with his Quakerism, made Crowley sensitive to the plight of his workers and their families. Many of Crowley's rules governed their care, including schooling, medical treatment and the disposition of injured workers who could no longer provide for their families. Law 97, for example, included a preamble that specified "the raising and continued supporting of a stock to relieve such of my workmen and their families as may be by sickness or other means reduced to that poverty as not to be able to support themselves".[14] In authoring the *Law Book*, Crowley imagined not just an efficient business but also a kind of utopia in which reason, when applied to the organization of society, would bring a better life to all.

It was Crowley's innovations in work time, however, as radical as they were effective, that established his place in time's history. In effect, he was taking a half step into the industrial revolution. The actual means of production (a term that would gain new importance with Karl Marx in the next century) remained enmeshed in the old ways—the ironworking itself was still done in small units or "shops". What had changed was each worker's "imperatives of time" as they now needed to fit their lives into the factory's time and the contours of the factory's day.

Crowley's employees laboured in the birth of a new kind of human time, one whose conception had occurred centuries before with the rapid diffusion of the mechanical clock. By the 1700s, during Crowley's lifetime, clock technology and manufacturing had become sufficiently advanced that the devices now appeared in the homes of merchant and lord alike.[15] It is notable, however, that the minute hand came into widespread use only in the late seventeenth century—the exact moment when Crowley was building his ironworks.[16]

It was the minute hand, in fact, that would propel human time into a new dimension. The minute was a small but workable unit of time. Minutes can be experienced; we can watch them pass and they can become the raw material of new engagements with time. The advent of the minute hand—destined to be watched by generations of students in school, factory workers on the shop floor and office workers at their desks—heralded an irrevocable change. It announced a new kind of time that would govern the home, the workplace and, most important, the laboratory. While Ambrose Crowley was constructing the first draft of industrial time, science was rebuilding its own definitions of time, and to complete that step it would need Isaac Newton.

THE DIVINE SENSORIUM: NEWTON'S
ABSOLUTE TIME AND ABSOLUTE SPACE

Born in 1643, Isaac Newton led a most unusual life that extended far beyond his scientific genius.[17] Beginning his work as a teenager, Newton would eventually reshape almost every branch of physics and mathematics. Yet he all but ended his scientific career by his late forties, when he left his position at the University of Cambridge and joined Parliament. Newton became master of the mint, a position meant to be something of an empty honour but which he took seriously, reforming English currency and actively pursuing counterfeiters.[18] A lifelong study of alchemy may have led to the mercury poisoning that some scholars claim accounted for his eccentric behaviour in his later years.[19] In his personal life Newton appeared to be as singular as he was in his professional efforts. Capable of both generous friendship and intolerant fury, he was an enigma to the people who knew him. By most accounts Newton died a virgin, having had no significant romantic attachments throughout his eighty-five-year life.[20]

One hundred years after his death, Enlightenment writers would canonize Newton as a hero of pure reason. But unbeknownst to them, Newton was a passionate Christian who devoted thousands of pages to biblical studies, including prophecy. Recently rediscovered documents show Newton to be a heretic, vehemently opposed to the doctrine of the Trinity at a time when such opposition carried a death sentence.[21] Throughout his life, Newton thought of himself as a "high priest of nature"; as an advocate of natural theology, he thought the study of nature revealed the creative hand of God.[22] His irrevocable belief in God's all-encompassing presence provided a principle that redefined space and time in his new physics.[23] Living in the midst of a culture that was building itself along newly imagined lines, Newton's conviction for a new natural theology manifested itself in new religious and scientific visions. And from those visions, the entire cosmos would be reworked.

Isaac Newton's most famous scientific work was his *Philosophiae Naturalis Principia Mathematica* (*Mathematical Principles of Natural Philosophy*), now usually known simply as the *Principia*. Published in 1687, the *Principia* established the science of classical mechanics—an all-embracing account of force, matter and motion. Newton's mechanics would transform the very nature of scientific endeavour, establishing a set of universal laws on the basis of which all phenomena could be described (or predicted). To construct this grand edifice, however, Newton first needed a new foundation for the stage on which physics is played—he needed to invent the absolutes of space and time.

For centuries, scholars struggled to understand the relationship between four critical concepts rooting the study of physics: time, space, matter and motion. Based on the long dominance of Aristotle, most scholars were sure that time and space held no separate reality.[24] For many of Newton's contemporaries, time held meaning only relative to changes in matter. Space held meaning only relative to the arrangement of matter. A chunk of material might be here at one moment in time and then move to there a moment later. But *here*, *there*, *now* and *later* held only relative meaning. They only made sense in relation to all the other matter, the other stuff, in the cosmos. Without a separate reality other than their reference to matter, time and space were, literally, nothing.

This perspective strongly impacted cosmological thinking. Aristotle continued to exert a strong influence even in Newton's age. According to the ancient philosophers, the universe was a *plenum*, a material continuum. In their view there could be no space without matter.[25] In an echo of Parmenides, a truly empty space was thought to be impossible. But in their demand that space and time be dependent on matter, scholars tied themselves into knots trying to formulate precise laws of motion. After all, motion is just an object's change in position (space) over some duration (time). If space and time did not exist without matter, how was matter supposed to move through them?

When the young Newton entered Cambridge he showed little interest in the study of Aristotle, immersing himself instead in the new perspectives of Kepler, Galileo and the French mathematician-philosopher René Descartes.[26] Descartes had gained attention with his own attempts to appropriately define space, time and motion just a few decades earlier.[27] Descartes' solution to the problem sidestepped the contentious possibility of a true void by imagining space continuously filled with primordial vortices. These vortices provided a kind of background to support matter's motion, carrying the planets through their Keplerian orbits. Though Newton would reject Descartes' views, they served as a foil to his own innovations.

At the beginning of the *Principia* comes a small *scholium* (Latin for "comment"). In just seven pages, the *scholium* rebuilds time, space and motion, clearing the way for Newton's mechanics and the new mechanical world.

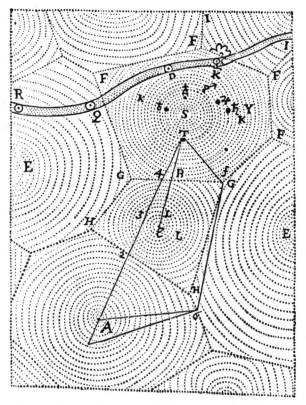

FIGURE 4.2. Descartes' cosmological physics imagined the solar system permeated by spinning vortices which carried the planets in their orbital motions.

In essence, Newton's innovation was to make space and time separate realities. Space is a "something" distinct from matter. It exists independent of any configuration of material. Space, for Newton, had properties independent of matter that were uniform and unchanging. Space, in other words, is the same everywhere in the universe. Just as important, time is a "something" too. It is distinct from matter and matter's changes. Time's flow is uniform everywhere and does not depend on the activity of matter. Newton's new perspective can be summarized as follows:[28]

- Absolute, true and mathematical time passes equally without relation to anything external and without reference to any change in matter or the way in which it is measured (e.g., the hour, day, month or year).
- Absolute, true and mathematical space is the same everywhere; its properties remain fixed without relation to changes in matter.
- Absolute motion is the movement of a body from one position in absolute space to another.

Thus Newton invented an *absolute space* and an *absolute time*. Real in and of themselves, both are constant, unchanging and independent of the relationships between matter and its changes. Newton's new vision for the space and time of physics would come to be called the "divine sensorium".[29] It was the domain of God's perfect perception of reality, an empty stage on which the play of the cosmos and the drama of physics would be enacted.

By defining an absolute space and time, Newton had successfully created his framework for defining motion. Then, with motion pinned down, he could define mechanics—the mathematical laws linking force and motion. The rest of the *Principia* would go on to articulate these mathematical laws; force, mass and acceleration would be clearly linked in a grand theoretical structure. Earth and heaven were tied together under this single set of laws, which came to be called Newtonian mechanics. Most important, these laws reworked both science and culture in ways still relevant today.

In what might be considered a marriage of the visions of Parmenides and Heraclitus, change and eternity were fused. Timeless, immutable laws governed the universe and the progress and nature of change.[30] By abstracting time and space from experience, Newton's theoretical machinery became the ideal to which all future cultural innovation would be held. The spirit of his mechanics—embodied as a faith in eternal, timeless and universal law—would soon be made concrete in the realms of government and other institutions as they rapidly attempted to master the globe.

MAPPING HEAVEN: ASTRONOMY AND COSMOLOGY AFTER NEWTON

It was a dark and stormy night on October 22, 1707, when a squadron of English warships approached the English Channel.[31] The twenty-one ships, commanded by Sir Cloudesley Shovell, were returning from an unsuccessful campaign against the French port of Toulon. Under a blanket of dense fog, the squadron's navigators believed themselves to be safely passing west of the last island outposts of Brittany and into clear Channel waters, but the weather made determinations of the squadron's east-west position difficult. The sailing masters had no way of knowing that the ships were plunging directly into the rocky outline of the Isles of Scilly. Within minutes of striking a sharp rock outcropping, the HMS *Association* disappeared below the waves, taking the entire eight-hundred-man crew and Admiral Shovell with it. Three other ships were lost that night, bringing the total number drowned to more than 1,200.

The Royal Navy's disaster of 1707 could be summed up in a single word: longitude. By the eighteenth century transoceanic commerce and military operations had grown to dominate geopolitics and geoeconomics. Maps and other navigational tools of the time were of limited use, as they couldn't provide accurate determinations of longitude—the east-west position on the map.

Knowing one's precise position on the globe (as in the middle of the featureless ocean) requires two numbers: latitude and longitude. Latitude is the position north or south of the equator and is easily determined with a simple observation of the sun's highest position in the sky. Longitude is another story entirely.

While latitude measurements are relative to the equator (or the poles), what special position should longitude be judged by? Nature did not inscribe a single giant arc running from one pole to the other for humans to judge east and west. Instead, longitude must be measured against arbitrary meridians—imaginary lines circling the globe and running through both poles. If, for example, we choose Greenwich, England, to be our reference, then longitude will be measured as east or west of that prime meridian running though Greenwich. But how do we make such measurements? Longitude measurements, it turns out, are always comparisons of time that are converted into space.

Traditional measurements of longitude required knowledge of astronomy, a calculation of local time and a very special book. The book gave times when the moon reached its highest point each night in Greenwich. If a sailor needed to find his longitude, he watched the moon and recorded the time when it reached the highest point in the sky that night (using a clock set to a local time determined by the sun's position). This was called "shooting" the moon.[32] The moon's maximum point on its nightly arc formed a kind of astronomical anchor in time. The sailor simply compared his local time measurement for the moon with the book's listing of the time the same event occurred in Greenwich. If the moon reached its zenith at 3:00 a.m. local time that day and the book told him the same event occurred in Greenwich at midnight, then he must be three hours ahead of Greenwich time. Three hours is one-eighth of the globe away, or 45 degrees of latitude east of Greenwich. By comparing different recorded times for a simultaneous event—the moon's highest point in the sky—comparisons of time were transformed into determinations of position (space).

But astronomical methods for determining longitude were complex and inaccurate. The stars rotated once a day and the moon moved against the fixed background of stars once a month. These celestial dance steps made astronomical determinations of longitude a time-consuming business. Worse still, typical errors in the method translated into uncertainties of tens of miles in position, more than enough to kill scores of sailors on ships in fog-bound seas.

The widespread introduction of mechanical clocks should have offered a solution. The idea was to carry a clock synchronized to Greenwich time and a clock that was regularly updated to local time via astronomical observations. Comparing the two clocks would yield a time difference easily converted into longitude. But the motion of ships, along with changes in humidity and temperature, prevented even the most accurate timepieces of the day from working well at sea. A new form of material engagement was needed.

In 1772 Captain Cook set out on his second voyage of discovery with a copy of clockmaker John Harrison's H4 shipboard chronometer.[33] Harrison had spent his life working to perfect his longitude-determining devices.[34] Returning three years later, after a voyage ranging from the tropics to the Antarctic, Cook reported that the clock's daily tally of seconds held steady. Translated into distance, its variation never exceeded eight seconds of longitude (two nautical miles at the equator).[35] Cook referred to Harrison's invention as "our faithful guide through all the vicissitudes of climates".[36] It is not known for certain whether Harrison knew of this success, but Cook's voyage proved beyond doubt that longitude could be accurately measured using timekeeping machines alone. The last epoch of sky-based time was ending, passing the baton to an emerging new machine era. In this way the contours of the world began to sharpen in human maps and minds.

As the globe was mapped with new precision during the seventeenth and eighteenth centuries, so too was the sky. Maps of the heavens were filled out by observations from ever more powerful telescopes as humankind took its first halting steps towards an astronomically based cosmology. As with so much else in the Age of Reason, the story begins with Newton.

In the *Principia* Newton laid out his vision of space and time, then used them as a background to formulate a new mechanics—a science that could accurately describe the relationship between force and motion. Newton's laws are embodied in the famous formula $F = ma$ (a force F applied to a mass m will produce an acceleration a). But this formula leaves the actual nature of the force undetermined. It is a general expression that applies equally to all forces—a cannon's blast, the friction of a rough surface or the pull of a mule team.

There was, however, one force Newton knew required special attention. In the *Principia* Newton provided a detailed description of a new agent he called gravity—an attractive force occurring naturally between any two objects. Its strength decreases with distance (specifically, the force of gravity diminishes with the square of the distance between the objects). Most important, Newton's gravity was universal, as applicable to apples falling from trees as it was to planets orbiting the sun. By giving a specific calculable form for gravity and making it universal, Newton wiped away the long-held distinction between the sublunar and celestial realms. With a single equation Newton cast two thousand years of Aristotelian physics aside and united the heavens and Earth.

The success of Newton's gravitation was quickly apparent to astronomers across Europe. Kepler's famous laws of planetary motion derive as easily from Newton's gravity law as an overripe fruit drops from a gently shaken tree. Astronomers soon seized on the universal law of gravitation, applying it to everything from planetary orbits to tides and the rotation of the moon. With Newton's precise formulation of the gravitational force, astronomers moved from simple descriptions of what they see in the sky to the first understanding of how these motions occurred. Through Newton's law of gravity, astronomy becomes a science of dynamics—an explanation of objects moving through space and time. The timing was fortuitous: movement was, for the first time in human history, emerging as a recognized property of the stars.

Telescope technology had been steadily improving throughout the seventeenth and eighteenth centuries, giving astronomers deeper and more resolved images of the heavens. The addition of the filar micrometer, a device that allowed precise telescopic measurements of celestial position, enhanced celestial cartography.[37] Just as nautical chronometers were producing accurate maps of the Earth, the new astronomical technologies were mapping the stars with enough precision to see their precise position in space *and* their movement in time.

In 1718 Edmund Halley came to question the orthodoxy that the stars were fixed in their positions relative to one another as he compared his observations of stellar positions with those found in Greek catalogues. Finding numerous discrepancies, he arrived at the astonishing conclusion that some stars had moved in the 1,500 years between ancient and modern observations. By 1738 his claims were verified for the star Arcturus.[38] Twenty years later, no fewer than eighty stars had been shown to exhibit so-called proper motion—movement across the sky. The term "fixed stars" became an anachronism.

These advances in astronomy renewed interest in cosmological questions. In some cases, new cosmological theorizing relied heavily on astronomical discoveries. In other cases, they remained heavily philosophical, with astronomy playing only a supporting role. Both forms of Enlightenment cosmology would play an important role in the development of the subject into our modern era.

The introduction of Newton's law of gravitation gave cosmological thinkers a tool they could use in asking specific questions about infinity and the evolution of the cosmos. Newton, in a series of letters to the

FIGURE 4.3. The Paris Observatory at the beginning of the eighteenth century. Newton's mechanics becomes the basis for understanding astronomical motions as astronomers begin charting the sky with ever greater accuracy.

theologian Richard Bentley, saw his own work pointing to a universe of infinite spatial extent. In a finite universe, the mutual gravitational attraction of each star upon the others would lead the entire system to collapse. Thus, an infinite universe composed of infinite stars was the more attractive alternative, but Newton understood that ultimately it too would be vulnerable to gravitational collapse. While a lattice of stars stretching to infinity would balance all gravitational attractions, the configuration was unstable; if one of the stars is nudged, gravitational forces would become unbalanced, forcing the stars to move towards one another and eventually collapse into giant clumps. Newton wrote that this instability made it hard to imagine how the stars could be "so accurately poised one among another, as to sit still in perfect Equilibrium".[39] The problem would come to be known as the gravitational paradox, and it would haunt cosmology well into the twentieth century. With his strong religious devotion, Newton "solved" the problem by assuming that God continued to play a role in the cosmos, periodically intervening to maintain order and prevent collapse.

While many astronomers remained bothered by this infinite number of stars (for example, how was such a number to be constructed?), the idea gradually gained acceptance. But with infinite distance came a renewed discussion of infinite time. In an era when Christian orthodoxy still required a universe only six thousand years old, most scholars were still unwilling to imagine an eternal universe. Newtonian astronomer James Ferguson saw how the gravitational paradox could be used to prove that the world's age was finite. He wrote:

> For, had it [the world] existed from eternity, and been left by the Deity to be governed by combined actions of the above [Newtonian] forces . . . it had been at an end long ago. . . . But we may be certain that it will last as long as was intended by its Author, who ought no more to be found fault with for framing so perishable a world, than for making men mortal.[40]

Adding to the debate in 1676, Danish astronomer Ole Romer discovered that light travelled with a finite speed. While many astronomers were "curiously reluctant" to recognize that looking out in space meant looking back in time, others grasped this powerful link between space and time.[41] As Francis Robert wrote in 1694, "Light takes up more time in Travelling from the Stars to us than we in making a West-India voyage (which is ordinarily performed in six weeks)."[42]

In the wake of Newton and the material engagement of new telescope technologies, all the ancient questions of cosmology had been reopened. Acceptance of an infinite Newtonian universe had renewed discussion of other universes. Newton opined that, given God's omnipotence, other cosmos with other laws of nature might well exist. The French mathematician Gottfried Leibniz (the co-inventor with Newton of calculus) also held that God could create other universes but chose not to since this one was the best of all possible choices. The Italian Jesuit astronomer Roger Boscovich even imagined that other "spaces" might exist causally disconnected from our universe.[43] Boscovich reasoned that nothing occurring in our universe could affect the behaviour and evolution of matter in these other spaces. Likewise, all events in these other universes could not have any effect on the course of cosmic history in our space. The idea was wild speculation for Boscovich's day but now appears as a premonition of ideas that play a pivotal role in our own historical moment.

Many modern alternatives to the Big Bang hinge on the concept of a multiverse. Thus these early discussions of its possibility within a Newtonian framework show us, once again, the resilience of the five great cosmological questions. Time and again, humanity returns to its same storehouse of cosmological possibilities even as new forms of material engagement alter culture, politics and economics.

THE BIRTH OF INDUSTRIAL TIME

As the eighteenth century came to a close, astronomers had triumphed in mapping the dynamics of the night sky. So complete was the victory of Newtonian mechanics in astronomy that even the role of God had diminished to insignificance. When Napoleon asked the renowned

French astronomer Pierre Laplace why God was never mentioned in his new book on celestial mechanics, Laplace famously answered, "I have no need for that hypothesis."[44]

It is, then, even more remarkable that at the very moment astronomy began to shed light on the movement of the stars, the industrial revolution would rob humanity of its most basic experience of night. As the foundations of Newton's mechanics gave birth to a new age of machines, industrial culture would rewire human time to change the boundaries of work and rest, day and dark.

By most accounts, the industrial revolution began in earnest at the dawn of the nineteenth century. The movement from cottage industries founded on small-scale, home-based operations to large-scale, machine-intensive industrial production is, once again, a story of material engagement facilitating the creation of new cultural and political structures. Unlike previous transformations, however, the coupling of idea and practice, theory and application, would be explicit and nearly simultaneous. The distinct processes of science would now shape how humans encountered the world's material properties. These would, in turn, shape the culture into which all future generations were born. As Newtonian mechanics became machines, both human time and cosmic time would be transformed.

The revolution began in England. New machines were the pivot points; textile production led the way as inventions such as John Kay's flying shuttle and James Hargreaves' spinning jenny allowed industrialists to turn cotton into thread on a massive scale.[45] Based on the efficacy of these machines, the factory system was born, and new methods of production came with it. Workers now focused their effort on these machines, carrying out only a few, repetitive tasks in the chain of steps needed to manufacture a product—allowing textile production to jump by a factor of three from 1796 to 1830.[46] Factories, originally located alongside rivers to tap the power of flowing water, were soon fuelled instead by James Watt's steam engine and could be sited anywhere. The creation of Robert Fulton's steamboat and George Stephenson's steam-powered train brought transportation into the industrial revolution, pushing aside the age of the sail and horse-drawn cart.

Human resources also fuelled the revolution. In order to create a ready pool of workers that could be drawn to the cities, the Enclosure Act of 1801 forced English peasant farmers off the land they had tended for generations.[47] While other countries would soon follow, England led the way in its invention of both machines and the cultural forms that would define the new age.

Many of the transformations in human time that arose during the industrial revolution are obvious enough that they need no reiterating. Suffice it to say that the factory system formed an entirely new time matrix for an ever-expanding fraction of Europeans. Time became compressed and abstracted for both worker and manager alike. For the first time in human history, minutes became a temporal unit of exchange— they counted and they could be counted. They became crucial as factory owners sought efficiency in production and workers resisted the push to turn their bodies into mere extensions of machines. For millennia, work had been limited by biology in terms of both human and animal exhaustion. Steam-powered machines freed work from these ancient biological constraints—at least in the eyes of those who owned the machines. The complete reordering of people's experience of time began with work but soon extended to every aspect of cultural and individual life. Often lost in discussions of punching the clock is the fate of that most basic of human temporal encounters, the experience of night.

THE END OF NIGHT

"No previous time in Western history experienced such a sustained assault upon the nocturnal realm as did the period from 1730 to 1830", wrote Roger Ekirch in his excellent *At Day's Close: Night in Times Past*. Living as we do in a world of twenty-four-hour illumination, we in the modern world find it difficult to recapture the extent of night's hold on human consciousness. "Night was man's first necessary evil", Ekirch begins, "our oldest and most haunting terror. Amid the gathering darkness and cold, our prehistoric forbears must have felt profound fear."[48]

Before industry reshaped human life, night was an elemental danger. Even the night air itself was thought to be thick and poisonous. The out of doors became hazardous as the fading of the sun's rays drained light and colour from the world. "Shepherds all, and maidens fair", wrote playwright John Fletcher in 1610, "fold your flocks up; for the air 'gins to thicken and the sun already his great course has run".[49] Doctors warned their patients to beware the "heavy vapours" of night. Physical health was not the only concern. In a prescientific age steeped in biblical lore and the certainty of demons and witchcraft, the night was a time of supernatural terrors. "Night belongs to the spirits", held a proverb; John Fletcher wrote that it belonged to the "blacke spawne of darknesse".[50]

Authorities, both religious and secular, had little power in the night and each home was counselled to become a small fortress. Cities and towns were fortified against night and its human dangers as well—many larger human habitations were surrounded by walls. The only way in and out of most cities at night was through fortified gates that were closed at sunset, watched over by guards and not reopened until dawn. But in the industrial age, these walls and gates were often torn down. What had once been protection from night terrors became a barrier to the efficient exchange of goods and materials that would need to continue, unabated, all night. "They are a relic of the past", said an eighteenth-century writer surveying Bordeaux's city walls and decrying them as a hindrance that must be "condemned by economic necessity".[51]

Before industrialization, the fall of night meant returning home and turning to sleep. But even sleep, that most intimate and personal arena, would be transformed by the industrial revolution. While candles and wood fires had long provided artificial lighting, the illumination they yielded was weak. If you wanted more light, you needed more candles, which were not cheap.[52] Thus, for much of our history, most human beings experienced night as a long round of sleep that began shortly after the fall of darkness. But their sleep was not our sleep. Instead it was a kind of slumber that has been all but lost to us.

"First sleep" was its common name in Europe before the industrial revolution. "Until the close of the early modern era, Western Europeans on most evenings experienced two major intervals of sleep bridged by up to an hour or more of quiet wakefulness", writes Ekirch.[53] Even the memory of this pattern is lost to us today. In our modern era we think being awake at night is a symptom of discord or difficulty—we have failed to get a "good night's sleep". But plays and diaries from Europe's preindustrial age tell a story of a different pattern, where two periods of sleep divided by a period of wakefulness was the norm. "I believe 'tis past midnight, for I have gotten my first sleep", wrote George Farquhar in his play *Love and a Bottle* (1698).[54] Thus the experience of night and sleep, so fundamental to the way we experience time, was fundamentally different before the industrial age.

Even before industry changed everything, as the era of science and reason began in the 1600s, attitudes about night had begun to shift. Fear of "night vapours" became the superstitions of a previous age, and people were more apt to be out after hours for social or commercial reasons. But as the industrial revolution took hold at the end of the 1700s, the millennia-old experience of night was turned fully on its head. Material engagement, in the form of new technologies for lighting, would drive the change, sweeping night away in a blaze of illumination. "Since the invention of gas light, our evening life has experienced an indescribable intensification", wrote a nineteenth-century diarist, "our pulse has accelerated, nervous excitation has been heightened; we have had to change our appearance, our behaviour and our customs, because they had to be accommodated to a different light."[55]

Beginning with burning wood, the long history of lighting was, until the mid-eighteenth century, a story of few innovations. During the preindustrial period, the candle and oil lamp represented the greatest inventions past pitch-covered torches. The first important departure in lighting technology came in Paris in 1760. In 1763, the Académie des Sciences held a competition for the development of a new light source to illuminate the city's crime-ridden streets.[56] The invention that followed was called the *réverbère* or reflector lantern. Using an oil-burning base with several wicks backed by two reflectors, the *réverbère* was many times brighter than ordinary lanterns of the time. As *réverbères* were deployed across Paris the results were hailed as a revolution. "Now the city is extremely bright lit", wrote one commentator. "The combined force of 1,200 *réverbères* creates an even, lively and lasting light."[57]

In truth, the lighting from the *réverbères* would seem pale by today's standards, but their connection with state authority via the police (they were often called police lanterns) made the populace feel as though the dangers of night had been pushed away. During the French Revolution the lanterns themselves would become symbols of the hated monarchical state. The practice of tearing down and smashing lanterns in political uprisings (vividly described in Victor Hugo's *Les Misérables*) became a favourite activity of would-be revolutionaries that continued up until the introduction of gas lighting.[58] The 1848 revolution against the Hapsburg monarchy is most strikingly remembered for the destruction of gaslight poles in Vienna and the ensuing pillars of fire erupting into the city's night of chaos.[59]

The widespread use of gas lighting marked the first real step into night's new day. Introduced at the turn of the nineteenth century, street lamps burning coal gas provided illumination ten times brighter than the *réverbères*. First appearing in 1807 on London's Pall Mall, gas lamps—more than forty thousand of them—would soon light two hundred miles of London's streets.[60] Other cities across Europe and America quickly adopted the new gas technology as a vast infrastructure of gasworks linking underground pipes to street lamps was set in place. Massive gasometers with capacities of thirty thousand cubic metres (a million cubic feet) or more were routine by the 1860s, becoming visibly potent symbols of industrial progress.[61] More visible, however, was the nightscape. Where the city at night once meant only danger to body and soul, gaslight engendered a new, more cheerful nightlife. As an 1829 visitor to London described the new face of night, "Thousands of lamps, in long chains of fire, stretch away to enormous distances. The display of the shops, lighted up with peculiar brilliancy, . . . is most striking in effect. The streets are thronged with people, and thousands of elegant equipages roll along to the appointed dinner-hour party."[62]

People were out, shops were open and night had been vanquished. But those shops needed merchandise, and here too, in manufacturing, gas lighting triumphed. The widespread use of gas lighting in factories allowed bosses to add night shifts. Production was now continuous and workers could clock in at midnight, keeping machines running on well-lit shop floors through the long hours till dawn.[63]

Electricity scored the final victory over night. When Thomas Edison presented his carbon filament lamp (lightbulb) at the Paris Electrical Exhibition in 1881, it was instantly hailed as the light of the future.[64] By 1882, the first central electricity-generating stations were operational in London and New York.[65] Electric lighting rapidly spread into factories, streets and, most important, homes. The last of these inroads was crucial in changing the human experience of night and time. The unpleasant smell and very real possibility of poisoning or explosion had made bourgeois households slow to admit gas lighting. But as one com-

FIGURE 4.4. Artifical lighting steals the stars. By the late 1800s, night was becoming day in the cities of Europe and America. Gas light, followed by electrical illumination, became commonplace, profoundly changing the experience of time by altering every aspect of human life including sleep.

mentator wrote, "All doors were open to electric light."[66] Electric illumination was hailed as not only superior to gaslight but also good for the health. "Electricity was regarded as positively beneficial, almost as a sort of vitamin."[67]

By the turn of the twentieth century, night and its ancient habits had vanished in the cities. Brightly lit factories ran twenty-four hours a day.[68] Brilliantly illuminated cities carried on their business under the unswerving arc of a thousand electric suns. Brightly lit homes allowed their inhabitants to substitute work or entertainment for rest. The quaint terms "first sleep" and "second sleep" disappeared. And, all the while, the great rush to turn darkness into day left growing multitudes of city dwellers unaware of what they had lost. The glare of artificial lighting blotted out all but the brightest stars and robbed even those beacons of their vividness and mystery. Scientists, on the other hand, had begun to claim the night sky for themselves, opening windows deeper into the universe and taking their first halting steps towards a truly scientific cosmology.

COSMIC HEAT DEATH AND THE ARROWS OF TIME: COSMOLOGY ADDS THERMODYNAMICS

Newton and his theories were a step ahead of the technologies that would define his age. Thermodynamics, the grand theoretical vision of the nineteenth century, operated in the other direction with practice leading theory. The sweeping concepts of energy, heat, work and entropy, which thermodynamics (and its later form, statistical mechanics) would embrace, began first on the shop floor. Originally the domain of engineers, thermodynamics emerged from their engagement with machines. Only later did this study of heat and its transformation rise to the heights of abstract physics and, finally, to a new cosmological vision.

Thermodynamics is a science of systems. If that term seems generic and abstract, it is meant to be. A system is any collection of interacting parts. The great beauty of thermodynamics is its applicability to any and all systems: mechanical, biological, celestial. Proposed as a universally valid theory of energy and its changes, thermodynamics would prove as useful for stars as it was for steam engines. With work extending across the nineteenth century, scientists such as Sadi Carnot in France, Rudolf Clausius in Germany, Willard Gibbs in the United States and Ludwig Boltzmann in Austria laid out a new theoretical framework for change and time. These new laws allowed scientists to see general principles of evolution at work even in the most complex systems. And evolution was the key—it became the watchword for the new century and its science. The core of thermodynamics was its ability to map the evolution of systems to the flow of energy. This emphasis on change and transformation allowed the new science to offer the first scientific explanation for the flow of both personal and cosmic time.

Thermodynamics added two new laws to the rule book of physics. Its first law tells us that the total amount of energy in a closed system is always conserved. The adjective *closed* simply means the system is isolated from the rest of the universe. Energy can take many forms: motion, gravity, magnetism and more. It can flow from one form to another just as a dropped stone converts gravitational energy into the energy of downward motion. But the first law of thermodynamics tells us that no matter what transformations occur, the total amount of energy you begin with must be the same quantity with which you end.

Of particular importance to nineteenth-century physicists was the recognition that heat was just another form of energy. In a return to the Greek ideas of Democritus and others, the atomic hypothesis was raised once again. Temperature was eventually seen as nothing more than a direct measure of the random motions of atoms. The dropped stone will feel slightly warmer after it impacts the ground because the energy of its "bulk" motion (the stone as a whole moving downwards) is converted into the random motions of its atoms when it slams into the pavement. By understanding and including heat in its inventory of energy, thermodynamics was born as science.

The first law of thermodynamics—"Energy is conserved"—was not a difficult principle to understand. The all-powerful second law, how-

FIGURE 4.5. The triumph of thermodynamics. A giant "Corliss" steam engine on display at the Philadelphia Exposition, 1876. Material engagement in the form of steam power became the basis for both the industrial revolution and the theoretical principles of thermodynamics.

ever, was something new indeed, and it added an entirely new conception of time and evolution to the framework of the cosmos.

The second law of thermodynamics told physicists that energy flowed from one form to another in a particular direction. It focused on transformations of energy that produce useful work: a burning fire creates steam, which turns a wheel; an exploding star sets a trillion trillion tonnes of gas into motion. According to the second law, transformations that move energy into a workable form must also create *un*usable energy. In other words, useful energy transformations always create waste.

For machines, the consequences of the second law are immediately apparent. A steam engine can never be 100 percent efficient; in burning coal to power it, some of the energy released will go into heating the engine itself rather than turning its gears. The same rules apply for the formation of stars. A cloud of interstellar gas can collapse under its own gravity to form a sun, but the collapse will generate heat, inflating the cloud and slowing the collapse. There will always be waste heat created. Physicists found a way to explicitly calculate this notion of waste heat in a new physical quantity called entropy.

Entropy can be thought of as the disorder in a system. All energy transformations that do work also create disorder. When you break eggs to make an omelette (the classic example), the tidy system of egg white and yolk gets irreversibly mixed. Thus in making an omelette you produce disorder and entropy. The second law tells us that all energy transformations that do work also create entropy and, most important, that the entropy in a closed system can never decrease.

This was the principal lesson of the second law: entropy for systems as a whole will always increase and can never decrease. At best, in a closed system, the entropy will reach a maximum value when equilibrium has been reached. At that point all evolution ceases. Place a hot cup of coffee in a box of cold air and heat will flow from the cup to the air until both reach the same temperature. In that final configuration of cup and air in equilibrium, the entropy of the total system will be at its maximum and all further evolution (in temperature at least) is ruled out.

Time and the second law of thermodynamics seem intimately connected. If you burn a tonne of coal to drive a steam train, the entropy of the system increases. Because you cannot reduce a system's entropy, you cannot unburn the coal. The transformation can only flow in one direction, and that direction appears to separate the past (low entropy) from the future (high entropy). In the nineteenth century, this so-called arrow of time—moving from past to future—and the entropy increase demanded by the second law appeared to many scientists to be equivalent.[69] Nineteenth-century cosmological thinkers quickly incorporated the relationship between entropy and time into new models of the universe's evolution.

The second law appeared to have implications for both the beginning and the end of cosmic time. "There is at present in the material world", wrote British physicist William Thompson, "a universal tendency to the dissipation of mechanical energy".[70] Dissipation meant entropy's generation of waste heat. Even in 1850 Thompson could see its consequences for the terrestrial future; he proposed that eventually the generation of this "waste" heat in the evolution of the Earth would render the planet "unfit for habitation of man". In the 1860s Rudolf Clausius took this thinking to cosmological heights, coining the term *heat death* for the universe. Clausius was sure that thermodynamics demanded a cosmic accumulation of entropy until a maximum was achieved and all further evolution ceased, leading to an eternal and universal stasis.

> The more the universe approaches this limiting condition in which the entropy is a maximum, the more do the occasions of further change diminish; and supposing this condition to be at last completely attained, no further change could evermore take place, and the universe would be in a state of unchanging death.[71]

It is noteworthy that these pronouncements were made by physicists rather than geologists or astronomers. What makes thermodynamics so powerful is that regardless of the system—the Earth, a star or the universe itself—the first and second laws will always hold true.

By the mid-1800s all scientific establishments had come to recognize that evolution was a fundamental principle. Darwin had already shown that life evolved, and geologists such as Charles Lyell had shown that the Earth evolved.[72] With its growing focus on dynamics, astronomy had shown that the heavens evolved as well. Could the universe, as a whole, be any different? Clausius and others argued that if cosmic evolution did occur, then its pathways must be no different from those of a steam engine.

Ready to go further than just predicting the heat death of the universe, Clausius was sure the principles of thermodynamics could be used to rule out some cosmological models as surely as it pointed to the veracity of others. Throughout the nineteenth century, a growing chorus of scientists and philosophers had begun to entertain cyclic, or oscillating, models of cosmic history in which creation was followed by destruction over and *over* again. According to Clausius, the second law ruled out such models.[73] Entropy generated in one cycle could not be disposed of at the beginning of the next. It would persist and eventually drive the whole system to equilibrium. Heat death could not be avoided.

Just as Clausius used the second law to argue against a cyclic geometry for the cosmic dimension of time, others would use it to argue against an infinite age for the universe. While the entropy of the world must increase towards an equilibrium state, it was easy to see that we have yet to reach that state. Thus, some writers argued, the universe cannot have existed forever and must have originated at some finite time in the past in a low-entropy state. For avowedly Christian scientists such as William Thompson, this apparent "proof" of a beginning held strong appeal. In an address to the British Association of Science, Peter Guthrie Tait, one of Thompson's Christian colleagues argued,

> The present order of things has *not* evolved through infinite past time by the agency of laws now at work, but must have had a distinctive beginning, a state beyond which we are totally unable to penetrate, a state, in fact, which must have been produced by other than the now acting causes.[74]

This other cause was, for Tait, the God of the Christian faith.

Infinite time versus finite age—the old debate had arisen again, but for the first time scientists had general scientific principles to use as tools in separating the possible from the improbable. Thermodynamics gave physicists some firmer ground from which to pose their questions and find their answers. When Tait asked, "What happened before?" he was able to set the question in the context of scientific principles of thermodynamics even while reaching to religion to provide an answer. Cosmological thinking was at the edge of a new age in which pure speculation based on purely philosophical reasoning was fading into the past and principles of mathematical physics were rising to the fore. The complete transformation would, however, take time.

It must be noted that there were also many scientists who rejected the use of thermodynamics for cosmology. Some, including Ernst Mach, claimed that no meaningful statements could be attached to the universe as a whole.[75] Others questioned how concepts such as entropy could be used in an infinite universe. It is also noteworthy that the first efforts to use thermodynamics as a cosmological principle remained divorced from detailed contact with astronomical data. So, in spite of thermodynamics, cosmology remained in a foetal stage during the 1800s and was still as much in the province of philosophy as it was in that of science. By the close of the nineteenth century, however, a thermodynamic vocabulary for cosmological debate had already been firmly established, and it would persist into our current era. Even today it shapes our conception of what might lie beyond the Big Bang.

ENDINGS AND BEGINNINGS: BRAIDING TIME IN COSMOS AND CULTURE

When considering purely mythological/religious narratives of the universe and time, it was easy to see how the braiding of cosmos and culture operated. Myths that suited the hunter-gatherer would not serve the farmer, and so they were replaced. But when entering the era dominated by science, the braiding between cosmic and human time becomes more subtle. Technologies don't simply appear fully formed from cloistered laboratories to do their work changing culture and fostering new cosmological ideas. Instead, the braiding of human time and narratives of cosmic time appear almost fractal—as if every fibre of each strand of the braid separates to form new braids with the fibres of the other strands.

Broad-brush accounts of cultural needs driving the next breakthrough in technology are too simple to reach the truth. And the truth is so much more interesting than a simple story of humanity simply discovering the objective account of cosmos and time. Cosmologist Edward Harrison, echoing Joseph Campbell in *Masks of God*, spoke of cultures creating "masks of the universe". Each mask is a kind of filter for the experience of the universe that cannot be removed cleanly in order to see the "objective" reality that lies behind it. Instead, each mask guides our investigations through the process of material engagement. With the maturation of science, human beings found powerful tools for entering into a new kind of dialogue with nature. The enigmatic entanglement between culture and scientific cosmologies shows us how, in peeling away layers of nature's behaviour, we also create new masks for what we call the universe.

As the pace of scientific and cultural change accelerated into the twentieth century, this enigma deepened. In the path to the present and the Big Bang (which now stands on its own precipice), the next step would see the threads of human and cosmic time become even more tightly woven.

Chapter 5

THE TELEGRAPH, THE ELECTRIC
CLOCK AND THE BLOCK UNIVERSE

The Imperatives of Simultaneity from Time
Zones to Einstein's Cosmos

SOMEWHERE BETWEEN NEW YORK AND PHILADELPHIA, USA ·
1881, 10:05 A.M. (GIVE OR TAKE)

He wants to stay calm but it's not working. Beneath his starched shirt the sweat is rising. So much depends on this interview; how is he supposed to stay calm? Through the train window he watches the landscape rumbling by. Usually travelling by train is exciting, a pleasure. But today he takes no pleasure in the ride. All that matters is that he make it to Philadelphia on time.

At least the train is moving again. More than an hour was lost as the train sat on the tracks. Oh, God, *he thinks*, I cannot be late.

For the twentieth time since he left Newark, New Jersey, he consults the gleaming pocket watch, a present from his new bride. This morning as he scraped stubble from his cheek with his straight razor she stood behind him, her generous smile reflecting in the mirror. "They are going to be so impressed with you," she said. "How could they pass on such a handsome and intelligent young accountant?" He had tried to smile. Then she showed him the box with the golden watch inside. "A good accountant keeps accurate time," she said. She threw her arms around him and whispered, "I love you" in his ear.

He needs this job, needs it badly. All their dreams of starting a family, buying a house—all his dreams for her—depend on his getting this job. But

117

FIGURE 5.1. A train on the Pennsylvania Railroad in West Philadelphia (c. 1874).

first the train has to arrive on time. He cannot show up late for the interview. He runs through the calculation again in his head. The train left Central Station at 8:25 a.m. and was scheduled to arrive in Philadelphia at 11:55 a.m.[1] His appointment was at 1:30 p.m. Now he had lost an hour as the train just sat there on the tracks. It was going to be so close. He looked at his watch again and a chill shot through him as remembered what she had said as he left. It hadn't mattered then. Now, suddenly, it meant everything.

"You have to reset your watch when you arrive," she'd said. "Remember, Philadelphia is far away. They have their own time."

A NEW NOW

Fifty years was all it took to partition the world's hours. After covering tens of millennia in our story of human and cosmic time, we now reach the boundaries of our modern life—a world of legalized, compressed and metered time. From the Palaeolithic to the Neolithic, from the first city-states to the Greeks' rational cosmos, from New-ton's mechanics to the industrial revolution, human time has changed and changed again. Cosmic time, in narratives of mythic creation or scientific evolution, has transformed as well. These changes reflected and refracted off each other as our engagement with the raw stuff of the world shaped the institutional facts of each human life. Our story of these changes has, up until now, run with the rhythm of centuries. For the most part, entire generations could pass with only small dif-ferences in their experience.

There are only forty or so years separating the American Civil War and Albert Einstein's theory of relativity. In those four decades, human time and cosmic time would undergo profound transforma-tions and influence each other as never before. In 1865 railways, pow-ered by the same steam technology that had driven the industrial revolution, were in the middle of their assault across the continents.[2] In that same year telegraph cables threading electrical impulses into instantaneous communication were just beginning to bind far-flung cities to each other.[3] As distance shrank, time became problematic in an entirely new way.

Simultaneity—the balance between your time in your location and my time in my location—suddenly moved from abstract physics into the realms of nation building and economic necessity. In 1865 simultane-ity was just beginning to become an issue of contention. In 1905 Albert Einstein would make it a cornerstone of his radical revision of physi-cal law that would lead the way to the first true scientific cosmology. The confluence of real-world and theoretical concerns with simultane-ity would prove to be no accident.

RAILWAYS AND TIME ZONES

It was a crisp autumn day in Chicago when the modern meaning of "now" was legislated into existence. On October 11, 1883, the first Gen-eral Time Convention convened in this great hub of the American Mid-west. Its mission was to rationalize the patchwork of hours that had spread like vines across a nation newly connected by transcontinental trains. Time reform was the order of the day.

In the United States and Europe, the growth of railways connecting city and village alike reshaped the human experience of distance and time. One hundred miles lie between New York and Philadelphia. In the 1770s, the fastest time in which that distance could be covered was two days (via carriage).[4] By the 1880s, regular train service cut the trip to a mere three and a half hours.[5] By linking distant cities in short trips, travellers were forced to confront the new and vexing issue of time standards. Each large city kept its own standard. Whose time was the traveller bound to? The clocks of each city were set according to a regional time standard often provided, remarkably, by astronomers working at the local university observatory. Thus noon in New York was not noon in Philadelphia.[6]

Before the advent of trains these local differences in time did not matter. If it takes a day and a half to travel from New York to Philadelphia, you are not likely to care about a five-minute difference in the definition of 2:00 p.m. If, however, you are planning to leave Philadelphia by catching the 5:05 back to New York after a day of business, those five minutes suddenly rise in importance. Thus the public meaning of time and the personal experience of "now" were reshaped in the forge of cultural innovation. As William Allen, secretary of the General Time Convention of Railroad Officials, put it, "Railroad trains are the great educators and monitors of the people in teaching and maintaining exact time."[7]

By 1880 a schizophrenic patchwork of local times had emerged. To deal with the mess, American railways adopted an ungainly convention. Inside the train, clocks were set according to specific major cities on a given railway line.[8] That meant it could be 1:00 p.m. inside the train and 12:00 p.m. in the towns the train was passing. By 1883 there were at least 47 lines on New York time, 36 taking time from Chicago and 33 with clocks in sync with Philadelphia. It was a mess of hours.[9]

Things were worse in France, as Paris became the standard for all railway lines. A traveller in Nice, almost six hundred miles from the capital, would experience three different times as he approached a railway station. First there would be the local time given by the clock in street; then he would find Paris time given by a clock in the railway waiting lounge; finally, he would encounter train platform time—set a few minutes different than Paris time to give the confused traveller "a margin of error" in catching his trains.[10]

The call to rationalize time rose across the world. The United States, with its continent-spanning jurisdiction, led the charge. The problem lay in the difference between the experience of local time— interpreted as it had always been by local celestial rhythms—and the new traveller's time, which had outpaced the planet.

The determination of noon, for example, depends on where you are on the rotating, spherical Earth. Define noon to be the moment when the sun reaches its highest point in the sky. Now imagine the meridian that runs along the Earth's surface from the North Pole to the South Pole and cuts right though where you are standing. Anyone standing anywhere along your median shares your understanding of noon and therefore shares your time. Their astronomically defined "now" will be the same as yours. But if you step just a little to the west or to the east, the astronomically defined time standard (noon, midnight, etc.) shifts. Time reformers at the Chicago meeting were demanding a global convention that would move people away from this local sun-centred interpretation of time.

Throughout the 1870s various schemes were submitted, some more radical than others. Taking reform to its extreme, Sandford Fleming, a powerful empire-building Canadian railway engineer, proposed a single worldwide system.[11] Under Fleming's plan it would be 3:00 a.m. everywhere on the planet at the same time regardless of local conditions of day or night.

But not everyone was interested in time reform. As plans like Sanford's one-world time became public, a growing chorus of protestors began vocally challenging any form of reworked hours. For those opposed to time reform, sun-bound time had to take precedence. As the superintendent of the U.S. Naval Observatory, John Rodgers, put it, "The Sun is the national clock. . . . No other clock can supersede it, as it is the one ordained by Nature to regulate man's life."[12]

The reformers came back with a hybrid system—a compromise between the economic needs of a nation stitched together by train lines and the natural local rhythms of day and night. They divided the nation up into time zones, each 15 degrees wide. The first, eastern zone began 75 degrees of longitude west of Greenwich, England (roughly the eastern edge of the East Coast). The second began 90 degrees of longitude west of Greenwich; the third time zone began at 105 degrees west, and the last 120 degrees west of Greenwich. Everyone 7½ degrees on either side of the meridian defining a zone's centre would use the time defined on that central meridian.[13]

Once the major railways began to sign on to this plan, its opponents quickly saw they were outgunned. In fact, the protestors were outdistanced. When the final vote at the General Time Convention was held,

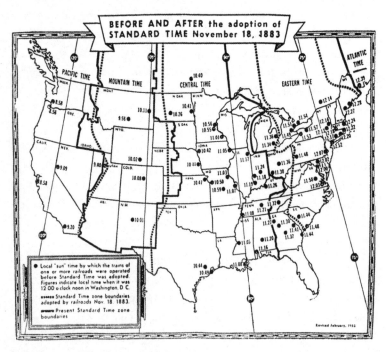

FIGURE 5.2. The divisions of U.S. time before and after the 1883 Standard Time Convention.

the tally was taken in miles of railway track. Reform got 79,041 miles, with only 1,714 miles against the plan.[14] The major railways had made their decision, forcing the major cities to fall in step. Understanding the need to tie their clocks to the economic lifelines of the railways, New York, Boston and Chicago agreed to give up their own local time standards and hitch their official rhythms to a centralized standard. Time had become nationalized by legislation.

Railways reshaped the human experience of time in more ways than simply providing short travel times between cities. The economic power inherent in the rapid transport of goods meant that oranges picked Monday in Florida could appear on the tables of middle-class families in New York a few days later.[15] All the raw materials of the furiously expanding industrial economy were circulating along rail-driven arteries. A new, communal pace of human culture was emerging from the human engagement with steam, coal and steel. Wires would soon be added to that arsenal.

THE TELEGRAPH AND THE POLITICS OF SPACE-TIME

While every steam train was a massive reminder of the power of industry to reshape life, a more ephemeral transformation contributed to new institutional facts changing human time. From one end to the other, the planet was being strung together with telegraph cable. If railway lines formed a circulation system for the new world order of time, telegraph lines carrying precisely metered electrical signals represented its rapidly growing nervous system.

The *click-click* of the telegraph provided people with their first direct experience of simultaneity at great distances. In 1867 transatlantic cables were drawn from Greenwich to the Harvard Observatory in Cambridge, Massachusetts.[16] Soon all the great cities of Europe and the United States were connected. A telegraph operator watching the electromechanical snaps of his receiver in Paris was tied to his fellow operator in New York tapping out the message. Time and space collapsed into the invisible domains of electric currents racing along wires lying on the ocean floor.

The consequences of establishing a common "now" for Paris and New York (or any other location linked by telegraph) would prove more far-reaching than the ability to send urgent news from one continent to another. Simultaneity in time translated into accuracy in space. It meant accurate maps. For the great land grab of late nineteenth-century empire building, mapmaking was a politically charged business.

The accurate measurement of global position had been an issue for centuries. As we saw in the last chapter, it was the determination of longitude—the east/west location from an agreed-upon meridian— that posed the problem. The accurate maritime clocks that John Harrison developed had gone a long way towards solving the problem, but carving empires out of Africa and the rest of the world demanded even greater precision. Determining if a rich copper deposit lay in a Belgian, Portuguese or British colony was more than an academic matter of cartography. Wealth and power would flow from exact maps.

Into this breach stepped the new technology of telegraphy, with accuracies that depended only on the engineering physics of tracking electric pulses through a very long wire. From the 1870s to 1900, the world powers went through a frenzy of cable laying. The Americans were wiring their western frontier. The French drew telegraph lines from the eastern shores of South America and draped them over the Andes.[17] Far ahead of everyone, the British strung lines across the entire globe, stitching India to Indonesia and Jordan to Johannesburg. By 1880 the Earth was crisscrossed by ninety thousand miles (140,000 kilometres) of cable on the ocean floor alone. It was, in the words of Peter Galison,

> a ninety-million-pound [forty-million-kilogram] machine binding every inhabited continent, cutting across to Japan, New Zealand, India, the West Indies, the East Indies, and the Aegean. Competing for colonies, for news, for shipping, for prestige, inevitably the major powers clashed over telegraphic networks. For through copper circuits flowed time, and through time the partition of the worldmap in an age of empires.[18]

With the new world-girdling network of telegraph cables, the message "It's 12:02 in Greenwich" or "It's 10:29 in Paris" could be (almost) instantaneously transmitted around the planet. Every node in the telegraph network could partake in simultaneity and share the same "now". Recall that longitude is determined through a simultaneous comparison of local time with some standard (such as Greenwich time). With an accurate measure of simultaneity, equally exact calculations of longitude could be established. Errors in the difference between local time and Greenwich time shrank to tiny fractions of a second. Prefiguring our own era of geosynchronous GPS systems, the web of telegraph cables netted the Earth together. Every location, each inch of soil, could now in principle be tallied precisely on the great and growing electric world map. Terra incognita was shrinking into a few jungles and Arctic wastes as synchronized time was transformed into precise maps.

The emerging era of electrically co-ordinated time drew the first outlines of the coming global civilization. Human life and its uses for time were transforming so rapidly that a generation could no longer recognize the one that preceded it. In the wake of this rapid reordering of the human temporal order, cosmic time would once again have to be re-imagined.

At the very moment when the world was desperate to construct networks of electromagnetic simultaneity, the young Albert Einstein was hard at work for the Swiss Patent Office. His day job was to evaluate designs for electromechanical time co-ordination devices—machines that could, for example, link a central clock in a factory office to hundreds of other satellite clocks distributed throughout a sprawling plant. His night job? To weave simultaneity into a theory of space, time, matter and energy that would change cosmology forever.

SIMULTANEITY NOW!
EINSTEIN, RELATIVITY AND THE
NEW REAL WORLD

A halo of grey hair standing on end as if charged at the scalp; a sage lost in thought standing before a chalkboard covered with indecipherable scribbled equations; the philosopher-saint standing before world leaders, arguing for universal peace. There are many iconic images of Albert Einstein that define the man and his legacy. His stature as one of the greatest scientists in human history, if not the greatest, makes him a central figure in our culture's mythology. The Albert Einstein we hold dear is the one who fits our needs. This is Einstein the über-theorist, the man who saw into the fundamental structure of reality, the man living in a world of pure thought and pure abstraction. Einstein's theory of relativity is held up as the epitome of pure science, an exploration of nature's essence far from the clamour of day-to-day life. But in the culture of his own day, the questions framing young Einstein's great achievement in relativity were not abstract. Instead they grew from bread-and-butter experiences that shaped his daily life. And within his own life, questions of time and simultaneity—manifest in the patent requests he saw every day—were explicitly paying for the young Einstein's bread and butter.

It was during his years in the patent office from 1902 to 1909, before he could secure an academic position, that Einstein developed the ideas that would become the theory of relativity. A separate myth has grown around these years of Einstein's life. In both popular accounts and in the stories we physicists hear in our training, Einstein's time at the patent office was an idyll, an aside from his great effort to construct the theory of relativity. The patent office might as well have been a fast-food restaurant, a place to kill time and make money while the real training was going on at home. The truth, however, is far more complex.

Simultaneity was a central issue for Einstein in the development of relativity. What does it mean to say that two events occur at the same time (as, for example, two trains pulling into two widely separated stations)? In physics textbooks the issue of simultaneity is usually presented as part of the general, abstract formalism of relativity theory. In the crucial years at the patent office, however, when the young Einstein was working out the key features of relativity, his concern with simultaneity was anything but abstract.

For these seven years, Einstein spent his days picking over the de-

FIGURE 5.3. The Zytglogge in Bern, Switzerland (c. 1905), with its fifteenth-century astronomical clock. All the clocks in this city where Einstein worked as a patent clerk were connected and synchronized via electrical signals.

tails of electromechanical time synchronization patents. It was the mechanics of synchronizing clocks via electromagnetic pulses that filled his mind during the workday. At home in the evenings, it was the theoretical mechanics of time, electromagnetic waves and simultaneity that consumed him, driving him forward towards a revolutionary new physics of time.

THE ELECTROMAGNETIC PARADOX:
RIDING THE FRAME OF REFERENCE

When Einstein first arrived at the patent office, he had a physics paradox already stuck in his mind like a burr. He could not know it then, but he would have to rebuild time itself to pull the burr free. With patent evaluation work keeping him enmeshed in the real world, Einstein struggled to see through the tangled theoretical machinery of his era's physics and gain purchase on a new vision of time and space.

The story begins not with grand abstractions such as time in and of itself but with the nuts and bolts of electromagnetic theory. Just eighteen years before Einstein was born, the British theoretical physicist James Clerk Maxwell gave physicists a set of equations linking all electric phenomena (charge, current, etc.) with magnetic phenomena (bar magnets, magnetic fields from moving charges, etc.). Maxwell had unified the domains of electricity and magnetism into a new field called electromagnetism. Considered a masterpiece, his consolidation of seemingly disparate phenomena into an underlying whole became an archetype for physics that persists even to this day.

Out of Maxwell's famous unifying equations came an explanation of light as an electromagnetic wave. Physicists were well versed in dealing with waves. They had studied their properties in everything from the windblown ripples of water in a pond to the periodic compressions of air molecules that form sound waves. But what Maxwell's equations made clear was that the visible light our eyes respond to is nothing more than waves of crossed electric and magnetic fields travelling through space at the tremendous speed of 300,000 kilometres per second (670 million miles an hour).

Just as water molecules slosh back and forth, supporting passing water waves, scientists expected that some medium existed to support electromagnetic (light) waves. Since light crossed vast distances between stars, they imagined an all-pervading "luminiferous aether" that filled the space between, supporting the light waves. While there was no direct evidence of the aether, physicists were certain it existed. Entire

FIGURE 5.4. Einstein at the Swiss patent office around 1905, the year he published his paper on the special theory of relativity. Rather than being simply a day job, Einstein's patent work dealt with core issues of time and synchronization.

textbooks of theory were even devoted to the luminiferous aether and its physical properties.

Maxwell, electromagnetic waves and the aether were all relatively new science when the teenage Einstein began his hungry exploration of physics. It was in these early years that he found an unanswerable question in Maxwell's account of light—an enigma that would follow the young man like a shadow all the way to the patent office each day.

"What would it look like", the young Einstein asked himself, "if I could ride a light wave?" A stationary observer watching a beam of light pass would see a train of wave crests rocket by, one after the other. But Einstein recognized that an observer riding the light beam, like a surfer catching the crest of an ocean wave, would look back and see nothing more than a wave frozen in space. No such frozen light wave had ever been observed. It was not part of the vocabulary of physics. More important, Einstein convinced himself that Maxwell's equations ruled out even the possible existence of such frozen light waves. Thus he found himself caught between a theoretical rock and mathematical hard place. He had a paradox on his hands.[19]

Einstein's dilemma rested on frames of reference. A frame of reference is the perspective, or platform, that constitutes an observer's account of the world. If you are standing in a field watching clouds pass by overhead, that is one frame of reference. If you are sitting in an aeroplane a mile in the air and moving at five hundred miles per hour as you watch the same clouds, you have a different frame of reference. Since the time of Galileo and Newton, physicists understood that the motion of a frame of reference could affect the outcome of experiments performed within that frame. Doing physics correctly meant being mindful of reference frames and their motion.

Slam your foot down on your car's accelerator and you will feel your body pressed backwards into the seat as the car surges forward. Blacken all the windows and you would still know that the car, your frame of reference, was accelerating. You would not need to see outside to confirm this because your body would feel the effects of the car's acceleration. Now imagine the car is moving at constant speed on a smooth road (with perfect shock absorbers—no bouncing, shimmying or road noise). Could you still tell if you were moving? If you couldn't look out the window, how would you determine your state of motion? Thus, some kinds of motion (acceleration) lead to effects you can measure with experiments, while others (constant velocity) do not affect experiments.

Over the years physicists learned how to shift their perspective back and forth between moving and stationary frames of reference and how to reconcile the different descriptions of physics each would observe. But Einstein's paradox hinged on the inability to shift between frames of reference. In the frame at rest relative to his light beam he "saw" a passing electromagnetic wave. In the frame riding with the light beam he "saw" a frozen wave. The two frames had two different descriptions of the light beam, and no physics could reconcile them.

While Einstein was developing his ideas on light and reference frames, other physicists were attacking the same problem for reasons of their own. By the end of the nineteenth century the all-pervasive luminiferous aether was in trouble and scientists across the world were struggling to save it.

In the late 1880s two American physicists, Albert Michelson and Edward Morley, tried to detect the aether in a new way. The idea was to use the Earth's motion through the aether as it orbited the sun. Their careful experiment was designed to pick up differences in the speed of light as the Earth moved through the background aether.[20] By bouncing light waves back and forth in directions aligned with, and perpendicular, to the Earth's motion, differences in travel times should have been detectable. The same effect occurs for motorboats on a choppy day when the wind drives strong waves on the water. Observers in the boat see the waves travelling with a different speed if they move with the wind (and the waves), or at some angle to the wind. But the Michelson-Morley experiment failed to detect any difference in light's speed no matter which way the Earth travelled.[21] It was as if the aether was not there, a conclusion few physicists in 1900 could conceive of without horror.

Some of the greatest physicists of the generation just ahead of Einstein had tackled the refusal of light to show any reference-frame-dependent changes in speed. Henri Poincaré, a giant of theoretical physics at the time, had taken the problem to heart. A mathematician of the highest calibre, Poincaré was deeply concerned with utilizing real-world applications of science in the French effort to create a globe-spanning grid of electrosynchronized longitude. Poincaré was also a strong proponent of the aether.[22] His efforts, along with those of other physicists such as Hendrik Lorentz of the Netherlands and George FitzGerald of Ireland, focused on saving the aether by allowing for a more flexible meaning of measurement when it came to light and its velocity.

To determine the speed of an object such as a bullet, you really need to measure two separate quantities: length and duration. Length would be the distance the bullet travels and duration would be the time it takes the bullet to cover that distance. Poincaré, Lorentz and FitzGerald developed new laws for lengths and durations between frames of reference. These laws allowed measurements for both to stretch or contract depending on the object's motion relative to the aether. If length and duration changed in just the right way, then the measured speed of light (length divided by duration) would always be the same. With this so-called length contraction and time dilation, the Michelson-Morley results were explained and the cherished luminiferous aether preserved.

DEATH OF THE AETHER, BIRTH OF SPECIAL RELATIVITY

Einstein did not care for the aether and so he did not care about saving it.

Plato's students had once spent centuries trying to solve the problem their master bequeathed to them: how to save the appearances of planetary motion. Einstein, like other students of his generation, was also bequeathed a homework problem by his elders: save the aether and explain how the speed of light appears constant even when frames of reference move relative to it.

But Einstein declined to play by these rules. He did not solve the problem—he changed it. While others spent careers trying to explain why the speed of light was constant, Einstein simply assumed the speed of light *was* constant and built his physics from that foundation.

In the crucial 1905 paper that launched relativity (published while he was still working at the patent office), Einstein reworked physics from the beginning, invoking kinematics—the study of motion—and its paired foundations of time and space. "I understood where the key to this problem lay", Einstein said to a friend. "An analysis of the concept of time was my solution."[23] Going back to his original thought experiment about light-wave surfing, Einstein recognized that the problem lay in describing phenomena from different frames of reference. Though he called his theory "relativity", what he was really after was invariants—he wanted to know which parts of physics did not change from one frame of reference to another. To resolve his paradox and find nature's true invariants, Einstein first abandoned the aether and then let go of Newton's space and time.

There are two postulates on which all of relativity rests. First, no special frame of reference exists from which the motions of all others can be judged. In other words, there is no "aether frame" for judging motion or rest. All motion is relative motion. Second, the speed of light must be the same for all observers, no matter what their state of motion. It was this second conjecture that opened the door to a strange new world of time and space.

If you are standing on the Earth watching light from a passing star whizz by, you will measure its speed to be 300,000 kilometres per second. A spaceship rocketing away from Earth at 270,000 kilometres per second (90 percent of the speed of light) looking at that same beam of starlight would also see it race by at 300,000 kilometres per second. According to the second postulate, people at every location, regardless of their state of motion, must measure the same speed for light.

To see how strange this behaviour of light is relative to our common experience, imagine two workers on a fast-moving postal delivery train. Both men work in the moving railway car with all doors and windows closed. The workman at the far end of the car heaves a bulky bag of mail to the other. The speed of the bag as it flies through the air is, from the perspective of the other worker, just the speed the first workman gave it when he let it go. The speed of the train does not affect their experience. But now imagine you are standing on a platform as the train roars by and the worker tosses the heavy mailbag to you from the open car. Would you want to catch the mailbag? Not likely. The speed of the bag would now be the speed at which the worker released it plus the speed of the train. From your frame of reference the velocity of the bag and the velocity of train have to be added together.

This addition of velocities is what physicists would have expected for light as well. Einstein, however, had a deeper vision. The second postulate of relativity is equivalent to demanding that the mailbag thrown from the high-speed train travels towards you at the same speed it left the railway worker's hands, as if the velocity of the train did not exist.

The deeper reason for this behaviour lies in the fact that Einstein made the speed of light an upper bound on all cosmic motion. Nothing can travel faster than light. It does not matter that this upper bound is the speed of something called light. It just matters that there *is* an upper bound on speed for everything in the universe. This fact alone changes the meaning of time and space.

Einstein realized that if the universe had a speed limit and light ran at the maximum, then something else had to give in order for light to achieve its constancy. As we have seen, every measurement of velocity is a mix of two other measurements, one for length and one for duration. This means that if the speed of light has to be constant, independent of its frame of reference, then measurements of length and duration can not also be independent in this way. Length (space) and duration (time) have to become flexible, changing from one frame of reference to another. In Einstein's relativity all time became local time and all space became local space.

Gone was Newton's divine sensorium. Time was *not* flowing smoothly everywhere through the universe. Instead of a single overarching Newtonian cosmic time, there was now a relativistic patchwork of times, each measured by observers moving relative to one another. Gone also was the metaphysical majesty of absolute space. In its place stood varying measurements of length made by many moving observers—onlookers will see different lengths for the same object. There is no one length for an object and there is no one time between events. It all depends on how you are moving relative to these events and objects. In just thirty-six pages of his 1905 article, Einstein unhooked time and space from their Newtonian mooring.

The flexibility of space and time in the new physics is cleanly displayed in the famous "twin paradox". Imagine a pair of identical twins both born at the same time. At age twenty, the more adventuresome twin rockets away in a spaceship. She travels to a distant star thirty light-years away at 99.9 percent of light speed. When she reaches the star she turns around and returns to Earth. The twin back on Earth has been waiting sixty years for her sister to return and is now eighty years old. The space-travelling twin, however, has only recorded about three birthdays during her round trip. That is her duration of the voyage. Her time was not her sister's time. For the sister who remained on Earth, the duration between the spaceship leaving and the spaceship's return was

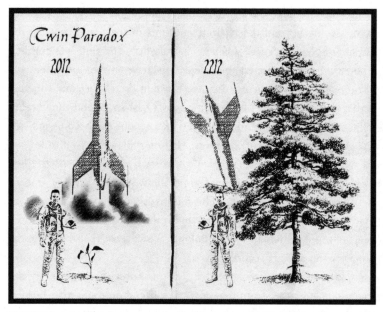

FIGURE 5.5. The "twin paradox" of special relativity. Time flows more slowly for a person who takes a round trip to a distant planet at near light speed than it does for everyone (and everything) on the home planet. There is no paradox, really. The different rates of ageing (the flow of time) are just a consequence of "relativistic time dilation".

far longer than it was for the astronaut twin. The beat of time flowed faster for the stay-at-home sister as measured by everything from clocks in the town square to the pulse in her chest.

The twin paradox demonstrates the relativity of space as well as time. The sister at home measures the distance between the sun and her sibling's destination as thirty light-years. But the space-travelling twin's odometer would click off only 1.8 light-years. The twins do not share the same time and they do not share the same space.

The important point to digest in thinking about relativity is that both twins are right. They have both made proper and accurate measurements of length and duration. Einstein's fundamental insight was that no "right" answer exists for questions about space and time because there is no absolute frame of reference with an absolute space and time from which to judge the answer. Understanding the correct physics requires seeing beyond the separate concepts of space and time. In the physics of Newton, space was one kind of entity and time was another. They were not connected and calculations never mixed measurements of time and measurements of space. Einstein drew space and time together to become part of a larger whole. Once space and time are no longer seen as separate and absolute, then each one can become individually flexible for different observers. My time will not be your time if we are moving relative to each other. My space will not be your space either.

Even the intuitive concept of simultaneity had to be revised in Einstein's new vision of physics. The intuitive idea—hardwired into our brains—that only one "now" exists and is shared by everyone lies at the heart of much of human social thinking. We all sense that we live in the same present and act accordingly.

Under the physics of relativity, all measures of simultaneity are frame-dependent. A claim that you and a friend on Mars were born at the exact same instant, the same "now", actually depends on the frame of reference of the person making that time measurement. In one frame of reference, the two events (your birth and your friend's birth) happened at the same click of a clock. In another frame of reference—where, perhaps, a clock-watching astronaut streamed through the solar system at 99 percent of the speed of light—you were born before your friend. In yet another frame of reference, where a different astronaut was passing through the solar system from the other direction, you were born after your friend. In relativity the simultaneous also becomes the local.

This result—that time flows more slowly for objects moving close to the speed of light, an effect called "relativistic time dilation"—is nothing less than shocking. That there can be no universally recognized, simultaneous present, no "now" for all creation does violence to our intuition because this is not the time we are born into. The problem, of course, is that the time we do recognize is the one our brains evolved us to see. Human bodies rarely moved faster than a few kilometres per hour before the late 1800s. Human minds never communicated with one another across globe-spanning distances using electrical signals before the late 1800s.

Thus we have no hardwired physics modules to p
tive understanding of relativity. Our brains evolved to int
time. Thousands of years of cultural evolution and materia ＿ ＿ment
have slowly taken us beyond that hardwiring. With relativity the pathways of our deepest physical reasoning, also born of material engagement, suddenly vaulted us past intuition and revealed an entirely new
form of time that would rework the cosmos.

FROM SPACE AND TIME TO SPACE-TIME: GENERAL RELATIVITY

Einstein's first paper on relativity, published in 1905, did not instantly
change the landscape of physics. As Peter Galison wrote, "There were
many choices open to a physicist wanting to understand . . . the electro-
dynamics of moving bodies . . . there were dozens of ideas vying for attention".[24] As physicists sorted through their options, one of Einstein's
early champions would also become his first serious re-interpreter.

Hermann Minkowski, a German mathematician and physicist, was
known for casting physics problems into the language of geometry, the
language of spatial relationship. In reviewing Einstein's early papers,
Minkowski saw a way to translate relativity into a powerful geometric
vocabulary that would alter all future descriptions of cosmology. Relativity, he discovered, was not simply concerned with objects extended in
space (the traditional study of geometry); instead, it described the structure of *events* in space and time taken as a whole.

Events were the real objects of concern in relativity. A light signal emitted from a spaceship was an event. The reception of that light
signal at a distant planet constituted a second event. The whole of creation was nothing more than a web of events situated in space and
time. Minkowski recognized that what mattered was not the location
of these events in three-dimensional space alone or their location in
time alone. Instead, relativity provided relationships between a cosmic web of events in something much larger. Minkowski cast relativity into the geometry of space-time, a new four-dimensional reality.

Space-time was the new stage on which the drama of physics would be enacted.

"Minkowski insisted that in the old physics of 'space' and 'time' scientists had been . . . . misled by appearance", wrote Peter Galison.[25] The philosophical implications of the new perspective were startling. Once again the ghost of Parmenides would hover behind a new development in theoretical physics. The future and past took on a different character in the so-called block universe of space-time. In this vision of relativity, next Tuesday, which we consider to be the future, already exists. The past and future are reduced to events that exist together in the totality of a timeless, eternal block of space-time.

While Einstein initially resisted Minkowski's geometric reworking of his relativity, other physicists saw the 4-D approach as more transparent, approachable and flexible. In fact, it was with the introduction of Minkowski's space-time geometry that the tide began turning towards Einstein's ideas.

Einstein himself would soon put space-time geometry to good use. His first efforts in 1905—now called the special theory of relativity—had focused solely on objects moving with constant velocity. By restricting himself to such a limited set of circumstances, Einstein had managed to tease out the mistaken notions of absolute time and absolute space that had remained at the base of physics since Newton. But velocities *do* change and the way they changed was the essence of Newton's physics. Newton had clearly shown how changes in velocity—accelerations— occur only through the presence of an imposed force. Forces produce accelerations. Einstein's next step was to understand the relativity of accelerating frames of reference. Making that leap meant dealing with Newton's other great achievement—gravity.

Like a film noir detective, Einstein looked at physics and tried to see just the facts. His approach was to find the most basic measurable effects, the elementary facts of experience, and build his theories from there. If two different situations yielded the same experimental results, then those situations were equivalent in the most basic sense of the word. By remorselessly insisting on this logic of equivalence, Einstein forged a link between accelerating frames of reference and gravity.

As he often did in his explorations, Einstein used a thought experiment to work out his next steps. Imagining himself as the sole occupant in a windowless capsule floating in space, Einstein asked, "How could I tell if the capsule was in motion?" Imagine for a moment that you are the capsule's occupant. The spacecraft is somewhere in deep space with a powerful rocket motor attached at one end. If the motor was turned off, there would be no way to know if the capsule was in motion. It might be stationary with respect to the stars or it might be moving at constant velocity. In either case, you and any equipment you had on board would just float freely inside your little enclosure. No experiment you could perform would reveal any difference between standing still and moving at constant velocity.

Now imagine the rocket motor is turned on. The capsule begins to accelerate. For the first few instants you and your equipment are floating freely, but now the "floor" of the capsule (where the rockets are located) comes rushing up to meet you as the entire enclosure accelerates in response to the rocket. The floor hits you and your equipment, sweeping you up and transmitting the rocket motor's incessant push. Pinned to the floor, you now feel as if you are being weighed down. You feel pulled towards the floor.

At this exact moment in the thought experiment, Einstein came to his realization: a person inside an accelerating rocket has the same experience as a person in a rocket at rest sitting on a planet's surface. Gravity and the rocket's acceleration produce the same effect. An experiment conducted in a closed capsule with the motor blasting cannot distinguish between the force of gravity and an imposed acceleration. The two situations could not be considered different; from a physics perspective, they were equivalent.

This principle allowed Einstein to make a bold conceptual leap. He did away with Newton's gravitational forces and, building on Minkowski's formulation of his own theory, substituted the geometry of space-time for Newton's gravity. The geometry of 4-D space-time now becomes malleable. Space-time is like a flexible fabric that can stretch and bend. The agent turning space-time into bubblegum is Einstein's relativistic merged matter-energy.

Drop a ball and it accelerates towards the floor. Using the principle of equivalence, Einstein translated acceleration into the ball's unforced movement through a curved space-time. The Earth, according to Einstein, does not create a gravitational force that pulls on the ball. Instead, it distorts geometry, the very shape of space-time, around it. Remove all imposed forces (like the support of your hand) and the ball is free to do what space-time wants it to do. It is free to fall along the curve of space-time the way water flows down a children's slide.

To construct his theory, Einstein had to extend Minkowski's mathematical insights. Ever since Euclid, the great Greek mathematian, scholars assumed that the geometry of space was flat. In flat space, for example, two parallel lines can be extended forever and never meet, like railway tracks running endlessly into space. It was not until a few decades before Einstein that mathematicians such as Bernhard Riemann began exploring geometry in curved spaces. You have more experience with curved spaces that you might imagine, since you live on one (the surface of the Earth is a curved 2-D space). While Minkowski's space-time for special relativity was flat, Einstein realized he would need to

FIGURE 5.6. Gravity and the flexible fabric of space-time. In Einstein's general theory of relativity, gravity is explained as the distortion, or "curvature", of space-time due to the presence of mass (mass-energy).

adapt Riemann's new mathematics for his general relativistic theory of curved 4-D space-time. The work was hard even for Einstein, who had to be tutored in the details of the new non-Euclidian geometry. After years of work, he published his definitive work on general relativity in 1916.[26]

The general theory of relativity cleanly extended Einstein's earlier special theory. Now, individual frames of reference moved on a flexible manifold of reality, a fabric of space-time that could be bent, stretched and folded. Mass-energy caused the distortions of space-time's fabric, and in turn, space-time guided the movement of mass-energy. Just as in special relativity, measurements of space (length) or time (duration) depended on individual frames of reference. In general relativity, an observer's location relative to a large body of matter could also affect space and time measurements.

Clocks close to a planet's surface run slower than those far away. Lengths measured close to a planet's surface yield smaller values than do measurements taken far out in space.[27] Time and space were still separately relative, but now it was the malleable, gumlike geometry of space-time that provided the framework in which to understand their totality. Space had been an empty stage in Newton's physics. Now, in Einstein's general relativity, space-time became a central player in the drama of physics and would soon become the platform from which modern cosmology would be launched.

HOW THOUGHTS AND THINGS MAKE A WORLD

Exploring the centrality of clock synchronization and simultaneity in Einstein's evolution towards relativity, Peter Galison asked the central question: "Was no one else in 1904–1905 *in fact* asking what it meant for an observer here to say that a distant observer was watching a train arrive at 7 o'clock?"[28]

The answer Galison provided—the answer we have explored in this chapter—is simple. In one way or another, the entire culture of the late nineteenth century was asking questions about simultaneity. Passengers riding swift trains from one city to another, families sending telegrams to distant relatives, mayors contemplating the co-ordination of town clocks, generals running military exercises synchronized by rail and wire—each in his or her own way was consciously, or unconsciously, asking this question.

Einstein grew up with the question. Electromagnetic clock co-ordination even appeared in one of the young man's favourite childhood science books.[29] By the time he reached the patent office, Einstein was in the thick of his culture's confrontation with simultaneity, clocks and time. As Galison wrote, "Reviewing one patent drawing after another in the Bern office, Einstein had a grandstand seat for the great march of modern technologies."[30] It is no accident that the technologies Einstein reviewed were aimed directly towards reworking our uses of, and material engagements with, time.

The theory of relativity revolutionized the meaning of temporality for physics and, as we shall explore in the next chapter, allowed the first true scientific cosmology to emerge. But the newly emerging relativistic time had its origins firmly rooted in the concrete concerns of human time. Train tracks and telegraph lines reshaped human culture. It was a radical new form of material engagement and from it would flow new institutional facts that radically, and rapidly, reshaped human culture. As Galison put it,

> Creating this standardized, procedural time was a monumental project that utilized creosote-soaked poles and undersea cables. It required a technology of metal and rubber, but also reams of paper, bearing, contesting, and sanctifying local ordinances, national laws, and international conventions. As a result, conventional, turn-of-the-century time synchronization never inhabited a place isolated from industrial policy, scientific lobbying, or political advocacy.[31]

Thus, from our material encounter with the world—our making of new things—new symbols and new ways of thinking emerged. These new cognitive possibilities flow both downwards, to transform everyday time, and upwards, to the realms of philosophy, physics and eventually cosmology.

The story of relativity's emergence is a pure example of human and cosmic time braiding together through the process of material engagement. While the myth speaks of Einstein wrestling alone with the realms of pure abstraction, the truth is far more interesting.

When we looked at the Neolithic revolution, we saw how symbols emerging from material engagement could transform human experience. What was true ten thousand years ago is true of the past century and is just as true now. Synchronized time was itself a potent symbol during Einstein's day. The cultural push for a single co-ordinated time invoked discussions of democracy and world citizenship. Relativity, with its emphasis on individual frames of reference and the primacy of each observer, became a symbol as well. "What all these symbols held in common", writes Galison, "was a sense that each clock signified the individual, so that clock co-ordination came to stand in for a logic of linkage among people and peoples that was always flickering between the literal and the metaphorical. Precisely because it was abstract-concrete (or concretely abstract), the project of time co-ordination for towns, regions, countries, and eventually the globe became one of the defining structures of modernity."[32]

In the next chapter we step forward into the twentieth century and the story of true scientific cosmology. It is a story that will lead us to our own precipice and potential revolutions in human and cosmic time. But before we take that step we must remember the critical role time has always played in the cultural transformations of the past. We have always made time and, in turn, our understanding of time in the universe has made us. Now we are at it once again.

Chapter 6

THE EXPANDING UNIVERSE, RADIO HOURS AND WASHING MACHINE TIME

Speed, Cosmology and Culture Between the World Wars

PITTSBURGH, PENNSYLVANIA, USA • 1935, 2:25 P.M.

The whistle would blow soon. There couldn't be much of her break left. The clatter from the assembly line wasn't making it much of a break anyhow. If she had a watch, she would know how much time she had, but of course she'd sold that long ago just to keep her and her boy in the house. The watch had been the last vestige of her earlier life, given to her by her husband during those years of wealth. At least those years felt like wealth now, with the whole country slipping to God knows where. But the factory clock would make sure she was back on the line in time.

The '29 crash had killed what was left of her husband's business and the resulting sorrow took him as well a year later. So now here she was, a single mother working the line assembling Westinghouse washing machines and doing piecework at night to keep her and her boy whole.

She was so tired. She had got up early to finish the wash and hang the laundry before she got the boy off to school. The irony of putting shiny new electric washing machines together all day when she was forced to do most of her own by hand was not lost on her. She was saving up to use her employee discount to buy one of the machines she spent each day putting together. But something always came up and the money always disappeared. Still she was grateful to have a job. Someday things would be better and she would have more time.

The shrill blast of the whistle jolted her out of the reverie. Back to the shift; back to the lines. Another two hours and forty-five minutes. She just had to get through two hours and forty-five minutes.

ELECTRIC SLAVES AND WASHING MACHINE TIME

Before electric appliances radically reshaped life and time at home, Monday was wash day, and it was called "blue Monday," for good reason: a bluing agent, added to rinse water to brighten clothes, gave the day its name. But the struggle involved in hand washing lent its own

FIGURE 6.1. Father Time puzzles over General Electric's "Monitor top" refrigerator. Introduced in 1927, the "Monitor top" was one of the first fridges to see widespread use. It was a powerful example of electric appliances changing day-to-day experience and the common experience of time.

meaning to the blues. In *Never Done*, a history of housework, Susan Strasser describes the process in the early days:

> Without running water, gas, or electricity, even the most sim-
> plified hand-laundry process consumed staggering amounts of
> time and labour. One wash, one boiling, and one rinse used
> about fifty gallons of water—or four hundred pounds [180 ki-
> lograms]—which had to be moved from pump or well or fau-
> cet to stove and tub, in buckets and wash boilers that might
> weigh as much as forty or fifty pounds. Rubbing, wringing,
> and lifting water-laden clothes and linens, including large ar-
> ticles like sheets, tablecloths, and men's heavy work clothes,
> wearied women's arms and wrists and exposed them to caus-
> tic substances. They lugged weighty tubs and baskets full of
> wet laundry outside, picked up each article, hung it on the line,
> and returned to take it all down; they ironed by heating several
> irons on the stove and alternating them as they cooled, never
> straying far from the hot stove.[1]

This was time and material engagement at its most fundamental level. Negotiating the human world of culture demanded clean clothes, but clothes required enormous amounts of time to clean. As electric currents began circulating through a world that was re-inventing itself from one year to the next, this most basic encounter with time would soon leave the realm of hands-on work and move into the domain of automation.

Machines for washing clothes have a long history. The earliest ver-sions were hand-driven contraptions that imitated the tedium of rub-bing clothes over a washboard. A lever, pulled by hand, drove two curved ribbed surfaces over each other with the clothes placed in between.[2] One such model, built in Manchester, debuted at the Great London Ex-hibition of 1862. The first electric clothes washers appeared in 1900.[3] These were basically motor-rotated tubs filled with water by hand, but the water would often run over the tub into the motor, delivering jolting shocks to its operator. The next thirty years were a story of manufacturers learning, step-by-step, how to transfer power from sufficiently muscular motors into enclosed washing mechanisms without electrocuting the customer.

Important steps were made in the 1920s, including U.S. appliance maker Maytag's adoption of the "agitator in a tub technology", which is still used, and the development of the modern look of the machine.[4] The familiar white enamelled sheet metal so recognizable today replaced the copper tubs and wrought-iron legs of previous washing machine incarnations. The sheet-metal skirt was eventually dropped below the level of the motor mount to encase the entire apparatus in one smooth-looking unit.

In 1937 the Bendix Home Appliances Corporation, founded in Indiana a year earlier by legendary salesman Judson Sayre, introduced the first fully automatic front-loading washing machine. The Bendix machine did it all—wash, rinse and spin. Women who just a generation before had spent days doing laundry could now simply load it, start it and leave it. The mind-numbing monotony and physical exhaustion associated with wash day was over. So powerful was the transformation in time wrought by this most mundane of technologies that more than one study has pointed to the electric washing machine as a precursor to the revolutionary women's liberation movement.[5]

Electrical appliances popped up like daisies in the 1920s and 1930s, altering the temporal landscape, especially in America. Electric vacuum cleaners, invented by a janitor who later loaned a version to his cousin Susan Hoover and her entrepreneur husband, shortened the daily task of sweeping and the more odious work of hauling carpets outside for beating.[6] Electric mixers, like the inexpensive Mixmaster, introduced in 1931, eliminated the hand kneading of dough.[7] With the advent of electric refrigerators, hauling heavy slabs of ice from the icehouse to the kitchen also faded to memory. Even intimate acts such as shaving gave way to electrification as the first mass-market shavers appeared in the early 1930s.[8]

In America, 60 percent of the 25 million homes wired for electricity in 1940 had a washing machine—a tremendous proportion for a nation still struggling with the aftermath of the Depression.[9] But the second part of this statistic, about the number of homes with electricity, is just as important as the first. From the 1920s to World War II the washing machine was just one of a host of home appliances reshaping the human experience of time by compressing housework, and these appliances needed electricity. It was electricity—the fruit of the previous century's scientific breakthroughs now becoming technology—that powered this radical, cultural reconfiguration.

America was well ahead of Britain in adopting electricity in the home. Immediately after World War I, in 1920, less than 10 percent of U.K. households had electricity, compared to 34 percent of all U.S. households.[10] By 1940, the U.S. proportion had doubled,[11] and in rural areas the number of households with electric power increased by a factor of almost thirty.[12] Electric power running to almost every home created a redistribution of work and time that was without precedent in all the long millennia that had preceded it. With electric power running appliances, each family had the equivalent of a small army of servants doing their vacuuming and their sewing, preserving their food, washing their dishes and of course cleaning and drying their clothes. Electric appliances reshaped the experience of daily time and created a new, mass-market vision of leisure.

Companies trying to sell these appliances to the lucrative American market were quick to find novel ways to advance the vision of life without housework. In 1928 the magazine *Modern Revelation*, under the auspices of the National Electric Light Association, offered a prize for the best essay on living with electrical appliances. In her winning essay, Wilma Cary compared the downtrodden, old-fashioned Joyce to her more modern neighbour, Mrs Stuart:

> Joyce believes that Mrs Stuart must be a terrible housekeeper
> [because she is often not home] until Joyce visits her neighbor
> one day. During this visit Joyce discovers that Mrs Stuart's se-
> cret is household appliances. With the assistance of these "elec-
> tric servants" Mrs Stuart is able to keep her house spotless, the
> laundry washed and ironed, and take her children on daily ex-
> cursions. After having completed all these chores, Mrs Stuart
> can still make a delicious dinner for her husband on her electric

stove. Amazing! Joyce sees the light and decides to persuade her husband to purchase these appliances for her to make her life easier and more enjoyable.[13]

This was the new electric world. What sane woman, man or child could want anything else?

Every aspect of the human universe was reshaped by the injection of this new current into daily life. In cities and town squares across America, electric lights blotted out the stars and turned night into day. Electrical appliances alleviated the drudgery of housework. Electricity even gave radio tubes their glow as they stitched the nation together through the voices of Rudy Vallee or President Franklin Delano Roosevelt.

In the cold night air of Mount Wilson, California, electricity was also at hand, running the motors that kept the giant 2.5-metre Hooker telescope turning gracefully in step with the wheeling stars, radically expanding the frontiers of astronomers' cosmic vision.

REALM OF THE NEBULA: ASTRONOMY STEPS INTO COSMOLOGY

Throughout the nineteenth century astronomers made steady progress mapping the sky. But it was not until the first decades of the twentieth century that the fundamental data of astronomy became relevant to cosmological debates. The transition was made possible because of the industrialization of telescopes (in scale if not in quantity).

A telescope is, for all intents and purposes, a light bucket. Stars appear fainter the farther they are from us.[14] Thus astronomers needed to build ever bigger telescopes to probe deeper into the night and further into the realms of cosmic architecture. The 2.5-metre Hooker telescope at Mount Wilson was a triumph of industrial-scale design and scientific precision.[15] The new technologies of electrification played their role in this domain as well: cables ran up the dangerous mountain roads to feed powerful motors that turned the dome and wound the precision timers for the telescope.

Weighing more than fifty-four tonnes, the telescope itself was a massive construction of steel and glass that pushed the limits of technology. The enormous mirror, polished to microscopic smoothness, allowed astronomers such as Edwin Hubble to see objects that were a thousand times fainter and millions of times more distant than ever before.

With so much power, the Hooker was the perfect instrument to answer a single question that had plagued astronomy for more than 150 years: What is the true nature of the Milky Way? Understanding the Milky Way was a critical first step in directly addressing the great questions of cosmology. Without a foothold on its size, shape and nature, it was impossible to form links between the yet unborn science of cosmology and the mature science of astronomy.

Before artificial light robbed us of our experience of the night sky, all humans were familiar with the Milky Way. This sky-spanning arch of diffuse light is the most striking feature of the Earth's night-time panorama. While some Greek astronomers had speculated about the nature of the Milky Way, there was little that could be done with the naked eye to explore it further. The first great leap in understanding occurred with Galileo. When he trained his small telescope on the Milky Way's dispersed band of light, it was instantly resolved into countless points. The Milky Way, Galileo discovered, was a vast system of stars.

Throughout the eighteenth and nineteenth centuries astronomers with ever-larger telescopes attempted to gaze farther and decipher the underlying geometry of the Milky Way as a stellar system. In 1784, after painstakingly counting the density of stars in different regions of the sky, William Herschel claimed that the Milky Way was shaped like a long thin bar with the sun positioned at the centre. Using similar methods a century and a half later, Dutch astrophysicist Jacobus Kapteyn concluded in 1922 that the Milky Way was like a squashed beach ball—an "oblate spheroid"—with the sun offset from the centre.[16]

The true shape of the Milky Way was really only one-half of a pair of questions astronomers were struggling to answer. The Milky Way's dimensions were of equal concern. How big was this ubiquitous stellar city? By Kapteyn's era in the early twentieth century, methods for determining astrophysical distance were becoming more sophisticated and more reliable. Kapteyn's best estimates set the Milky Way's girth from one end to the other at more than 100,000 light-years (a light-year spans almost 10 trillion kilometres).

Size mattered to these astronomers for a simple reason. As the twentieth century got under way, they wondered if the Milky Way and the universe were synonymous. It was quite possible that all the stars in the universe belonged to the Milky Way. If that was true, then the Milky Way, surrounded by an infinite void, was the material cosmos.

It was also possible that other star systems similar to the Milky Way in size and character existed as distant "island universes". The veracity of this island universe theory was a point of contentious debate among astronomers at the time. As the giant Hooker telescope was being constructed, the astronomical community had splintered into different sides of what was called the Great Debate. The heart of the question revolved around a class of astronomical objects known as spiral nebulae.

When astronomical surveys in the eighteenth and nineteenth centuries carried out an exhaustive census of the sky's inhabitants, they found more than just stars and planets. Diffuse, cloudy objects called nebulae (Latin for "clouds") also showed up. Some nebulae were round and smooth. Some were irregular and spiky. But it was the enigmatic pinwheel-armed spiral nebula, first discovered in 1845, that garnered the most attention.[17] When seen edge-on, these spiral nebula often had a clear disc-like appearance.[18]

The disc-like geometry of the spiral nebula hit a strong note of recognition for some astronomers. Discs of stars had already made a significant appearance in the realms of cosmological conjecture. As far back as 1785, the philosopher Immanuel Kant had offered an influential model of cosmic history by predicting that the Milky Way was formed from a giant spinning cloud of gas that had collapsed under its own gravity. Kant hypothesized that a disc-shaped Milky Way would be a natural result of such a collapse.[19] By the early twentieth century, a growing body of observations led many astronomers who recalled Kant's theories to imagine spiral nebulae to be separate systems of stars like our own Milky Way. If they were the same size as the Milky Way, these island universes, called galaxies, must be at great distances to appear so

FIGURE 6.2. Astronomer Edwin Hubble and the powerful Hooker telescope. Hubble's work with the electrically powered Hooker allowed him to greatly expand the scale of the universe by showing galaxies as distant star systems and to discover cosmic expansion.

small in the sky. Opponents of the island universe hypothesis dismissed this interpretation of spiral nebulae, arguing that the required distances were so great as to be unimaginable. The universe, in their vision, could not be that big. The spiral nebulae, they argued, must be nothing more than interestingly shaped clouds of gas and stars embedded much closer to home within the Milky Way.

The argument came to a head in the confrontation between two well-known astronomers, Harlow Shapley and Heber Curtis. The debate occurred in a packed auditorium at the 1920 meeting of the National Academy of Sciences in Washington, D.C.[20] Shapley, formerly the director of the Mount Wilson Observatory, opened the argument by sharply attacking the island universe theory. He had developed his own model in which the universe was one "enormous all-comprehending galactic system".[21] Curtis, a prominent astronomer from the Lick Observatory, defended the island universe theory by presenting multiple lines of evidence that spiral nebulae were distant, separate galaxies.[22]

The debate ended without a knockout. Both men were struggling with biases and incorrect assumptions that no one understood at the time. What is truly remarkable about this Great Debate, however, is its historical moment. As late as 1920, a time when aeroplanes and radios were becoming commonplace, science had yet to determine the nature of our own galaxy or prove the existence of others. The true dimensions of galactic space, and of the universe itself, would not be resolved until the appearance of Edwin Hubble.

At the time of the Great Debate, Hubble was not yet the towering figure in astronomy he would one day become, but he had already made impressions on the community. Tall and handsome, the young Hubble arrived in Pasadena (the home of Caltech and the offices of the Mount Wilson Observatory) in 1919.[23] He was fresh from serving in World War I and, after spending time in Oxford on a Rhodes Scholarship, had taken somewhat unsuccessfully to English affectations.[24] While Shapley was firm in his conviction that spiral nebulae lay inside the Milky Way, Hubble favoured the island universe theory.[25] Hubble decided to mount his own attack on the issue, but to solve the riddle of the spirals he would need a reliable means of measuring their distance.

Distance vexes astronomers. There is no way to run a tape measure to the stars. Astronomers must find proxy measurements to convert into distance. Measurements of brightness can, in special circumstances, fit the bill. The apparent brightness of any light source decreases with distance. That is both basic physics and common experience. Car headlights are painfully bright up close but those same headlights will appear faint from a mile away on a dark night. Thus if you know the intrinsic brightness of the source—like knowing a lightbulb burns at one hundred watts—you can use this dimming effect to find its distance.

By comparing how bright an object appears to be with how bright you know it to be intrinsically, its distance can be directly computed. The problem for astronomers is that stars, and other celestial sources, do not come with "100 watts" printed on the side. Luckily, certain classes of celestial objects have properties that let astronomers deduce their all-important intrinsic brightness. These are called "standard candles" and they are worth their weight in gold. Once identified, a standard candle makes determining distance as simple as measuring brightness.

By the 1920s a special class of pulsating star called Cepheid variables had already been firmly identified as a standard candle. Cepheids are pulsating stars. They brighten and dim in cycles over a period of days or weeks. In 1908 the astronomer Henrietta Leavitt discovered a direct relationship between a Cepheid's pulsation period and its average intrinsic brightness.[26] In essence, Leavitt found a way to read the "100 watt" label printed on stars. Because of her work, once an astronomer found a Cepheid variable and measured its period, he or she could quickly compute the star's distance (and the distance to any objects around it).[27]

The Hooker telescope was big enough to let astronomers see individual stars in the bigger spiral nebulae. On October 5, 1923, Hubble spent the night probing the great spiral nebula in Andromeda for signposts he could use for gauging its distance.[28] The next day while comparing his night's work with previous observations he found what he was looking for on his photographic plates. To his surprise and delight he had come across a Cepheid variable in Andromeda. With a few simple lines of maths Hubble used the newly discovered standard-candle star to sweep away one hundred years of debate.

Using the Cepheid variable, Hubble calculated that Andromeda was almost a million light-years from the Earth. This was far larger than any astronomer's estimate for the Milky Way's outer boundary and implied that the spiral nebula in Andromeda could not live within the Milky Way; it was without doubt a spiral galaxy. Harlow Shapley had by this time moved on to Harvard but he had not given up on his belief that spiral nebulae were part of the Milky Way. On receiving news of Hubble's discovery, Shapley (who nursed an intense dislike of the younger man) told a student, "Here is the letter that has destroyed my universe."[29]

Hubble's result showed astronomers that spirals were indeed galaxies and, more important, that the universe was far larger than anyone had imagined. From Hubble's discovery onward the measure of space would be taken in terms of galaxies and their measurable cosmic distribution. Cosmology was leaving its era of philosophical speculation and entering its astrophysical age.

BUILDING UNIVERSES:
THE GAME BEGINS

True scientific cosmology demanded a theory of the universe as a whole, a complete and all-encompassing mathematical description of space and time. Such a theory would be a model, a mathematical representation capable of describing everything that happened in the cosmos. The model must also tell scientists what to expect while making their observations and allow the raw data of astronomical investigation to be compared with theoretical expectations.

If cosmology was finally to move from the realm of quasi-philosophical speculation to the firmer ground of science, it would need a testable account of the universe. It would have to become a branch of physics and astrophysics. The universe would have to be treated like everything else physics studies: an atom, a rock, a cow. But the universe contains all atoms, all rocks, all cows and all physicists. It not only contains everything like a giant box, it *is* the box. How could scientists describe the totality of existence from the inside?

In many ways cosmology was waiting for Einstein. He and his general theory of relativity found a way to make the theoretical description possible.

All earlier attempts to build models of the universe were hobbled because they lacked Einstein's great insight into the nature of space-time. The first step towards a successful model was the special theory of relativity, as it swept aside Newton's divine sensorium of absolute time and space and replaced it with a unified 4-D space-time. The task was completed when general relativity linked the flexible fabric of this space-time to the large-scale distribution of mass-energy. In this way, gravity was recognized as nothing more than the malleable fabric of space-time, with mass-energy acting as the agent driving space-time's gravitational distortions.

It did not take long for Einstein to begin using his field equations, linking space-time to mass-energy, to construct cosmological models. But the effort required one critical assumption that all his work, and the work of others that followed, would have to accept. To derive a mathematical description for the universe as a whole, Einstein had to assume the universe was perfect in one key sense of the word—it had to be very smooth and very symmetric. The technical terms are *homogeneity* and *isotropy*, but the meaning is simple: on cosmic scales everything has to look the same from every perspective. Einstein could not use his equations to derive cosmological models unless he assumed that the universe was perfectly symmetric on the largest scales.

A perfectly smooth ping-pong ball looks the same no matter what angle you view it from. Thus physicists say that perfect spheres, like ideal ping-pong balls, are maximally symmetric. Likewise, if you examine a perfect sphere by standing on its surface and wandering around, your location should not matter for the description—every location looks the same. By assuming that the real universe was maximally symmetric, a description of a small region of space-time in Einstein's equations could become a description of all space-time—a mathematical description of the entire universe.

With this assumption in hand, cosmic history and cosmic architecture became accessible for Einstein to explore. The ancient questions of cosmology still remained. Did the universe have a beginning or had it always existed? Was space infinite or was it somehow bounded? But with his newly formulated relativistic cosmology, Einstein had the conceptual tools needed to provide mathematically definite answers to many of the great questions.

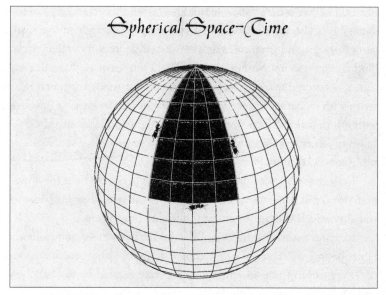

FIGURE 6.3. Einstein's first universe. In Einstein's first cosmological model, the whole of cosmic space was spherical. 3-D space wrapped back on itself the way the 2-D surface of a balloon is without edges or boundaries. Objects (such as the depicted ants) are constrained to move through space according to its shape.

The limits of cosmic time and space were tackled first. Newton had accepted a spatially infinite universe but, with his biblical bias, he could not accept one that was eternal in time. Unlike Newton, Einstein wanted an eternal universe. Like most scientists of his day, Einstein believed the universe had always existed and always would exist, which led him to search for solutions to his equations that would represent a finite, closed and eternal universe. *Finite* means there is only so much space to go around, only so many cubic centimetres existing in the cosmos. In this context, *closed* means that space has no edges, no brick walls for starships to bang into when they arrive at the end of creation.

To understand what this looks like, imagine the surface of our own planet. The Earth has a finite surface area (as we are becoming painfully aware). It is also unbounded. Head west for long enough and you will come back around from the east after travelling in an extended circle. This familiar example shows us that the 2-D surface of a sphere, like the Earth, is a curved space that is both finite and unbounded. General relativity, with its curved space-time, allowed Einstein to create a universe with just these kinds of properties, only in higher dimensions. The 3-D space of Einstein's first model of the universe was hyperspherical—it was the 3-D version of a 2-D spherical surface.

Let's stay with the analogy of the 2-D spherical surface for a moment longer. It will give us some insight into the critical issue of dimensionality, which will be important in the chapters to come.

Imagine a spherical balloon. Like the Earth's surface, the balloon's fabric defines a 2-D space (the technical term for space here is a *manifold*). We can imagine an entire 2-D universe defined by the balloon's surface. This universe might include 2-D creatures cheerfully unaware that there are any higher dimensions extending beyond their world. Thinking cosmologically, we can see that for these creatures there is no "inside" or "outside" of the balloon. As higher-dimensional creatures, we 3-D beings can see the balloon is curved. We can see that it separates an inside from an outside. But that privileged distinction exists only in a space with more than two dimensions. For the 2-D creatures, no such extra space exists or needs to exist.

Remember that the total 4-D space-time of general relativity *is* reality. It is all there is. In Einstein's first cosmological model, the 3-D spatial part of space-time was curved like the balloon's surface. There was no 3-D "inside" or "outside". If you had a starship, you could head off in any direction and, after a very long time, return to where you started from the opposite compass point. In this way the long-standing paradox of boundaries in space—"brick walls" at the edge of the universe—had been resolved. In thinking about the standard idea of cosmic edges, or boundaries, Einstein wrote to a friend, "If it were possible to regard the universe as a continuum which is finite (closed) with respect to its spatial dimension, we should have no need at all of any such boundary conditions."[30] With curved space Einstein could build a finite but unbounded universe as a solution to the equations of general relativity. Time presented a different problem.

Einstein was searching for solutions to his equations that described a static universe. When he looked more closely at his solution—the model universe predicted by his equations—Einstein saw that it was unstable. A little nudge and his closed, hyperspherical universe began to contract or expand, just like a balloon withering under deflation or stretching under inflation.

This gravitational instability was somewhat similar to the one Newton had discovered two hundred years before for his smooth, infinite distribution of stars.[31] Einstein was convinced that both contraction and expansion were absurd possibilities. To protect his universe from any kind of evolution, the great scientist fudged. He added an extra term to his equations called a cosmological constant. The cosmological constant filled all space with a kind of antigravity that locked the universe into rigidity. It was an act of cosmic meddling he would soon come to regret.

Within just a few years Einstein had company in his cosmological sandpit. Equations in mathematical physics are like Lego sets. Just because you built a tractor with your box of Lego doesn't mean someone else can't use the same bricks to make an aeroplane. Almost immediately after Einstein published his model universe, Willem de Sitter, a professor of physics in the Netherlands, found an entirely different cosmological solution to the equations of general relativity. The de Sitter universe also appeared static, stable and closed. When confronted with de Sitter's work Einstein saw that it too represented a valid solution to his equations. But there were aspects of de Sitter's model that struck Einstein as deeply flawed. Most important was that de Sitter assumed a universe was empty of matter. When Einstein chided de Sitter on this point, the Dutchman responded that "empty" could be interpreted simply as an approximation to a very low density of matter.

De Sitter's universe had another strange property that would prove far more important for history. In his solution, time ran slower for distant observers than for those nearby. One consequence of this cosmic time dilation was that light emitted by distant sources would stretch as it travelled through space-time. Wavelengths would elongate, making light from distant sources shift from the shorter, bluer end of the spectrum to the longer wavelengths at its red end. It was puzzling behaviour. Eventually the true reason for the redshift would emerge as de Sitter's cosmological solution was recognized as *a universe that moved*.[32] It took some time for astronomers to recognize this point, but once they did, it was clear that de Sitter's space represented an expanding space. Expansion was an idea that would soon be on everyone's mind.

THE EXPANDING UNIVERSE IN LIGHT AND MIND

"Nobody who hasn't done it could ever realize how cold it was", Milton Humason, the mule driver turned astronomer, later said of his long nights of astronomical observation.[33] Humason had arrived at Mount Wilson as a teamster when the Hooker telescope was still under construction. Eventually he found work as an electrician at the observatory and when his skills guiding the telescope were recognized he eventually was made a full-fledged member of the astronomical staff. Together, Milton Humason and Edwin Hubble would spend countless hours in the tiny cage perched atop the 2.5-metre telescope. But it took many nights of training the giant instrument on individual galaxies to get accurate readings of their light and extract a measure of their motion. The effort would prove worthy of frozen hands, as motion had become the central question of cosmology circa 1930.[34]

Even before Hubble discovered that galaxies were separate star systems, the motion of spiral nebulae had been a contentious subject. Astronomers can measure a celestial object's motion towards or away from the Earth by looking for changes in the light the object emits. Changes in the wavelength of light are like cosmic speed guns for astronomers, allowing them to chart cosmic motions and map out cosmic architecture. The secret lies in the fingerprints of the universe's elements.

Heat any element, such as a tube of hydrogen gas, and it will glow with a few very narrow and precisely defined bands of colour (this is the physics behind colourful neon lights). These emission lines, as the bands of light are called, form a unique elemental fingerprint of colours (i.e., wavelengths). When astronomers look at a distant object they use a spectrograph to break light up into its component colours. These spectra allow astronomers to see exactly how much energy arrives at each different wavelength. In the early years of the twentieth century, astronomers began collecting a menagerie of galaxy spectra, which would open a new door on cosmic evolution.

Anyone who has heard the shift in pitch of a passing ambulance siren knows that motion and frequency are related. A stationary ambulance siren emits sound waves with a definite frequency (and a definite wavelength, which is directly related to frequency). When an ambulance moves towards you the sound waves get crowded together. The crowding forces the wavelength to get shorter and the frequency to rise (so that the pitch you hear is higher). When the ambulance passes and is moving away from you the siren's sound waves are stretched out. The wavelength increases and the frequency drops (so that the pitch you hear is lower). This is the famous Doppler effect and it holds just as true for the light waves transmitted by galaxies as for the sound waves emitted by ambulances on Earth.

In the early 1900s astronomers began searching spectra from spiral nebulae for Doppler shifts. The goal was to find the motions of the nebulae (not yet recognized as galaxies) through space. Doppler shifts towards shorter wavelengths—or the blue end of the spectrum—would mean a galaxy was moving towards us. And Doppler shifts towards the longer wavelengths—or the red end of the spectrum—would mean a galaxy was moving away from us.

In 1912 astronomer Vesto Slipher, working at the Lowell Observatory in Arizona, had painstakingly accumulated enough light from a few distant galaxies to record a decent spectra.[35] Most of Slipher's galaxies showed redshifts. Only a few galaxies like Andromeda, the Milky Way's neighbour, appeared to be moving towards us. With so many redshifts, Slipher's data were intriguing, but they gave astronomers no purchase on an underlying pattern. Once spiral nebulae were recognized as galaxies, new questions emerged. Were all galaxies redshifted? Did they all have the same velocity? If not, which galaxies had the highest speeds? Was a galaxy's redshift determined by its position in the sky or distance from Earth?

To unravel the enigma of the galaxy redshifts Edwin Hubble returned to the 2.5-metre Hooker telescope in 1928 with Milton Humason acting as assistant. Humason's extraordinary patience and steady hands at the controls on long, cold nights made him the ideal partner. Through the exacting work of recording spectra from galaxies across the night sky, the pair hoped to find the relationship between velocity and distance for galaxies and pick up on a pattern that would make sense of Slipher's initial discovery. Doppler shifts in the spectra would

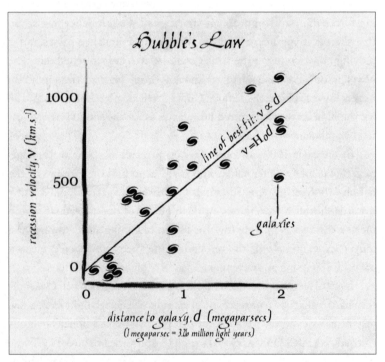

FIGURE 6.4. The Hubble Law. Hubble discovered that all galaxies were receding from one another. In particular, the recession followed a simple law with galaxy velocities increasing with their distance from us. This is what would be expected for an expanding space-time.

give them the galaxy's velocity, while Cepheid variables would have to be identified in each galaxy to determine its distance from Earth. It was painstaking work but slowly a pattern did emerge.

In an epochal paper published in 1930, Hubble and Humason demonstrated that nearly all galaxies were redshifted, receding at tremendous velocities and, as the great English astrophysicist Arthur Eddington put it, were shunning us like the plague.[36] More important, however, was the relation between recession velocity and distance. Hubble and Humason found that galaxy recession velocities increased as their distance increased. In a beautifully simple linear relationship, distant galaxies were moving away from us faster than those nearby. Hubble refused to interpret his own results, allowing theorists the prerogative of explanation. But he and most everyone else knew the message in the data. Einstein had been wrong. The universe was not static. It was moving.

Einstein's equations had proven more prescient than their creator. By adding the cosmological constant, Einstein had prevented the mathematics from leading him to a conclusion that was now clear to scientists around the world: the universe as a whole was expanding. "It was my greatest blunder", the chastened Einstein would later say about the cosmological constant.[37] In our own era, Einstein's blunder has returned in a different guise, reshaping modern cosmological debates away from the Big Bang. But in 1930 the cosmological constant was seen as nothing more than a mistake.

Einstein may have been wrong about the universe's motion but so too was de Sitter. Hubble had found redshifts, but they were not the ones predicted by de Sitter's expanding universe. The simple linear relationship between expansion and distance, so clear in Hubble's data, did not match what de Sitter's model predicted. If relativistic cosmology was to yield a theoretical complement to Hubble's astonishing discovery of an expanding universe, it would have to come from someone other than Einstein or de Sitter. Remarkably, by 1930 the linear pattern of velocity and distance expansion had already been predicted by scientists not once but twice.[38] More remarkable still, both solutions had been forgotten.

Alexander Friedmann, a young Russian theorist, was the first to discover the correct version of an expanding relativistic universe. The son of a ballet dancer (his father) and a piano teacher (his mother), Friedmann had survived the horrors of World War I and the communist revolution with his scientific ambitions intact.[39] A mathematical physicist, he was deeply interested in the mathematical properties of general relativity's cosmological models. In a 1922 paper published in the famous German journal *Zeitschrift für Physik*, Friedmann found a new cosmological solution to the equations of general relativity that included both matter (unlike de Sitter's) and expansion (unlike Einstein's).[40]

Exploring the field equations of general relativity, Friedmann found a broad solution with three possible versions of cosmic history, each of which predicted the form of expansion Hubble had discovered. Two of Friedmann's solutions had a space that was infinite and unbounded. In these universes any two randomly chosen points would move farther away from each other as cosmic time marched forward. The third solution began with expansion but eventually turned around and started to contract. Space was finite and bounded in this version of the universe and two randomly chosen points would first move away from each other before eventually being swept back together. The trajectory the universe took in reality depended on the balance between cosmic expansion and the force of gravity pulling everything back together—a balance that could be described in a single number.

If the average amount of mass-energy in the universe (its mass-energy density) was above a critical value, then gravity would eventually beat expansion and the universe would slow down, change direction and collapse on itself. But if the mass-energy density in the universe was below this critical density, the universe would expand forever, thinning out until it was all but empty. If the mass-energy density was exactly equal to the critical density, the universe's expansion would be slowed by the force of gravity but only enough to halt its stretching infinitely far into the future.

The critical density parameter could be cast in a form called omega (Ω): the ratio of the actual density in the universe to Friedmann's critical value. It was omega that set both cosmic fate and cosmic geometry in his models. An omega larger than 1 meant a closed and finite universe,

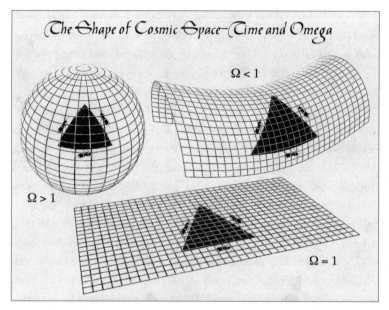

FIGURE 6.5. Cosmic geometry and the omega parameter. In the expanding cosmological models of Friedmann (and Lemaître), the global geometry of space-time depends on omega, a number describing the mass-energy content of the universe.

a moving version of Einstein's 3-D hyperspheres; an omega equal to or less than 1 meant a space that was infinite and unbounded. The search for omega's real value would come to dominate observational cosmology for decades to come.

Reading over Friedmann's paper, Einstein at first rejected the work, warning that the results appeared "suspicious". After a series of exchanges with Friedmann, the young Russian proved his case, and Einstein withdrew his objections. But, as sometimes happens, Friedmann's work failed to reach the small band of cosmological cognoscenti and within a few years it was forgotten.

Eight years later, in 1930, Hubble's discovery of the expanding universe was sinking into the minds of theoretical astrophysicists around the globe. So completely had Friedmann's work disappeared, however, that even Arthur Eddington, one of the first astrophysicists to embrace Einstein's relativity, could wonder out loud at a conference why no cosmological solutions could match the observed expansion. Ironically, Eddington's mistake lay buried not just in the journals but, in a way, right under his own nose.

Georges Lemaître, a professor of astronomy in Belgium, was a former student of Eddington's. Lemaître was not just a scientist working at the frontiers of theoretical cosmology; he was also a Catholic priest. Like Friedmann, he too had seen the horrors of World War I through close-range fighting and the use of poison gas.[41] The experience propelled the quiet young man to a life of both science and spirituality. After first studying general relativity with Eddington and then getting his PhD at MIT, Lemaître began exploring new cosmological solutions. In 1927 he derived the same expanding-universe model that Friedmann had found a few years earlier. Lemaître published his work in a little-known Belgian journal, where it languished, but he did manage to show Einstein his paper at a conference. Once again, however, the great scientist missed the chance to see the truth in his own equations. Einstein told Lemaître: "Your calculations are correct but your physics is abominable."[42]

Once Hubble's results made news, Lemaître immediately contacted Eddington. The quiet priest informed Eddington of his own solution and its nearly perfect prediction of the observed form of cosmic expansion. With Eddington's help, Lemaître's results quickly gained attention from the scientific community. Hubble's data now had their stunning interpretation. Galaxies were not flying through space away from one another; space, or rather space-time itself, was expanding. The linear pattern of galaxy recession velocities increasing with distance was to be expected if all of space were being uniformly stretched, carrying the galaxies along with it like pennies glued to an expanding rubber sheet. Without a doubt, the universe itself was expanding.

The discovery of the expanding universe was big news. From Europe to Japan to America, Hubble's discovery made for breathless headlines in newspapers and thrilling announcements in the radical new medium of the age: radio. It would be no small irony that at the moment

Hubble discovered the expanding universe of space-time, radio was rebuilding the human universe by contracting space and constructing an entirely new form of experienced time.

EVERY HOME SHOULD HAVE A RADIO!

The twentieth century's most intimate and enduring experiment in experienced time was born at an invitation-only white-tie party at New York's Waldorf-Astoria Hotel.[43] The invitations had gone out to the city's elite: celebrities, politicians and pundits. But the most honoured guests that night were an entire nation of radio listeners. It was November 15, 1926, the night the National Broadcasting Company, a true radio broadcasting *network*, was born. From 8:00 to 10:00 p.m. the same music accompanying the elegant dancers at the Waldorf was also emanating from radio sets across the continental United States. The music was carried on waves of electromagnetic radiation broadcast by a network of antennas spanning the country from New York to Los Angeles.[44] A different station owned each antenna tower, and each station had just been brought under the programming supervision of NBC.[45] Though the BBC had managed to make its first near-national broadcast four months earlier, via the long-wave antenna of station 5XX, the Daventry-based station had far less ground to cover and remained an experiment. NBC's Waldorf broadcast announced the beginning of true synchronous programming, and the impact of that night on Americans' experience of time would echo through the decades. The first true culturally shared sense of simultaneity—the first American experience of a nationwide "now"—was born.

As a commercial venture, broadcast radio began inauspiciously. In 1889 physicist Heinrich Hertz discovered how to generate and receive long-wavelength electromagnetic (radio) waves by driving oscillating electric currents though metal wires.[46] The business and political worlds took notice but were unsure what to do with the new invention. With telegraph and telephone wires becoming ubiquitous across the world, the first uses for the new wireless technology were imagined to be point-to-point communication. In other words, the killer application for radio was expected to be radiotelegraphy.

The Marconi Wireless Telegraph Company of America, founded by the brilliant Italian pioneer Guglielmo Marconi, started with a simple business plan.[47] American Telephone and Telegraph dominated the American telecommunications business of the day. Marconi wanted to upend AT&T's dominance of telegraphy by replacing wires with waves. By the end of the twentieth century's first decade, radiotelegraphy had made inroads in at least one domain: ship-to-shore communications. The potential for this form of radiotelegraphy was made concrete in 1912 when Morse-coded radio bursts broke news of the *Titanic*'s sad fate. The story of the *Titanic*'s sinking spread by radio from the *Carpathia*, the first ship to reach survivors, over to Newfoundland and down the East Coast, relayed from one Marconi wireless station to the next.[48]

In spite of its importance for naval messaging, point-to-point radio telecommunication had obvious limitations. Any hack with a radio set could pick up what were intended to be confidential messages. Its inability to offer privacy made radio a poor replacement for the telephone. Taking radio beyond radiotelegraphy would require entrepreneurs who could imagine it as something more than a wireless telephone.

David Sarnoff was one of those visionaries.[49] Born in 1891 in Uzlian, a Russian shtetl, Sarnoff immigrated with his family to the United States in 1900.[50] A quick study, the young Sarnoff rapidly turned his fascination with the telegraph into a working knowledge of Morse code and a job as a telegraph operator. He was an ambitious man and rose quickly within the Marconi Company. When General Electric purchased the business in 1919 to create the Radio Corporation of America (RCA), Sarnoff was elevated once again. He would soon prove his worth to the new company. In a prescient memorandum on the future of radio, Sarnoff wrote,

> I have in mind a plan of development which would make radio a "household utility." . . . The receiver can be designed in the form of a simple "Radio Music Box" and arranged for several

different wave lengths. . . . The box can be placed in the parlor or living room, the switch set accordingly and the transmitted music received.[51]

But this plan faced a classic technological chicken-and-egg dilemma. In 1919 RCA might imagine the public buying up a fortune's worth of radio sets, but what would they listen to? There was no one broadcasting in 1919 and, therefore, no reason to buy a receiver.

By 1920, the broadcast problem began to resolve itself. In that year Frank Conrad, a Westinghouse employee in Pittsburgh, set up a transmitter in his home. Pulling a Victrola up to the microphone, Conrad occasionally played opera records for the smattering of radio enthusiasts who owned receivers.[52] Conrad's broadcasting exploits caught the attention of local newspapers and, soon enough, the attention of his bosses. On October 27, 1920, the first U.S. licenced radio station, KDKA, started transmitting from atop the Westinghouse building.[53]

Other stations appeared at a brisk pace. By the autumn of 1921 six to twelve new operations were setting up shop each month. As broadcasters appeared, so did their audience. By 1922 Americans were spending more than $60 million a year on home radio sets.[54] The early growth of radio into the public consciousness was explosive. But early broadcasting bore little resemblance to what we experience today. In particular, the idea of programme time was yet to be invented.

At the start, American radio stations were entirely haphazard in their programming. They transmitted for at most a few hours each day, often appearing on the air only when their owners had time or opportunity. When a station in Schenectady, New York, began featuring regularly scheduled music, it was an important enough innovation to warrant a feature story in the *Washington Post*.[55] One reason the *Post* ran the story was that Washington, D.C., listeners were picking up the Schenectady station. The ability of radio waves to bounce off electrically charged upper layers of the atmosphere meant that, in an era with few active transmitters, stations could travel hundreds of miles from their source. Radio was breaching distances for Americans in an entirely new way, introducing a single "now" to widely separate populations hungry

FIGURE 6.6. Family time around the radio circa 1940. Regularly scheduled programming by radio networks redefined the common experience of time for most Americans beginning in the 1920s. Television would deepen the experience of "programme time."

for programming they could count on. In a few years, radio would begin to set time by changing the behaviour of its listeners.

The early days of radio had its share of cranks and innovators. Sometimes these were rolled into the same personality. The first form of regularly scheduled programming, destined to reshape the American experience of communal time, came not from RCA but from a medical quack in Milford, Kansas.

John R. Brinkley was a doctor of dubious credentials.[56] He had made his fortune offering sexual restoration to men through, of all things, goat testicle implants. The wealth he gained in this pre-Viagra age allowed him to buy the radio station KFKB (which stands for "Kansas First, Kansas Best" or, alternatively "Kansas Folks Know Best"). With its powerful three-thousand-watt transmitter, KFKB reached across almost a thousand miles.[57] While using the station for his own brand of hucksterism, Brinkley saw the power of the new medium. Only a few years after its inception in 1923, Brinkley's KFKB provided radio with its first full day of scheduled programmes. A typical morning on KFKB treated listeners to a range of music and quackery. From 7:00 to 7:30 listeners heard hints to good health, followed by a half hour of Bob Larkin and His Music Makers. The half hour from 8:00 to 8:30 was lecture time with Professor Bert.[58] And so the radio day unfolded as a mix of music, news and sales pitches from Dr Brinkley. Filling the day—that is, filling time—with continuous radio programming was a brilliant innovation that would outlive the good Dr Brinkley (by 1930 he had been exposed as a fraud and had his licence revoked).[59]

With the onset of the Great Depression in 1929, radio was one of the few businesses that continued to grow. As regular programming seeped into the everyday experience of Depression-era Americans, a new breed of celebrity was born. *Amos & Andy*, a comedy about two black taxicab owners, debuted in 1928.[60] For years it was one of radio's most popular programmes, settling into its nightly fifteen-minute broadcast on NBC at 7:00 p.m. Eastern time and then rebroadcast at 10:30.[61] The two white actors, Freeman Gosden and Charles Correll, and their racially insensitive creations became the "presiding deities of the twilight time", and they earned huge salaries. Like *Amos & Andy*, other radio shows and their stars drew large, consistent audiences each night as Americans began shaping their evenings around favourite programmes. In essence, people became trained to live on radio time through the popularity of shows such as *Amos & Andy*. But while other stars grew in the radio firmament, none shone as brightly during this first phase of broadcast culture as Rudy Vallee.

Born Hubert Prior Vallée, the handsome Yale graduate was a consummate performer.[62] With his saxophone, crooning voice and meticulous eye for detail, he went from a job as a small-time bandleader to host of the longest-running and most beloved radio programme of the era. His break came in 1928. WABC lacked funds to hire an announcer during broadcasts of Vallee's performances at the Heigh-Ho Club in New York.[63] With no other options, the station gave Vallee the microphone and told him to announce his own shows. Thus came the introduction, destined to be repeated a thousand times, that would lead the great expansion of America's broadcast universe: "Heigh-ho everybody, this is Rudy Vallee announcing and directing the Yale Collegians from the Heigh-Ho club at thirty-five east Fifty-third Street New York City."[64]

By the early 1930s, Vallee owned the 8:00 p.m. Thursday slot on NBC as presenter of *The Fleischmann's Yeast Hour*.[65] It was Vallee who single-handedly created the concept of the variety programme, featuring music, comedy and even scenes from Shakespeare played by leading actors including Claude Rains and Jimmy Cagney. Like clockwork, every Thursday evening at eight o'clock, Americans (especially American women) turned their radio dials to Vallee's show. The importance of Vallee to the American psyche is amply demonstrated by the infamous case of a husband in the Midwest who snidely asked his Vallee-obsessed wife, "Why don't you get something worth listening to?" She promptly shot him dead and continued listening.[66]

By the mid-1930s, the country was deep into the golden age of radio. The medium was as ubiquitous in the lives of people then as the Internet is in ours now. Surveying transformations in the human experience of time across fifty thousand years of evolution, scholars often give emphasis to calendars and clocks—technologies that directly measure and parse time. The growth of radio demonstrates, once again, an entirely different movement in our use of time.

The development of radio broadcasting networks created two entirely new cultural experiences of time. The first was a shared simultaneity—a communal "now"—that manifested itself in sports, breaking news and hugely successful popular entertainment. Along with the possibility of a single "now" for Washington and New York, Boston and Chicago, Denver and Seattle, a new encounter with time was also established as people structured their lives around programming. Radio time was something new, the abstraction of programming schedules made concrete in the warp and woof of daily life. Seven o'clock was *Amos & Andy* time. Eight o'clock was Rudy Vallee time. The human movement through life directly implicated in the human movement through time had shifted once again.

Within just a few decades, radio would play a defining role in establishing the Big Bang and the universe with a beginning. But at the moments of its first inroads into culture, radio was not yet ready to be used by science as a tool for cosmos building. Instead, during the decades before World War II, researchers were hard at work organizing the framework for relativistic cosmology and taking their first steps towards a cosmology that answered fully to astronomical data.

UNIVERSES IN MOTION

At first it might seem strange to compare washing machines and radios to expanding universes. Washing machines are the stuff of day-to-day life; they are objects at hand, fashioned from enamelled steel, rubber and copper tubing. Radio sets are also material objects whose weight we can feel in our hands, even though their purpose is to capture invisible electromagnetic waves. The expanding universe, however, lives somewhere further afield, its reality lying at a location between ideas and facts proven by scientific practice. But as we have already seen, the proven facts of relativity could not be separated from the creosote-soaked telegraph poles that launched the first nascent experiences of dispersed simultaneity at the turn of the twentieth century. As astronomy and cosmology took their first steps towards each other in Hubble's discovery of an expanding universe, the upwards and downwards movements of material engagement were seen everywhere. The human world and its image in cosmic science were both in the midst of rapid, profound and dizzying change.

What makes the first steps towards Big Bang cosmology so compelling was how fast scientists had to accommodate a new range of results and theories. In little more than a decade, the universe went from a single collection of stars to a vast and perhaps uncountable array of galaxies, each composed of billions of stars. From a cosmos composed of a single galaxy extending across, at most, a few hundred thousand light-years, the dimensions of the universe suddenly expanded a thousandfold and more. In that same time the universe, thought to be the very model of eternal quietude and repose, became an exploding manifold. The empty space of Newton's universe was replaced by a dynamic space-time best imagined as an elastic fabric sweeping galaxies apart. In the midst of this new data, a tumult of ideas and explanations bubbled up like foam in a swiftly moving stream. The great questions of cosmology were being transformed from philosophical speculation into data points. No one was sure what to believe, what to expect or what would come next.

Scientists live in the real world no matter how abstract and removed their professional concerns might be. The enigmatic entanglement between cosmic and social time wove the two concerns together with the electric-powered threads of machines and instruments. Within a single decade, radio redefined what it meant to "be there" at events as diverse as boxing matches and ballroom dances. Current-driven machines compressed time and effort from days to hours and shifted both roles and expectations in the most intimate human frontier, the home. The skies began to fill with planes. Ships could stay in continuous contact with land. Cars became ubiquitous on an ever-expanding network of tarmac-covered roads. Everywhere and in every way, space and time were being redefined by culture through its machines. And the narratives of cosmology were changing just as completely and just as rapidly.

Chapter 7

THE BIG BANG, TELSTAR AND
A NEW ARMAGEDDON

The Nuclear Big Bang's Triumph in a Televised Space Age

BIKINI ATOLL, THE SOUTH PACIFIC • MARCH 1, 1953, 7:54 A.M.,
H+16 MINUTES

Jesus Christ! he thought. The water blew straight out of the toilet!

He was scared but that was to be expected. This was his first test shot, after all. It was the other guys in from the bunker who worried him, the ones who had seen atomic explosions before. They were looking pretty scared now too and that was really terrifying.

He was outside the bunker, looking straight up at the blue Pacific sky and the mushroom cloud, pure white, towering twenty miles overhead.[1] My God, he thought, what power! For months he had been looking forward to this very moment, but now, as the minutes ticked off and the radiation level ticked upwards, he just wanted to get away from this terrible place.

Everything had gone smoothly in the run-up to the detonation. This was "Castle Bravo," the first in the Castle series of U.S. hydrogen bomb tests. They were using a different design for the nuclear detonator this time, but that seemed like just a detail to him. All the previous tests out here in the Pacific had gone off fine. After the final inspection of the "package" out on the atoll the head scientists had helicoptered the miles back to the control bunker. The countdown was tense but that was no surprise either. Countdowns were supposed to be tense.

175

It was his job to call out the time: "H minus 15 seconds, H minus 10 seconds . . ." At preciously 6:45 a.m. the switches were thrown and they waited. With no windows in the bunker, they were blind to the blast. From a ship many miles away the test results were radioed in: "Detonation achieved."

Everyone cheered. Then, quickly, they braced themselves for the ground shock that might or might not come.

It came all right, that was for goddamn sure. And it was nothing like what they'd told him to expect. The building pitched like a seesaw. It lasted only seconds but it felt like an eternity. Then another one hit. He had to drop to the floor to keep from being tossed over. A few seconds later, the blast wave from the multimegatonne explosion swept over them. The concrete bunker trembled like an old wooden frame house. That was when the water shot straight up out

FIGURE 7.1. The Castle Romeo nuclear test. Castle Romeo was part of a series of explosions in the Operation Castle series. Romeo was conducted on March 27, 1954, and yielded an eleven-megatonne explosion, the largest up to that point. The nuclear science behind these weapons was the same as that used in Big Bang "nucleosynthesis" cosmology.

of the toilet. It terrified him but the others just laughed it off like it was some kind of amusement ride. To the experienced guys, it seemed like it was all big fun. But that was before they started measuring the radiation.

At first everything seemed okay. The numbers on the Geiger counter were rising but the level was in the safe zone. Then the chief got worried. The radiation levels jumped so fast he had to change settings on the Geiger counter even to get a reading. Something was going wrong, really wrong. One of the military guys said rad levels in the bunker were past dangerous and the rooms in the other facilities were much worse. There was no place they could go.

Now everyone was being called inside. Christ, they were all going to be trapped.[2]

NUCLEAR AGES IN COSMOLOGY AND CULTURE

The Castle Bravo test at Bikini Atoll was the worst radiation contamination event in U.S. history.[3] The twenty-two-megatonne explosion was almost three times more powerful than expected and a change in wind direction dropped radioactive fallout on U.S. scientists and sailors scattered on ships across hundreds of miles.[4] It is estimated that three hundred people received dangerous levels of radiation exposure.[5] A Japanese fishing vessel with the unfortunate name of *Lucky Dragon No. 5* was caught at the edge of the "safe zone". Fallout rained down on the unprotected and unwarned fishermen, and soon they fell severely ill. A week later, one of the *Lucky Dragon's* crew died. Fallout from Castle Bravo drifted to Australia, India, Japan and even the United States and parts of Europe. While the test was supposed to be secret, it quickly became an embarrassing international incident, driving calls for atmospheric test bans and providing a palpable reminder that any thermonuclear exchange would have global consequences. For many, the end of human time was feeling like a distinct possibility.

Atomic and nuclear weapons were the direct material manifestations of the two great scientific revolutions of the twentieth century. The first revolution was Einstein's relativity. It was his unification of matter and energy, in the ubiquitous equation $E=mc^2$, that gave the weapons their godlike powers of destruction. Only a gram or so of matter need be fully transformed to energy and a mighty city could be vaporized.[6] But what is matter? What, truly, is its nature and its constituent parts? Those questions lay at the heart of the century's other great scientific overturning—quantum physics.

By the end of the nineteenth century, physicists began constructing a new generation of experimental devices that allowed them access to the world on increasingly smaller scales. Atomism, the Greek doctrine of a universe composed of minute but fundamental flecks of matter, made its return. It was reincarnated not as philosophy but as the end point of experimental investigation.

As experiments reached down to regions measured in billionths of a metre, physicists were suddenly able to probe a staggering array of new phenomena: the atomic basis of heat, the nature and behaviour of atoms and their constituent parts, the subtle relationships between light, energy and matter.[7] These were domains of nature that scientists had never accessed before and bewildering new behaviours were unveiled by the experiments. As physicists confronted the world through their new instruments, they were forced to radically alter their approach to, and conception of, physical reality. They often ran headlong into their own conceptual biases while working to reveal the properties of atoms.

Attempts to make sense of the experiments using the physics of the day, what we now call classical physics, failed entirely. Classical physics contains a heavy dose of gut-level intuition, the fruit of millions of years of evolution and the hardwired physics modules working in our brains. This intuition rises from our direct experience of the world at our own scale: the resistance of a heavy stone being pushed aside, the lurch of our stomach when we fall. The new physics did not rise from this kind of evolved experience. Instead the rules governing atoms seemed to mock the imperatives of our intuitive physics.

During the first three decades of the twentieth century, physicists such as Niels Bohr in Denmark, Werner Heisenberg in Germany and Paul Dirac in England responded to the new data and new questions with bold conceptual leaps. In a phenomenal display of human creativity, these scientists reached beyond their own training and created an entirely new branch of physics we now call quantum mechanics.[8]

The quantum mechanical articulation of atomic behaviour was stunning in its speed and completeness. By the late 1920s, physicists had worked out the structure of elements down to the exquisite details. Democritus had got it right two thousand years earlier: every element, from the lightest hydrogen gas to the heaviest lump of uranium, could be decomposed into tiny atoms. But atoms, the physicists found, were not the lowest level of structure. Every atom was made up smaller fundamental particles. At the atom's centre, carrying most of its mass, was a nucleus composed of nucleons: electrically charged protons and electrically neutral neutrons. Surrounding the nucleus was a swarm of orbiting electrons. All atoms were electrically neutral, with equal numbers of negatively charged electrons balancing out the positively charged protons.

A classical physicist hears that atoms are made up of smaller particles and she imagines they look and behave much like microscopic billiard balls. These little spheres should bounce into each other, spin, hold an electron charge and react to gravity or magnetic fields. Most important, she would imagine the tiny specks of matter to possess definite properties. Those properties should be measurable to any degree of accuracy (just as you could precisely measure the position of a cue ball on a pool table). The problem with this kind of common sense was that it did not hold up on the microscopic level. Physicists quickly found that it was impossible to build working, predictive theories—that is, mathematical models—using the billiard ball model of atoms. Nature, it seemed, was just not built that way. As Werner Heisenberg once said, "Atoms are not things."[9]

There were many surprises in the new quantum description of subatomic reality but two aspects would come to stand out. The first was embodied in the word *quantum*, which is Latin for "how much".[10] Classical physics imagines the world to be composed of objects whose properties appear in a smooth continuum. A bicycle can travel at five miles an hour, ten miles an hour or any speed in between. A ball can be dropped from five feet above the floor, ten feet above the floor or any height in between. In the quantum world all critical facets of physical reality are quantized—they appear only in discrete divisions. The energy of an electron orbiting a nucleus can take on only specific quantized values and nothing in between. When the electron changes its physical state, it jumps from one discrete value of energy to the other without ever having the intermediate values. It's like climbing a staircase by only appearing and disappearing on the steps.

Along with quantum jumps in physical properties, the new physics also forced scientists to give up their cherished notion of pure causation. When Thales began the Greeks' inquiry into nature 2,500 years earlier he bequeathed to his students the idea that all physical effects were assumed to have direct physical causes. Likewise, every cause had to have a definite and well-defined effect. If randomness did exist in the world, it was really only the result of our ignorance. We think that rolling dice and probabilities go together only because we do not have perfect knowledge of every atom in the dice. If we did, the classical Newtonian picture of the world tells us, we could predict with certainty how the dice would roll every time (and make ourselves a lot of money in the process). But at the atomic level even this idealization falls away.

Quantum events are subject to an inherent uncertainty built into nature. Radioactive decay is the archetypal quantum mechanical event. When the nucleus of a radioactive element decays, it sheds some of its constituents and transmutes into a different isotope (the same element with fewer neutrons) or another element altogether (with a different number of protons). In a clear affront to classical sensibilities, there is no way to predict exactly when an individual radioactive nucleus will decay. It is an event that is fundamentally random. The technical term for this is *acausal* (without cause). Radioactive decay is an acausal process.[11] Quantum mechanics can give exquisitely accurate probabilities that describe large collections of nuclei and their behaviour, but predicting the fate of an individual nuclei simply cannot be done. On the atomic and subatomic levels, nature does not behave that way. Quantum mechanics raised uncertainty to the level of a fundamental physical principle.

As the 1920s ended, physicists turned from atoms as a whole to the inner workings of the nucleus. A basic understanding of what made a nucleus of helium (two protons and two neutrons) different from a nucleus of carbon (six protons and six neutrons) was slowly fleshed out. As the nuclear nature of the elements was teased apart, an important question made its first appearance: How had these nuclei been born and what set their abundances? A causal survey of the elements made it clear that some, including hydrogen, were common and others, such as gold, were not. It was through this question of elemental abundances that nuclear physics and quantum mechanics stood poised to make their entrance in cosmological theory.[12]

BEGINNING THE BEGINNING: THE PRIMEVAL ATOM

The story of modern cosmology is often told as a straight line from Hubble's expanding universe to the glorious confirmation of what in the 1960s would be called "hot" Big Bang theory. History, however, is far more complex and interesting. The Big Bang was proposed not once but three separate times over a span of thirty years. Until its final incarnation, it was resisted by a sizeable faction of physicists and astronomers with rival, alternative cosmologies gaining considerable attention and followers. But like any good story, the triumph of the Big Bang has its share of characters, obstacles and fortuitous accidents.

The Big Bang's first incarnation originated with Lemaître, the same shy Catholic priest-scientist who provided a proper general relativistic account for Hubble's expanding universe. Lemaître's solution had come to be called the Eddington-Lemaître model because together the pair had successfully reintroduced it to the astrophysical community. But Eddington was a strong advocate for an eternally expanding universe and would brook no discussion of cosmic origins.[13] "Philosophically, the notion of a beginning of the present order of Nature is repugnant to me", Eddington said.[14] Lemaître disagreed. Motivated by issues of science and not religion, he felt discussions of the universe at its earliest moments could be fruitful and productive, yielding real, measurable predictions.

In exploring the universe's origins Lemaître took his cosmological cue from quantum physics. Knowing radioactivity to be fundamental to the process of splitting large nuclei into smaller ones, Lemaître imagined the same process might be writ large on the cosmos. Following his own expanding universe models backwards, Lemaître imagined an early stage of cosmic history in which all mass-energy was packed into a single primeval nucleus.

In Lemaître's theory, first proposed in the mid-1930s, the history of the universe became a process of super-radioactivity. The primeval atom split and split again, with smaller and smaller "atoms" dividing out until, ultimately, all the particles present today were born. Once again the ancient questions of cosmology would return to be recast into a new era's language. The most important aspect of Lemaître's theory was its use of quantum physics' infamous indeterminacy to avoid questions of what caused the universe (and time) to just begin. Lemaître was leaning hard on quantum physics to vault over a cosmological dilemma of origins known for centuries and often referred to as Kant's First Antimony.[15]

Three hundred years before Lemaître, Immanuel Kant asked how the universe could be explained through a deterministic cause when it must embrace all causes. Since the universe encompasses all things and, therefore, all causes, what can exist outside to set it in motion? In essence the First Antimony states, "That which causes all effects cannot itself have a cause." Quantum mechanics gave Lemaître a way around Kant's dilemma. As if responding directly to Kant, Lemaître wrote,

> Clearly the initial quantum could not conceal in itself the whole course of [cosmic] evolution; but according to the principle of indeterminacy, this is not necessary . . . the whole story of the world need not have been written down in the first quantum like a song on the disc of a phonograph. The whole matter of the world must have been present at the beginning, but the story it has to tell may be written step by step.[16]

Using the phonograph metaphor of his own culture's material engagement, Lemaître saw that quantum mechanics allowed cosmic history to unfold in the moment, without a predetermined track set by Newtonian physics. In other words, the primeval atom by itself is enough. Uncertainty, built into the very foundations of quantum mechanics, let Lemaître off the hook in terms of specifying a cause for subsequent evolution. The primeval atom eventually would decay without a cause, leading to cosmic transformation and evolution, just as an atom of plutonium will eventually decay without a cause. Lemaître was able to use quantum mechanics' inherent indeterminacy—its claim that randomness is intrinsic to nature—as a foundation to build a fully scientific narrative of cosmic evolution.

It is, however, important to see that this early version of the Big Bang is not a theory of creation ex nihilo. The primeval atom exists already, and no explanation is given for this brute fact. Lemaître's theory of cosmic evolution begins *after* creation in the sense that the primeval atom already *is*. Like all versions of the Big Bang, including the modern one, Lemaître invented not a theory of creation but a theory of after creation.

The nuts and bolts of after creation meant tracking general relativistic solutions for the universe as far back in time as the physics allowed. Using Einstein's equations, physicists had learned to conceive of cosmic evolution in terms of what they called the universe's radius. In general, the cosmic radius could be thought of as the distance between any two arbitrarily chosen points in the space-time fabric. If space expanded, then the radius of the universe increased and all points were carried away from one another. If space contracted, then the radius decreased and all points were carried towards one another. Lemaître knew that if he followed his cherished expanding universe solution backwards in time to its extreme, then at $t = 0$ the radius was zero. Thus, if the models were to be believed, at the beginning of time the universe had no volume at all. All its mass would be compressed into a single geometric point—an obvious absurdity. This singularity, as it was called, of infinite mass-energy density and zero size would haunt Big Bang cosmologies into our own era. For his part, Lemaître recoiled at the infinities of the singularity and denied it could have any physical meaning. He was sure that *something*, some other kind of physics, must come into play, allowing the universe to avoid the infinities of a singularity.

It is worth noting that Lemaître flirted with notions of a cyclic universe as one way of avoiding the singularity at the beginning of time. In 1922, Friedmann had been the first to find closed trajectories of cosmic expansion followed by contraction, and Lemaître was willing to consider these solutions as one turn in an ever-repeating cycle. He wrote that cyclic models possessed "an indisputable poetic charm and make one think of the phoenix of legend".[17] In the end, however, he abandoned the cyclic cosmos, believing it was ruled out by astronomical observations. This led to his primeval atom and a cosmology with the *appearance* of a beginning in the first atom's primal decay.

The birth of the beginning did not go smoothly. Without firm supporting evidence, Lemaître's primeval atom was simply too far on the hairy edge of conjecture for most scientists. John Plaskett, a Canadian physicist, called Lemaître's theory "the wildest speculation of all . . . speculation run mad without a shred of evidence to support it".[18] Many cosmologists were wary of taking too seriously any mathematical models beginning with a singularity. The noted American cosmologist Richard Tolman warned of the "evils of autistic and wishfulfilling thinking" when it came to mathematics, reality and the origin of spacetime.[19] Others, including Eddington, continued their philosophical objections: "The beginning seems to present insuperable difficulties unless we agree to look on it as frankly supernatural", quipped Eddington in a lecture.[20] Lemaître did have an ally in Einstein, at least, who "was enthusiastic about the idea".

The first attempt at Big Bang theory also faced a more serious difficulty in terms of astronomical data. Hubble's simple, linear relation between galaxy recession velocity and distance (called Hubble's law) could be used as a clock. By telling astronomers how rapidly the universe was expanding, Hubble's law could be turned upside down to infer how long it had been since all the galaxies were piled on top of one another. In this way, Hubble's law provided an "age of the universe" of about two billion years. While that may seem like an incomparably long time, it was, in fact, far too short. Astronomers had already found good reason to believe that the sun was ten billion years old. Geologists, working on their own, were ready to put the Earth at five billion years old.[21] Thus, the notion that Hubble's expanding universe implied a beginning ran straight into a paradox: the universe was younger than the objects in it. This discrepancy became known as the "age problem" and neither Lemaître nor any of his supporters had a good answer for it. With no solution to the age problem in sight, opposition to Lemaître's version of the Big Bang remained strong.[22]

By the beginning of the 1940s efforts to include quantum mechanics in accounts of cosmic origins all but stopped. As the battles of World War II swept across Europe and Asia most scientists were engaged in their own war-related work. Quantum mechanics would move from the realm of abstraction and experiment into the domain of material engagement as physicists worked round the clock to deploy the nucleus as a tool of war.

THE DOOMSDAY CLOCK: TURNING HOURS TO MINUTES IN NUCLEAR WAR

The clock face was a potent symbol. The hour hand was set at twelve. The minute hand was set a few minutes before the hour—midnight, the end of the world.

The Doomsday Clock was the 1947 brainchild of the editors of the *Bulletin of the Atomic Scientists*.[23] Many of them had helped develop the first atomic bomb, and now they hoped to alert the public to the dangers of atomic weapons. The clock was a graphic representation of their best estimate of nuclear war's proximity. Originally set at seven minutes to midnight, the clock would be reset nineteen times up to the present day. The invention of the intercontinental ballistic missile in the 1950s would be one critical pressure point pushing the hands of the clock closer to midnight.

The weapon dropped over Hiroshima, Japan, on August 6, 1945, was an atomic bomb, meaning it relied on nuclear fission: the splitting of a large atomic nucleus into smaller daughter nuclei with an ensuing release of energy. "Little Boy" incinerated Hiroshima in seconds, killing more than a hundred thousand people. It had an explosive yield of twelve thousand tonnes of TNT.[24] In the aftermath of the war and the instant rivalry with the Soviet Union, the United States began planning for the possibility of atomic war with its former ally. When the Soviets exploded their own device in August 1949 (using plans stolen from the United States by scientist-spy Klaus Fuchs) the atomic arms race was on.[25]

During the Manhattan Project scientists had discussed the possibility of a "superbomb" using nuclear fusion rather than fission. By fusing light elements together, Edward Teller, the project's chief advocate, calculated possible explosive yields that were a hundred times higher than those of atomic weapons. After extensive debate, including condemnation of the project as immoral, the United States began work on the superbomb. In 1952 the first U.S. fusion device, code-named "Mike", was exploded in the Pacific.[26]

The destructive power of the new weapon was staggering. One physicist involved with tests remembers watching his first fusion explosion: "Now, on kiloton shots it's a flash and it's over, but on those big shots it's really terrifying. . . . I never will forget that experience of the thermal effects from the very high yield shots."[27]

One year later the Soviet Union detonated its own thermonuclear weapon. Watching newsreels of the hydrogen bomb tests, it was easy to feel the old myth of Armageddon—the end of time—had become real and manifest in this device.

With nuclear weapons in hand, both the United States and the Soviet Union soon turned their attention to the mechanics of weapon delivery. In the United States, the Strategic Air Command (SAC) was given the task of waging nuclear war. With the cigar-chomping General Curtis LeMay leading the way, SAC built a strategy relying on long-range atomic bombers. For General LeMay the totality of a nuclear weapon's violence led to strategy focused on lethal "country-killing" blows early in a conflict. Beginning in the late 1940s and into the 1950s LeMay and SAC developed the technological capacity to keep bombers "orbiting" on the edges of Soviet airspace, always ready to deliver their small Armageddons.[28]

The introduction in 1954 of the KC-135 Stratotanker meant that

FIGURE 7.2. XB-47 Bomber prototype, circa 1957. Long-range bombers and in-flight refueling kept nuclear armed planes "orbiting" close to their targets, reducing the time to attack down to hours. The development of intercontinental missiles would reduce that time even further to tens of minutes.

American nuclear bombers could refuel in the air, essentially giving them unlimited range. This was part of LeMay's strategy of counter-force. If war came, SAC would destroy, or "counter", Soviet nuclear forces on the ground (as opposed to targeting Soviet industrial capacity). The Soviets, for their part, adopted a first-strike strategy, stressing "the importance of landing the first, preemptive nuclear blow". The Soviet emphasis on surprise and the continual presence of U.S. bombers near the Soviet landmass gave rise to a sense that both nations were on a nuclear hair trigger. The "time to target" in the early 1950s, for both U.S. and Soviet bombers, was only a few hours.[29]

The public became increasingly aware of this nuclear hair trigger as the 1950s progressed. The development of civil defence programmes in these years was meant simultaneously to prepare the population for a nuclear war and convince them they might survive. Massive, multicity nuclear preparedness drills—Operation Alert—were run once a year. The lengths to which the drills were made realistic were as impressive as their expected results were dubious. Newspapers ran fake stories the day after each Operation Alert recounting the numbers of civilians killed

FIGURE 7.3. Operation Alert wipes out Buffalo, New York. "Fake" front page prepared by the *Buffalo Evening News* as part of national civil defence's Operation Alert.

and wounded. On July 20, 1956, the *Buffalo Evening News* published a special "emergency" edition as part of that year's operation. The headline screamed, "125,000 Known Dead, Downtown in Ruins".

The culture of atomic age nuclear preparedness—from a civil defence film called *Duck and Cover*, featuring an animated character called Bert the Turtle, to backyard bomb shelters—seems both contrived and bizarre. For those who lived through the era, however, the sense of angst was both real and pervasive. A nuclear sword hung over everyone's head, swaying like a pendulum ticking off final moments.

By the end of the 1950s, the introduction of a radical new technology would heighten tensions. Both the United States and the Soviet Union had taken notice of Germany's V-2 rocket bombs in the closing days of World War II. The fourteen-metre-tall liquid-fuelled rockets were the world's first long-range ballistic missiles.[30] While the V-2s did little to affect the outcome of the war, their range and accuracy were astonishing. When the war ended, both U.S. and Soviet forces did their best to accumulate as many of the unused German rockets as possible, along with German rocket scientists. With Wernher von Braun leading the Americans and the legendary Sergey Korolyov driving the Soviets, the race was on for continent-bridging nuclear-tipped missiles.

The Soviets won the race. In August 1957, under Korolyov's direction, the Russians tested their R7 rocket. It was truly an intercontinental ballistic missile (ICBM), soaring across 3,700 miles in its first successful flight. Two months later another R7 hurled a sixty-centimetre silver ball packed with electronics into orbit. The payload was called Sputnik, and its incessant chirps from more than two hundred miles overhead let the world know that even the stars were now a frontier in the Cold War. Two years later the United States joined the ICBM club when its Atlas D missile was successfully tested. The era of the nuclear bomber's superiority was ending; the future belonged to the missile.[31]

The ICBM changed both nuclear war strategies and the public perception of its fate. After launch a typical ICBM will spend just three to five minutes in its boost phase, rising into space. The midcourse phase, when the missile reaches heights of 625 miles or more, lasts only twenty-five minutes.[32] The re-entry phase, during which the missile dives back through the Earth's atmospheric blanket, lasts a little more than two minutes. All told, from launch to delivery, a population targeted by an ICBM would have less than thirty minutes' warning before their annihilation.[33]

In 1986, the United States, the United Kingdom and the Soviet Union had more than seventy thousand nuclear weapons targeted at one another.[34] The addition of multiple independent re-entry vehicles (MIRVs) in this final phase of the arms race meant that a single missile could vaporize up to ten separate targets. Had these full arsenals ever been unleashed, the apocalyptic conclusion of human civilization would have been fast and horrifically efficient, with the last missiles simply "bouncing the rubble around". Just as the Bendix had done for wash day, the ICBM compressed the apocalypse into just under an hour.[35]

A NUCLEAR UNIVERSE: THE BIG BANG, TAKE TWO

The nuclear revolution transformed cosmology just as surely as it changed world politics. However, it would take just a little more time for the shift to be recognized in cosmology. The Big Bang—that is, a universe evolving from a primal state—would be re-invented theoretically once more after World War II. Unlike Lemaître's first primeval atom incarnation, the Big Bang's second "discovery" would take the form of a full-blown nuclear cosmology.

A few years before the start of World War II, the geochemist Victor Goldschmidt completed the first full census of elements.[36] Goldschmidt's work showed that hydrogen and helium, the lightest elements with the simplest nuclei, made up the bulk of matter. The profusion of heavier elements generally dropped with weight and nuclear complexity. Thus iron, with 26 protons and 30 neutrons, was less abundant than oxygen, with 8 protons and 8 neutrons, but more bountiful than uranium, with 92 protons and 143 neutrons.[37]

Out of Goldschmidt's census came a riddle: Why and how had the universe ended up with just this distribution of elements? When World War II ended, one nuclear physicist set out to use his new science to create a cosmology that could provide an answer.

George Gamow had never been conservative in his habits or his science. Born in Russia and trained under the tutelage of Friedmann, he chafed under the confines of Stalinism. In 1928 Gamow found a powerful explanation for one ubiquitous form of radioactivity called "alpha" decay.[38] Using the fundamental uncertainty that was built into quantum mechanics, Gamow showed how nuclear particles (2 protons and 2 neutrons) could "tunnel" out of a larger nucleus, disappearing from inside the nuclear interior only to reappear outside and escape. The discovery gained him enough fame to allow his emigration to the United States in 1934.[39] Irreverent, impatient and and a heavy drinker, in the United States he quickly enhanced his reputation for theoretical brilliance. For a scientist such as Gamow, Goldschmidt's elemental abundances were too tempting to leave to someone else's explanation. With typical insight and audacity, he leapt to the conclusion that Goldschmidt's data were nothing less than a fossil record of cosmic history—a history that must have begun with a bang. To explain the elemental abundances, Gamow imagined an early universe that was a billion times denser and hotter than today's. This ultradense, ultrahot starting point would be Gamow's Big Bang. It would be the starting point for his calculation of the origin of the elements—a theory of cosmic nucleosynthesis.

Much had changed in the years between Lemaître's primeval atom and the atomic bomb. The weapons developed in Los Alamos, New Mexico, were the products of a robust nuclear science. When Gamow began his cosmological work in 1946 a great deal of nuclear physics had already been discovered that went far beyond just the existence of protons and neutrons. A growing menagerie of new particles was being catalogued, particles with names such as *muons* and *mesons*. Physicists charted reactions between these subatomic particles to create a detailed working knowledge of nuclear chemistry, where quantum mechanical transformations turned one set of particles into another. Gamow's audacious goal was to use this growing body of knowledge to map out the first moments of cosmic history and explain the elemental abundances we see today.

His first paper on the subject in 1946 outlined the basic idea.[40] The relativistic cosmology of Friedmann and Lemaître would be the space-time container into which Gamow poured his nuclear physics. Just as in Lemaître's primeval atom theory, Gamow began immediately after the beginning. Rather than a single all-encompassing primeval nucleus, Gamow imagined the early universe to be a soup—a primordial soup—of neutrons.[41]

Gamow proposed that well-timed nuclear reactions in this primordial soup were the explanation for Goldschmidt's elemental abundances. Starting a fraction of a second after the dreaded singularity, Gamow's theory had a cosmic timer built into it. As space-time expanded, according to the Friedmann-Lemaître solution, the temperature and density of the universe smoothly declined. As the temperature and density fell, a brief window in time would occur when conditions were just right to turn the entire universe into a nuclear reactor. All Gamow had to do was follow the details of the reactions and their dependence on the cosmic background of density and temperature and he was sure all the known elemental abundances could be recovered.

But details were not Gamow's strong suit. He was a big-picture guy and putting meat on the skeleton of his grand idea required the heavy lifting of hours and hours of laborious calculations. To carry the idea to the next level Gamow asked his graduate student Ralph Alpher to take on the project. Alpher, the son of Jewish immigrants, had exactly the right balance of patience, technical competence and physical insight to take Gamow's proposal and turn it into a theory complete with predictions.[42]

In 1948 Alpher and Gamow fleshed out important details in the model. The entire process of "nuclear cooking" began just 0.01 second after the universe began from the singularity (whatever that meant) and took less than a half hour to complete.[43] At first the primeval soup of neutrons (which Alpher dubbed "Ylem") would decay (meaning quantum mechanically transmute or transform) into protons and electrons. This transmutation created the first hydrogen nuclei (protons). Colli-

FIGURE 7.4. Ralph Alpher with a slide rule, circa 1950.

sions between protons and neutrons would then lead to deuterons, a heavy form of hydrogen. Continued collisions would add neutrons and protons until helium, with its two protons and two neutrons, became abundant. Then, like a chain being built one link at a time, the other elements would follow until, just as Gamow had originally proposed, all the elements in the universe were built up with the abundances shown in Goldschmidt's data.

As the pair prepared their manuscript for submission to the journals, Gamow, at the last minute, saw the opportunity for a joke he couldn't resist. Referring back to the first letters of the Greek alphabet—*alpha*, *beta* and *gamma*—Gamow inserted the name of his friend Hans Bethe, a Nobel Prize–winning nuclear physicist, between his and Ralph Alpher's in the authors list. Writing later, Gamow recalled,

Dr Bethe, who received a copy of the manuscript, did not object, and, as a matter of fact, was quite helpful in subsequent discussions. There was, however, a rumor that later, when the α, β, γ theory went temporarily on the rocks, Dr Bethe seriously considered changing his name to Zacharias.[44]

Ralph Alpher was not amused. He worried that having the names of two famous physicists on the paper would devalue his own (enormous) contribution to the work. History would, unfortunately, prove him right.

Gamow's joke aside, the initial calculations in the Alpher, Bethe and Gamow paper seemed promising. A second paper in 1948 improved on the first by including parameters for key reactions only recently published by Argonne National Laboratory in Illinois (the lab was the fruit of the wartime nuclear physics effort). From 1948 to 1951, Alpher was joined by another young physicist, Robert Herman. Together they published a series of detailed papers uniting nuclear physics and its understanding of the very small with general relativity's understanding of the very large. The synthesis created a new kind of cosmological theory describing the detailed evolution of space, time *and* matter.

The calculations of Alpher, Herman and Gamow were a tour de force of theoretical physics. Like an exquisitely choreographed ballet, all the known particles played their role on a space-time stage described by an expanding universe of general relativity. Each species in the subatomic zoo mixed with others and with particles of light (photons) in the ultrahot, ultradense primordial soup. Particle populations remained in perfect equilibrium, changing back and forth from one form to the other, until the cosmic expansion cooled and thinned the soup. Nuclear reactions depend on temperature and density. As the cosmic temperature dropped, some reactions slowed to a crawl, like a cake that stops baking as the oven cools. In this way the universe's expansion carried the primordial soup through successive stages where some particle species stopped transmuting and were "frozen out" of the ongoing reactions. Their abundances became fixed for all the cosmic history to follow, becoming the nuclear fossils we find in our own era. By tracking these

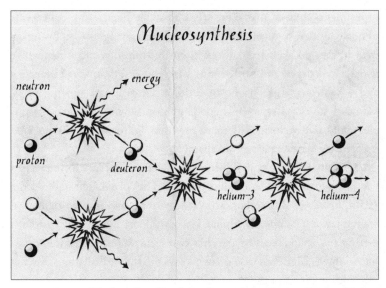

FIGURE 7.5. Nuclear "cooking" in the early universe. In the Alpher, Gamow theory of Big Bang nucleosynthesis, light atomic nuclei combine to form heavier nuclei. Here hydrogen nuclei (a single proton) combine to form helium nuclei (two protons and two neutrons). The universe is hot and dense enough for nucleosynthesis only during a brief period. All nuclear cooking ends a few minutes after the Big Bang.

reactions and the successive freeze-outs, Alpher and his collaborators followed the nucleonic dance of the early universe through the age of the nuclear cooking. Within half an hour after $t = 0$ the nuclear show was over and the elemental abundances were set.

Alpher, Gamow and Herman's calculations would form the basis for all modern theories of the early universe. But when the work was published, the authors had a difficult time convincing scientists of its worth. The detailed nuclear calculations were alien to most astronomers. More important, for the physicists who could follow the calculations, there was a hole in the theory that tamped enthusiasm for the Gamow-Alpher "hot Big Bang".

Building heavy elements from light elements in a step-by-step fashion was the fundamental idea behind Gamow's nuclear cooking. But experiments showed that there were no stable nuclei with collections of nucleons that summed to 5 or 8. If nuclear cooking, as Gamow imagined it, created a nucleus of beryllium-8 (4 protons and 4 neutrons), it quickly decayed and turned into a nucleus of lithium-7 (3 neutrons and 4 neutrons). Like trying to climb a staircase with missing steps, Gamow's step-by-step nuclear cooking faced a "mass gap". Heavier nuclei could not be formed because unstable lighter nuclei disappeared before they became part of the chain.

Gamow, with his typical disregard of details, was sure this mass gap was just a matter of details and would soon be solved. Throughout the 1950s, however, all attempts to cross the mass gap in a cosmological setting failed.[45] While the Alpher-Gamow theory did a superb job predicting the abundances of the first two elements—hydrogen and helium—the heavier ones would not give way to their calculations, leaving other scientists sceptical of the entire enterprise.[46]

Alpher eventually left research physics to take a position with General Electric.[47] There he worked on a number of projects including the physics of atmospheric reentry for ICBMs and other spacecraft. In one of cosmology's most unfortunate episodes, Alpher went to his grave bitter that he had never received his full measure of credit for shaping the century's dominant cosmological model. Herman also left academia. Gamow, for his part, shifted his emphasis to biology and left cosmological theorizing behind. Thus, the second discovery of Big Bang theory ended with a whimper. By the mid-1950s, most researchers were unaware of their work, but the researchers left a gift before departing the cosmological stage. It was a kind of Easter egg that would one day crack open to revolutionize the field. Buried within their papers was a prediction that would remain unnoticed for another decade until fate, time and telecommunications satellites intervened.

LONDON • JULY 23, 1962, 8 P.M.

"Ladies and gentlemen, we have just been informed that this baseball game is being seen in Europe, right now, over the Telstar satellite. Let's give all the baseball fans in Europe a big hello from Chicago."

FIGURE 7.6. Telecommunications satellite as rock star. Original album art for the Joe Meek song "Telstar," performed by the Tornados.

Joe Meek was so excited he couldn't keep himself on the piano bench. Baseball! All the way from America as it was happening! It was too much.

From the first moment of the telecast, when the American flag flickered on the screen, Meek was almost giddy. He might be a record producer with number-one hits now, but fifteen years ago he'd been a radar operator in the RAF. In all that time he'd never lost his passion for electronics, space exploration and the dream of what they both meant for the future.

He kept coming back to the image of Telstar sailing gracefully overhead in orbit, miles above the atmosphere. He wanted to make people see what he saw, a new world of wonders that was just beginning.

Within an hour Meek had most of the song, a tribute to Telstar, composed. He could feel it was going to be a hit. He knew it in his bones, just like that first time he'd produced Humphrey Lyttleton's "Bad Penny Blues" five years before. On that song he'd used his expertise in electronics to compress the sound of Lyttleton's piano. Hump had been so angry with him for "tinkering" with the sound—until the song went to number one. Meek knew his Telstar anthem would need a lot more electronic effects than that first effort. The song was about the future and it would have to sound like the future.

It only took another hour to finish and as he walked away from the piano Meek had to smile. This was going to work. Tomorrow he would call that band the Tornados and see if they wanted to make history with him.

DEPLOYING A GLOBAL NOW: THE BIRTH
OF TELECOMMUNICATION SATELLITES

"Telstar", a three-minute pop masterpiece, was rushed into production after Decca records heard the Tornados' rendition of Meek's composition. Overdubbed with a clavioline, an early version of an electronic keyboard, Meek gave the recording a signature eerie spacelike sound that hovered above the infectious triumphant melody. It was an instant hit, rocketing to the number one spot in the U.K. charts and remaining there for five weeks. But, like its namesake, "Telstar" would prove to be a global phenomenon. Within two weeks it crossed the ocean and the Tornados become the first British band to score a spot on the U.S. *Billboard* pop charts.[48] What began with a grainy image of the American flag beamed from a passing satellite turned into the first salvo of the British pop invasion.

Just fourteen days before the historic broadcast, Telstar 1 blasted into orbit atop a Thor-Delta rocket.[49] The launcher was, essentially, a modified ICBM.[50] Like the continent-spanning nuclear missiles reshaping the Cold War, Telstar and the telecommunications satellites that followed represented another revolutionary form of material engagement which, like the missiles, had its roots in World War II.

Radar (an acronym for "radio detection and ranging") was one of the most powerful technologies to emerge from the war. The ability to bounce radio waves off objects to detect, or target, distant enemies had obvious advantages. The development of radar accelerated in the late years of the conflict, but the atmosphere's radio-sensitive upper layers (the ionosphere) posed a challenge to radar engineers. It was not clear at the time if the technology would be useful for targets at very high altitudes, such as orbiting missiles. As the war ended and the conflict with the Soviets began, the U.S. military knew it had to determine if radar could be utilized in space.

The U.S. Army quickly formed Project Diana with a single goal: to push radio signals through the ionosphere and bounce them off the moon. The generals did not have to wait long for results. On the morning of January 10, 1946, Project Diana's Belmar, New Jersey, transmitting station was warmed up and ready to go. As the moon rose over the horizon the transmitter shot out a powerful blast of radio emissions. Just 2.5 seconds later (the amount of time it takes light to travel to the moon and back) the station's oscilloscopes flickered. The dancing traces on the scope marked the echoed signal's return. Humans had, for the first time, "touched" a celestial object and begun their space age.[51]

Another milestone was crossed eight years later when James Trexler, an engineer at the U.S. Naval Research Laboratory, spoke slowly into a microphone at the lab's Maryland radio facility. After a 2.5-second delay, Trexler heard his words bounced back to him from the moon. With a human voice transmitted from the Earth to the moon and back again, communications and the high frontier of space were joined in scientists' imaginations.[52] The public's imagination would come next.

It did not take long for both sides of the Cold War to lay their plans. In 1955 J. R. Pierce of Bell Labs wrote an influential article making the case for "orbital radio relays"—satellites that either passively reflected signals or more complex craft that received and rebroadcast signals. Pierce soon became a stalwart advocate for satellite communications and led many pioneering Bell Labs efforts, including the Telstar project.[53]

The development of ICBMs and work on communication satellite launch vehicles were a paired effort. By the late 1950s the United States was lifting payloads into low Earth orbit.[54] In 1960 the newly created National Aeronautics and Space Administration (NASA) launched its first experimental communication satellite—a giant thirty-metre Mylar balloon called Echo 1. Echo was essentially an orbiting mirror, allowing telephone, radio and television signals to be bounced from the Earth to the rapidly moving satellite and back again. Such passive telecommunications platforms would be no substitute for an active re-broadcast of signals with greater power and accuracy. Active signal re-broadcast was Telstar's job.[55]

New forms of material engagement require new forms of institutional behaviour. Just as time synchronization in the era of the train and telegraph would demand conventions between nations, telecommunications satellites would require even higher levels of co-operation among institutions. Telstar was not a NASA project but a commercial venture of AT&T through Bell Labs, their extraordinary research wing. At the heart of the project was an array of multinational agreements between AT&T, NASA, Britain's General Post Office and France's PTT (Postes, Télégraphes et Téléphones), all designed to develop transatlantic satellite communications.[56]

With its first broadcast in 1962, Telstar caused a cultural sensation. In truth, however, the Telstar satellites were never destined to be more than a way station towards satellite communications. Two Telstars were put into operation but their low Earth orbit meant each spent only twenty "broadcast-serviceable" minutes above the horizon during their two-and-a-half-hour orbits. Geosynchronous satellites were the real solution for space-based instant telecommunications. Orbiting the Earth once every twenty-four hours, a geosynchronous satellite could be "parked" high above the equator, continuously broadcasting to almost any position below it. Just two years after Telstar, the 1964 Tokyo Olympics were broadcast live to the United States via the newly launched Syncom 3 satellite. One year later, Early Bird, the first commercial geosynchronous satellite, achieved orbit, and the world began its new experiment in culture and cultural time in earnest.[57]

For all the impact of television in the 1950s, the images it conveyed were essentially local. Film of distant events was shipped by plane back to TV studios. "The film had to be processed, edited and flown across oceans ... before it could be broadcast", said Walter Cronkite, recalling his own experience of Telstar's impact (he shared anchoring duties of the first broadcast with Chet Huntley). Telstar transformed TV's grasp of the world. "We had no idea", said Cronkite, "that a technology so young would become master of events so quickly".[58] Just days before the 1969 moon landing, the last geosynchronous satellite needed to create a twenty-four-hour global network was eased into orbit. The world was linked together in a single simultaneous present just in time to watch the first human steps on the moon.[59]

While the cultural and political impact of telecommunications satellites was immediate (consider how nightly newscasts of the Vietnam War affected U.S. policy), the broader changes in human experience, and the experience of time, would play out over longer periods. The creation of twenty-four-hour cable news networks such as Sky News and CNN would not be possible without telecom satellites serving as their broadcast backbone. From the Olympics to the indelible images of the World Trade Center Towers collapsing in ruin, space-based telecommunications satellites altered the human experience of space and time by creating that all-important global instant. Just as telegraph cables had crudely stitched the world together one hundred years earlier, the swarm of satellites inhabiting Earth's orbit played their own unique role in reshaping human time in ways appropriate to the space age.

It should not be seen as a coincidence, therefore, that at the very moment this new human time was emerging, the same technologies that made it possible allowed scientists to stumble across the final act in the discovery of the Big Bang.

THE BIG BANG TRIUMPHANT: AN ACCIDENTAL UNIVERSE

The 1950s were not kind to Big Bang cosmology. Gamow and Alpher's theory remained unable to cross the mass gap and explain the creation of elements heavier than helium.[60] In addition, continuing concerns about the time scale of cosmic evolution led many scientists to conclude that Big Bang cosmology still predicted a universe younger than its stars. While each of these problems would eventually be worked out, many scientists still harboured a deep suspicion of cosmologies that began with a bang. Beginning time was never a popular option for scientists.

The reasons for this bias are an important part of cosmology's narrative. While we like to believe that science is a dispassionate search for the truth, in reality scientists come to their investigations saturated with beliefs that can turn their investigations in certain directions and set their minds with particular convictions. What makes science unique is the ability for the data eventually to speak for itself and trump personal convictions. But as we explore the braided evolution of human and cosmic time, we would be in error if we neglected the ways individual and institutional paradigms shape both research and its interpretation.

Nowhere is this more apparent than in cosmology, with its close links to the domains of religion and mythology.

It was religion, or the reaction to it, that spurred work on a popular alternative to the Big Bang. In November 1951, Pope Pius XII gave endorsement to the combined work of Gamow, Alpher and Lemaître.[61] In an address to the Pontifical Academy of Sciences the pope proclaimed that "present-day science, with one sweep back across the centuries, has succeeded in bearing witness to the august instant of the primordial *Fiat Lux* [Let there be light]". He said, "Hence, creation took place. We say, therefore, there is a Creator. Therefore God exists."[62]

Lemaître, the Catholic priest and scientist, was horrified. He knew that his or any other theory could always be disproved. Travelling to Rome, he counselled the pope against linking the faith to any contingent scientific hypothesis. Others, however, were just as clear as the pope on links between the Big Bang and biblical Genesis, and they had their own, very different reasons to be hostile.

In the Soviet Union, official communist doctrine's disgust for religion meant that any cosmological theorizing with traces of Christian dogma was suspect. Big Bang cosmology was officially viewed as "astronomical idealism, which helps clericalism."[63] It did not help that at least two scientists who had supported Friedmann's original relativistic models of the expanding universe had died in Stalin's purges. In response, Soviet astronomers in the 1950s largely gave up "the study of the universe as a whole".[64]

In England, a more rational objection—and alternative—to the beginning of time was born. Fred Hoyle, a Yorkshireman of considerable energy and wit, deeply distrusted religion and found the idea of a "moment of creation" untenable.[65] With fellow Cambridge physicists Hermann Bondi and Thomas Gold, Hoyle proposed a new model, a steady-state alternative to Gamow's ideas. Steady-state cosmology accepted the universe's expansion but sought to make it eternal and unchanging. By allowing matter to be continuously created throughout the universe, new galaxies could form in the empty voids that opened up by cosmic expansion. Thus, the universe could be dynamic and expanding and still always appear the same. Any region of space viewed at any time in cosmic history would look like any other: a few new galaxies forming while the rest expand away from one another. The steady-state model was parsimonious with continuous creation. Just a single hydrogen atom popping into existence each year "in a volume equal to St Paul's Cathedral" was enough to do the trick.[66]

While many scientists scoffed at this "magic" creation of matter as a feature of the steady-state model, Hoyle responded that it was no stranger than having all the mass-energy appear at once at the beginning of time. It was Hoyle who, during a BBC lecture series on the nature of the universe, coined "Big Bang"—as a term of derision. Unfortunately for Hoyle, both the term and the theory it described stuck. But the BBC programme did make Hoyle a household name in 1950s England and it gave the professor and his wife the opportunity to purchase a cherished electrical appliance—a fridge.[67]

By the beginning of the 1960s, cosmology seemed stuck in a ditch. The steady-state model had both strong supporters (mostly in Britain) and strong critics. Along with the other cosmological theories still in circulation, such as the work of Lemaître and Gamow, it remained a viable explanation for cosmic history that could embrace the one piece of data everyone agreed on—cosmic expansion.

There had, however, been some good news for the Big Bang during these years. Its two major empirical objections were overthrown. Astronomers had come to understand that all elements heavier than lithium were built at the centre of stars rather than at the origin of the universe. With stellar nucleosynthesis working so well for heavy elements, Big Bang nucleosynthesis could focus on what it did best—explain the origin of light elements. Astronomers had also revisited Hubble's early estimates of the cosmic expansion rate. The age of the universe in Big Bang models was moved closer to ten billion years, which put it in line with the age of stars. But in spite of these advances, astronomers remained sceptical. With so much conflict and so little hard data, the field of cosmology was largely ignored, making the cosmological surprise waiting in New Jersey all the more remarkable.

In 1964 Bell Laboratories had facilities scattered across the country. These labs were scientific powerhouses pushing forward the frontiers of everything from transistors and lasers to computers and, of course, satellite telecommunications. In 1959, in a field just outside of Holmdel, New Jersey, Bell Labs' Crawford Hill facility had constructed a giant ultrasensitive antenna for bouncing signals off the Echo 1 satellite. Once Telstar was launched, the fifteen-metre, horn-shaped microwave receiver was freed up for other uses. Two Bell Labs employees with PhDs in astronomy, Arno A. Penzias and Robert W. Wilson, jumped at the chance to put the antenna to astrophysical use. The two astronomers planned an ambitious research programme examining distant objects using microwave light. But as they prepared the horn to take data they encountered what appeared to be a technical problem that brought their ambitions to a screeching halt.

There was a noise in the system that would not go away. A steady microwave hiss persisted in the horn regardless of the direction the astronomers pointed their antenna. At first Wilson and Penzias were sure the signal was nothing more than an electronic artefact and they struggled for weeks to root out the problem. The electronics were rebuilt but nothing changed. Layers of pigeon guano were scrubbed from the antenna surface, to no avail. Nothing worked. It was as if the entire sky was saturated with microwave radiation that peaked at a wavelength of 7.35 centimetres. The pair realized the truth of their situation only when a friend handed Penzias an article written by astronomers at nearby Princeton University.

The sky *was* saturated with microwave radiation. More important, the 7.35-centimetre radiation was exactly the consequence of what Alpher, Herman and Gamow predicted in their groundbreaking, and now all but forgotten, papers fifteen years before. Penzias and Wilson had found an all-important relic of the hot Big Bang, an electromagnetic fossil of the universe's earliest epochs.

THE COSMIC MICROWAVE BACKGROUND AND THE BIG BANG'S TRIUMPH

Gamow's Big Bang was always a kind of cosmic nuclear archaeology. The first few moments of the universe left traces on the world we see

FIGURE 7.7. From telecommunications to cosmology. Wilson and Penzias stand in front of the horn-shaped receiver (designed for satellite communications) they used to discover microwave "echoes" of the Big Bang.

today. If we were clever enough, we could read those traces. In his earliest work, Gamow considered the cosmic abundance of elements to be just such an imprint. As he, Alpher and Herman sharpened their conception of the hot Big Bang with detailed calculations they found other relics that should have survived billions of years of cosmic evolution. Seven different times in their various papers, Alpher and collaborators predicted that a Big Bang would leave the universe saturated with a particular kind of fossil electromagnetic radiation.

While the universe would only be hot and dense enough to forge elements for a few minutes, the end of the nuclear era did not end the universe's particle alchemy. As space continued to expand, the temperature and density of the cosmic soup continued to drop. Matter, in the form of protons, electrons, helium nuclei and other particles, was well mixed with photons (quanta of light). The physics of this mix imprinted the photons with a fixed signature—a fossil imprint of history—that was discovered by Penzias and Wilson's curious probing.

In the late 1800s, physicists learned that any hot, dense object emits light with a characteristic pattern called a *blackbody spectrum*. The red glow of an iron rod in a fireplace is a common example of blackbody light. Within the heated iron rod matter and photons are strongly coupled together exchanging energy in rapid-fire reactions. Physicists also learned that most of the light a blackbody emits comes at a wavelength strongly dependent on the blackbody's temperature. Place the same iron rod in a blast furnace and it will glow white-hot as the peak blackbody emission shifts to include more visible wavelengths (colours). Astronomers routinely exploit this property of blackbody emission. Hot, dense objects (such as a star) can have their temperatures taken from millions of light-years away simply by recording their spectra, confirming its blackbody character and noting where the peak emission occurs.

As exotic as the early universe might have been, it was still a collection of hot, dense stuff. Thus, the entire early universe was a blackbody. Matter particles and photons jostled together. Matter absorbed the photons and then spat them out again in endless reactions. In this way, the young universe was saturated with blackbody light. Then, about three hundred thousand years after the birth of time, the blackbody photons were frozen out of the party.

The key event was the capture of electrons by protons to form the first hydrogen atoms. The newly formed hydrogen could not absorb the blackbody photons. The physics of hydrogen's internal constitution made interactions with the blackbody photons all but impossible. A "decoupling" of matter (the now ubiquitous hydrogen atoms) and the bath of blackbody photons occurred relatively swiftly. The photons were left without dance partners. Just a few thousand years before, they could not travel more than a millimetre through the universe without being absorbed, but after decoupling they were left orphaned, free to wander the universe unimpeded. As time marched forward, the only change this fossil light would experience was a stretching in wavelength directly tied to the expansion of space itself.

Through their calculations Alpher and Gamow saw how the nuclear thermodynamics of a hot Big Bang implied a universe full of fossil blackbody radiation. They even knew its temperature. In one of the great acts of scientific prescience they predicted a cosmic background of "thermal photons" with an average temperature of about five degrees kelvin (five degrees above absolute zero). This is exactly what Penzias and Wilson stumbled upon.

The Bell Labs scientists did not make this monumental third "discovery" of Big Bang cosmology (after Lemaître and then Gamow, Alpher and Herman) alone. The paper Penzias had been given was from Princeton physics professor Robert Dicke and his collaborators. Dicke had unknowingly re-derived Alpher's results. Just as Penzias and Wilson were calling to seek a consultation with the Princeton physicist, Dicke was in the process of building his own microwave detector to look for the background radiation. After listening to Penzias explain their findings, Dicke reportedly put down the phone and said, "We've been scooped."[68]

Alongside Hubble's recognition of the expanding universe and the abundance of light elements, Alpher and Gamow's prediction of the cosmic microwave background (CMB) ranks as one of the greatest cosmological advances of the twentieth century. By unwittingly confirming such a specific prediction of hot Big Bang cosmology, Arno and Penzias found the lever and fulcrum needed to upend all competing models of the universe. The microwave photons filling the sky with their perfect blackbody signature were direct proof that the universe had once been far hotter and denser than the cold, empty darkness we see today. The steady-state model offered no ready explanation for the fossil photons filling space, and it soon disappeared as a viable alternative.

By the end of the 1960s the hot Big Bang, with its marriage of subatomic scale, quantum physics and universal-scale general relativity, *was* cosmology. The Big Bang had triumphed. Cosmic time had a beginning even if no one understood what that really meant.

As the decade came to a close an appropriate resonance in popular imagery and theoretical ideas had emerged, linking cosmological beginnings and human endings. New cosmologies always require imaginative interpretations. From Egyptian murals of sky gods to angels on the ceiling of the Sistine Chapel to Depression-era WPA murals invoking the march of science, cosmologies are always ultimately a public endeavour. Attempts to imagine the first instants of the Big Bang—the blinding light and the rush of superheated matter—required their own set of visual metaphors. The necessary images were already close at hand, emanating from the same nuclear science that gave the hot Big Bang cosmology its success. Thermonuclear bomb explosions were a Big Bang of truly awesome power that everyone had already seen.

Chapter 8

INFLATION, MOBILE PHONES AND THE OUTLOOK UNIVERSE

Information Revolutions and the Big Bang Gets in Trouble

BIRMINGHAM, ENGLAND · DECEMBER 2002, 11:39 A.M.

Maybe she could just drop out entirely, tell people to call her on the phone or send a letter by the post.

She popped open the Outlook window on her screen. There were forty-two new messages. Tami in PR wanted to add her to Thursday's meeting. Thomas in HR needed a consult on the new hire by day's end. The Manchester team was asking for specs on the new modules. And that new wanker from management had sent another motivational message. She sagged backwards in her seat. Forty-two new messages in an inbox that already had 206 unread e-mails, and she hadn't even had lunch yet.

She had come in early this morning just to clean out her e-mail inbox. A full hour and a half wasted sorting through requests, reviews, jokes, chain letters and spam.

Reply, file, delete, reply, reply, delete, delete, delete.

By 9:00 a.m. she'd got down to a mere 168 messages. Hit delete all, she'd thought, smiling to herself. But she resisted and tried moving on, focusing on the PowerPoint slides for tomorrow's tech review (neatly blocked out as a forty-five-minute blue rectangle on her Outlook calendar). She'd resolved to not even glance at her e-mail inbox.

But she'd failed.

By 9:50 a.m. there'd been eighteen new messages waiting for her atten-tion. Another ten new e-mails had appeared by the time she was ready for a cuppa at 10:45. When she dropped her bag back on her desk at 11:01, nine more messages had shown up. Why even bother responding when that would just generate more replies?

She looked at the Outlook calendar again. Twenty-one minutes till the lunchtime sales briefing. Then what? More blue blocks partitioning the after-noon into teleconferences, interviews and meetings. Another day of wasting time being just-in-time. Another day spent getting nothing substantial done. Maybe she should just browse for images of cute kittens.

ACCELERATION

The last decades of the twentieth century were witness to yet another rev-olution in material engagement, institutional facts and human temporal experience. This time, however, we were the witnesses to the revolution. Everyone over the age of twenty has a foot on either side of the divide as human culture and human time stepped from its analogue era into the digital domain. Many of us can still dig up memories of a world unme-diated by material engagement in the form of silicon microcircuit chips.

Remember when you had to be home to get a call?

Remember when you had to order a map from The AA to plan a trip?

Remember when your calendar was something that hung on the wall?

We have been fully present during this transformation in material engagement, institutional facts and cultural time. We have seen the pro-cess, still very much ongoing, emerge from below our direct conscious-ness to seep into every aspect of culture, reshaping our most intimate experiences of life through time.

Before the advent of mobile phones, for example, we were funda-mentally more alone with our thoughts and more present in the lives directly before us. There was nowhere else to be. Urgent or nonurgent communication had to wait until we could find a device connected by a wire to the wall. Now we reach out to communicate at a whim, fill-ing time simply because we are bored or responding to a thought that crossed our mind. The simple act of walking down the street and calling a friend to check in is as profound and radical a shift in the experience, use and conception of time as anything that passed before, from the Neolithic to the industrial revolution.

Given our proximity to this tectonic surge of cultural re-creation, it is impossible to know how long the digital age will last. The agricultural revolution reshaped human experience for many thousands of years. The industrial revolution created cultural forms that extended across centuries. Will the digital revolution simply lead us to a dead end or will it drive sustainable forms of culture that cross generations? While we cannot answer that question now, we are close enough to the dawning of the electronic era to see firsthand how new forms of technology—that is, new forms of material engagement—directly reshape the human experience of time. Reflecting on the last fifty thousand years of cultural transformation, we should expect that the changes in human time manifested through digital technology would mirror changes in cosmological time. That expectation has been fulfilled.

The introduction of mechanical clocks shifted the organization of the European day and eventually provided a new metaphor for the heavens—a precise, cosmic clockwork set in motion by God's hand. Centuries later, the introduction of steam power set the industrial revolution's new machine age in motion and drove the rhythms of its workers' punch-clock lives. The science of thermodynamics, emerging from those steam-powered machines, advanced a new understanding of time and transformation in terms of energy, entropy and evolution. Thermodynamics yielded its own metaphors and conceptual tools that reshaped cosmological thinking. Then, just before the dawn of the twentieth century, trains and telegraph wires created new experiences of simultaneity across vast distances. Einstein's theory of relativity used its own new vision of simultaneity as a pivot point for merging space and time into space-time. Once a fully relativistic account of space-time and its flexible geometry was available, cosmology was given its first complete language. Always and again, transformations in cosmic and human time surged back and forth, each one supporting the other in metaphorical and material realms.

By the last decades of the twentieth century, silicon technology dominated our material engagement with the world. Machines made possible by silicon microcircuits—computers, personal digital assistants, mobile phones and GPS devices—were accelerating the immediate and very personal movement through daily life. These silicon "machines" moved at speeds so fast their cadence was far more native to atoms than to humans. By building culture timed to their clock cycles, our own time and experience were compressed in ways both thrilling and exhausting. In both our working and personal lives we were expected to do more because these machines would make it possible. And so we entered a new time whose contours were as closely felt and intimately lived as the tick-tock world of our great-grandparents or the sun-parsed days of our more distant ancestors.

At the same time, the scientific capacities unleashed in the computer age pushed our cosmic narrative of the Big Bang to its limits. Computer simulations, massive data-gathering projects and space-based telescope platforms revealed new challenges to any cosmology that would begin with a beginning. In the closing years of the twentieth century, the pace of life, time and cosmic evolution all were set in a permanent state of acceleration.

YOU'VE GOT TOO MUCH MAIL: ENTERING THE OUTLOOK UNIVERSE

Time is not simply what we read off a clock. Instead, time as it is lived can be defined as what we do and how we go about doing it. "There is no such time as 'this afternoon' or 'one-o'clock' or 'two o'clock'", said Suzuki Roshi. "At one o'clock you will eat your lunch. To eat lunch is itself one o'clock."[1] What emerged in the last decades of the twentieth century was a new form of movement through the day. New, silicon-based forms of material engagement were rapidly embraced and rapidly used to re-imagine the institutional facts of culture. Those facts were expressed most intimately by what we did and how those behaviours shaped what we thought. Thus the revolution in time occurred in our

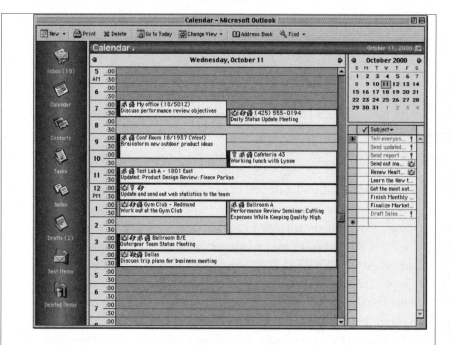

FIGURE 8.1. A screenshot of Microsoft Outlook on a Macintosh personal computer, circa 2001.

hands and between our ears. This pairing of action and thought would become explicit in the relationship between communication and time management as e-mail led us all into the Outlook universe.

Distraction is expensive. According to Basex, a company that studies knowledge workers and their use of technology, unnecessary electronic interruptions cost the world economy $650 billion per year in lost productivity. In Basex's research the black hole sucking in all that money is "information overload" driven by that most basic of electronic applications: e-mail.[2]

There were many digital technologies that transformed our sense of time during the 1980s and 1990s. Desktop computers reworked the very notion of work. The World Wide Web brought instant access to everything from the Louvre's collections of fine art to bottomless pits of porn. Online day trading amped up financial time and made markets as jittery as a coked-up comedian. But of all these changes the most pervasive and insidious was the introduction of e-mail. Few lives in the developed world have been untouched by e-mail and its demands. It was the early killer app for computer networks, spreading like a virus through the ecologies of personal and business life. But this greatest of all applications began humbly enough as a kind of afterthought in a Massachusetts Institute of Technology computer lab.

The history of technological revolutions is full of dramatic first lines. Samuel Morse's first message pulsing across telegraph lines was "What hath God wrought."[3] Alexander Graham Bell's famous first message spoken into the telephone was "Mr. Watson—come here—I want to see you."[4] E-mail's origins, unfortunately, fail to rise to such a dramatic standard.[5] "QWERTYUIOP" was, by all accounts, the first modern cross-computer e-mail. Ray Tomlinson, the man credited with creating the first networked version of e-mail, has rather vague memories of the event: "I sent a number of test messages to myself from one machine to the other. The test messages were entirely forgettable. . . . Most likely the first message was QWERTYUIOP or something similar."[6]

In 1971 Tomlinson was working as a computer engineer with a contractor for the U.S. Defense Department. The company's project was to help develop the Advanced Research Projects Agency Network (ARPANET), the original version of what we now call the Internet. ARPANET was a work in progress at the time, consisting of only fifteen nodes at places such as UCLA, Stanford and the University of Utah.[7] Rather than the laptops we now carry in briefcases and the smart phones we carry in pockets, in those days the term *computer* referred only to mainframes that looked very much like the megaliths of the Neolithic era. Mainframes were housed in vast air-conditioned "machine rooms", attended to by teams of classically nerdy acolytes. Most important, these machines were solipsistic lords of their own domains. Users worked within the confines of a single computer system, and the systems did not talk to one another. The system's users were, however, always interested in talking.

As early as 1964, computer programmers had developed ways to leave one another messages on files that could be shared within each stand-alone system.[8] Tomlinson recognized that the users of ARPANET, which was intended to be a network of many distinct and geographically remote computers, needed something similar but more sophisticated. His achievement was the development of the all-important protocols for file transfer.[9] These protocols were coded rules that allowed computers to talk to one another. In particular, Tomlinson developed rules for sending and receiving messages between linked machines. Adding the now ubiquitous @ between the computer user's name and the computer system's name was the final piece of the puzzle. "The @ sign seemed to make sense", Tomlinson later recalled. "I used the @ sign to indicate that the user was 'at' some other host rather than being local."[10]

It did indeed make sense. For the growing population of scientists using ARPANET for daily research, Thompson's creation seemed like a natural outgrowth of their activity. A 1973 ARPA report on the new electronic messaging service concluded, "A surprising aspect of the message service is the unplanned, unanticipated, and unsupported nature of its birth and early growth. It just happened, and its early history has seemed more like the discovery of a natural phenomenon than the deliberate development of a new technology."[11]

ARPANET users liked e-mail and did not have to be encouraged to use it. Their acceptance of the new tool offers an object lesson in the realms of time and culture. Here, at the very dawn of a new form of material engagement, we can see the process begin. E-mail was not dreamed up as a means to change human communication and the human experience of time. In 1976, just a few years after ARPANET's developers wrote their report, the rest of the world still experienced telecommunication as the bright buzzing ring of the phone on the kitchen wall or the nightstand. Mail meant a paper envelope whose weight could be judged by its heft in the hand. But for the small group of ARPANET's users, a new kind of activity had become possible. By typing a message on a keypad that instantly appeared to its receiver somewhere else in the country, a new kind of behaviour and a new way of spending time had been imagined into existence. It was the seed of new material engagement cracking open beneath the culture's surface. The rest of the world, for good or ill, would soon nurture that seed into something entirely novel for the human experience of time.

In the 1980s, personal computers became both a wildly successful business and a ubiquitous "appliance" in homes and businesses. The IBM PC was introduced in 1983 and Apple's groundbreaking Macintosh first went on sale in 1984. These machines, connected through dial-up services such as CompuServe and AppleLink, brought e-mail to a broad population of users. Meanwhile, dial-up "bulletin board systems" (BBSs) were growing in popularity, allowing people with common interests to exchange messages. Though each system had its limitations, together they formed the arena for a period of "training", in which a vanguard of home-based users established new behaviours on the electronic frontiers of culture. Most people, however, knew of e-mail as only something they read about in *Newsweek* or heard about through a more tech-savvy friend. Most important, e-mail was still simply a means of communication; it had yet to become merged with other functions such as time management or social networking. Pencilling in an appointment on your calendar was still done with a pencil.

The adoption of e-mail in the business world was the most important step in e-mail's march to world domination and role as master of a new form of lived time. Hosted on local area networks (LANs), e-mail was enthusiastically embraced by corporations that had already developed internal computer cultures. Almost overnight, or so it would seem, working in an office meant dealing with the imperatives and politics of electronic communications. Gossip and the rumour mill had long been the staples of office social behaviour, though they were largely limited to water cooler conversations and coffee break dialogue. Now, electronic communication, with all its potential for misinterpretations, would be added to the daily social dance, including dealing with the disaster of hitting SEND ALL on an e-mail meant only for a select few.

The 1998 romantic comedy *You've Got Mail* marks a turning point in e-mail's novelty and intrusion into social time. The film follows the online romance between Kathleen Kelly (Meg Ryan) and Joe Fox (Tom Hanks). Though competitors in the real world of business (she runs a small bookstore, he runs a bookstore chain), in the anonymous world of e-mail they bare their souls and fall in love. The timing of e-mail conversations, the expectations of a sent message's reply and the "wow" quality of using this new medium to carry out the age-old romantic dance make *You've Got Mail* a clear example of a culture redefining its norms of behaviour (and hence time) around a new form of material engagement. *You've Got Mail* should be compared with another Ryan-Hanks romantic comedy vehicle from just five years earlier. In *Sleepless in Seattle* the two lovers find each other thanks to an old-fashioned snail-mail letter. A decade or so after *You've Got Mail* a film centred on e-mail would make little sense, as the culture had already absorbed it into the background of daily time. In 2010, films about Facebook in which e-mail was just one component would be relevant, but not e-mail alone. In 1998, however, e-mail was still new and had just started taking us into the seamless integration of digital information, life and time.

An important feature of the 1980s desktop LAN system was the development of ever more intuitive user interfaces. E-mail hovered in the background for many of these applications. Even the simple ability to append attachments to an electronic message created a vast new terrain of possibilities for workers to navigate. The era of the ever-circulating, never-finished spreadsheet had arrived.

By the early 1990s, the true Internet, born of U.S. federal projects such as ARPANET, Usenet, Milnet and others, had begun its own conquest of the world. Here, the convergence of human time and cosmic time becomes clear. The first successful Web browser, Mosaic, was developed specifically for astrophysics research.[12] The brain behind Mosaic, Marc Andreessen, was an undergraduate working at the National Center for Supercomputing Applications at the University of Illinois.[13] Andreessen developed Mosaic as a tool for his astronomer bosses to share research-related files, and he put hypertext—onscreen text referencing immediately accessible files—centre stage in his development.[14] The now ubiquitous hypertext was a radical innovation at the time and emerged directly from the realms of particle physics research. Tim Berners-Lee was the programmer who first developed hypertext transfer protocol (HTTP), the protocol on which the World Wide Web is built, while working as a programmer at CERN, the principal European particle physics research institute.[15] The fact that Berners-Lee created a digital tool to aid particle physicists in their research—which, almost overnight, became an essential element of a cultural transformation—marks a remarkably explicit example of material engagement changing both human time and the cosmic vision of time. The mid-1990s saw technologies such as hypertext and Web browsers exploding into culture through a newly commercialized Internet. It was through these new forms of material engagement that human life collapsed onto the two dimensions of our computer screens.

As companies rushed to the digital frontier, instant communication by e-mail was recognized as a service of such fundamental importance that it could serve as a backbone for other digital products and services. As search engines grew in importance, the offer of free e-mail accounts became a standard means of drawing users into the accelerating universe of electronic content. In just a few years, addresses with @yahoo.com, @hotmail.com and @gmail.com would become the standard for a world moving to new electronic clock cycles.

The ubiquity of e-mail allowed software companies to build integrated platforms using electronic communications as their foundation. These tools would soon come to dictate and shape our days. Microsoft's Outlook was introduced in the late 1990s, and by the early 2000s, its linked e-mail, calendar, contacts list and to-do list functionality became omnipresent. Outlook had its roots in the 1980s with a Canadian company called Consumers Software and their Network Courier application. The program was ported to Windows in 1986, morphing into Microsoft Mail 3.0 as part of the phenomenally successful Windows 3.0. When the Windows stand-alone calendar program Schedule+ was merged with Microsoft Mail, the plate tectonics of culture shifted in an earthquake known as Outlook's personal information management (PIM) system.[16]

Personal information management, the all-encompassing ever-updating electronic repository for your life, existed in gadgets before Outlook. Personal digital assistants such as the Palm Pilot had already pushed boundaries in bundling the shifting facts of daily life into an application. But Microsoft's ubiquity meant Outlook would become, for a time, the de facto standard in training millions in how to move along the contours of a new time.

The day, that most intimate experience of temporality, now took on a new symbolic form. In the neatly ordered patchwork of precisely timed, colour-coded appointments, each linked to its contacts and to-do bullets, human life became an exercise in information management. Time and communication had been paired before in older technological shifts from the telegraph to the telephone, but as digital technologies came to fully rule over the shape of our days, the two became seamlessly interwoven. Efficiency in communication became the driver for efficiency in all domains of the day. With e-mail acting as the universal solvent, our personal lives were broken down and alchemically transformed into personal information. This became the fluid pushing through the body of our new streamlined lives.

In the Outlook universe, time became a kind of flexible geometry (metaphorically in spirit reminiscent of Einstein's relativity). The coloured blocks on the calendar screen denoting meetings, teleconferences, children's playdates and gym workouts could be stretched and squeezed endlessly as each of us worked to "manage" our time. The neatly stacked rectangles gave the illusion that time *could* be managed with a precision that mirrored a universe whose own story was told in billionths or trillionths of a second. We began to expect that life could be parsed with the same accuracy as the boundaries of those coloured blocks on our electronic calendars.

A 2002 announcement for a lunchtime programme at my city's central library tells visitors that the talk will begin at 12:12 p.m. It is as if the announcement's author imagined the speaker could somehow be "dragged and dropped" into this kind of precision. No doubt a global-send e-mail carried the news of this sharply timed event to thousands of recipients, where just a click on the screen would add it to their linked calendar program. In 1972, would a talk beginning at 12:12 have made sense at all?

By 2002, it was an Outlook universe and we were all living in it even if we owned a Mac or remained Linux rebels. Personal information management was bound to appeal to us regardless of what platform our machines used. It formed the basis of time management. In a hyperconnected world, efficiency in our use of time was something that could now be managed with a totality that included whom we knew, where we went to meet them and what we planned to do with them once we arrived. From playdates to speed dates, our days accelerated in the name of efficiency. We were moving so fast that time had been crowded out of our minds and onto our devices. The most intimate internal experiences of time had been outsourced due to overload.

Life was moving much, much faster. The electronic datebook living on desktops, laptops and, soon enough, mobile phones would become the new medium for standardizing the personal universe and synchronizing it with the social network beyond. The personal universe, relentlessly networked to the social and cultural cosmos we were embedded within, was accelerating. And all the while, the collective universe of cosmological science was running headlong into its own changes and its own new meaning for the term *acceleration*.

SETTING STANDARDS FOR THE UNIVERSE: STANDARD MODELS IN COSMOLOGY AND PARTICLE PHYSICS

By the mid-1970s, the triumph of the hot Big Bang was complete. Three unassailable pillars of evidence now supported the Big Bang: the expansion of the universe, the abundance of light elements and the astonishing presence of the cosmic microwave background. To deny the Big Bang, an opponent needed to topple each and every pillar. As the years went on and supporting data mounted, critics faced a Herculean task. The Big Bang, with its implied beginning for space and time, had become the standard model for cosmology. Everything else became a kind of crazy speculation, a heresy against a growing mountain of observational facts.

But two of these pillars (the abundance of light elements and the CMB) rest on the quantum mechanical description of subatomic physics. The Big Bang's triumph was just as much a victory for particle physics—the study of subatomic structure—as it was for astrophysics. In this fundamental domain of science, another standard model had also emerged.

Throughout the 1950s and 1960s, physicists had poured enormous effort and resources into the development of giant particle accelerators, machines designed to bring the smallest specks of matter to the highest possible speeds (which really means the highest energies).[17] The goal was to smash particles into each other to probe their inner structure. Scanning the remains gave physicists hints of the internal constitution of particles such as the proton and neutron. Physicist Richard Feynman described the effort this way, roughly comparing atoms to watches: "All we can do is smash them together and see all the funny pieces (gears, wheels and springs) which fly out. Then we have to guess how the watch is put together."[18] By the mid-1960s, accelerator-based studies produced a remarkably complete and successful description of the subatomic menagerie known as the *standard model of particle physics.* Its successes as well as its limitations remain relevant forty years later and are still shaping options for the cosmological revolution we stand poised before today.

The spray of particles emerging from countless accelerator collisions allowed physicists to see the bedrock on which all forms of mass-energy rest. In their efforts, they found that only two fundamental classes of matter exist: leptons and quarks. The electron is the most familiar of the leptons, but two other lepton "generations"—the muon and tau—fill out the family. Quarks constitute the other classification of matter and are the building blocks of the ubiquitous protons and neutrons inside every atomic nucleus. Each and every particle, quark or lepton, comes with an antimatter twin. The anti- and normal particles have opposite electrical charges. Antimatter and matter are mortal enemies of a kind. When matter and antimatter particles collide, they completely annihilate one another in a spray of energy.

Taking a census of all the universe's particles and antiparticles was not the only job of the standard model. Particles "feel" one another by exerting forces. As far as we know there are only four forces at work in the universe: gravity, electromagnetism, the strong nuclear force and

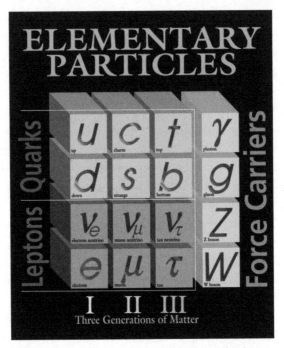

FIGURE 8.2. The standard model of particle physics. This table shows all the fundamental elementary particles and force carriers of the standard model. There are three "generations" of quarks and leptons and four "force bosons", one for each of the fundamental forces.

the weak nuclear force. Gravity, known since Newton, is remarkably weak and requires large collections of matter to exert significant forces. Electromagnetism is the domain of electric charges and electric currents. The strong nuclear force binds nuclei together, while the weak nuclear force is intimately connected to radioactive decay. Each, in its own way, has been part of our material engagement with the world. We have used knowledge of gravity in everything from harnessing the energy of falling water to accurately tracking the trajectory of an artillery shell. Our material engagement with the electromagnetic force is directly responsible for all the electronic miracles we have come to live with, from mobile phones to the Internet. Our material engagement with the nuclear forces led directly to nuclear weapons and all the changes those devices entailed.

The standard model provides a description of all these forces and their effects on subatomic particles. The forces are mediated by a separate class of particles called force bosons. The photon, for example, is the particle that carries electromagnetic force. Exchanging photons is the way charged particles "feel" the electromagnetic force. The magnets hanging on your fridge are kept in place by unseen photons dashing back and forth between magnet and metal refrigerator door. Force bosons called *gluons* mediate the strong nuclear force, while so-called W and Z bosons carry the weak nuclear force between particles.[19]

This exquisitely articulated web of particles, antiparticles and forces was the triumph of the standard model. With it, physicists had a grand map of matter at a fundamental level. But it was not *the* fundamental level. The standard model could not answer a long and important list of questions. Why were there just four forces? Why did each force have a different strength? Why were there two different families of particles, the quarks and the leptons?

Of particular importance was the set of numbers that had to be fed into the standard model from experiments. These numbers were the so-called constants of the standard model and, among other things, they set the strength of the forces. There were about twenty of these constants scattered across the theory. They were not predicted by the standard model but had to be directly measured. It was like getting a job where you know you will be paid hourly but don't know the pay rate until you look at your paycheck. Having twenty unpredicted numbers running around your best theory for matter was something of a letdown for physicists. In their hearts, they were hoping for an ultimate and final theory of reality that predicted everything that could and did happen, including all the values for their constants.

The standard model also seemed to be delicately tuned to the values of these constants. The constants could take only very specific values or else the universe would have evolved in an entirely different direction, including making the development of life impossible. This "fine-tuning problem" was an anathema to physicists, because they not only had to find a way to predict the values but also had to predict why only these exact values were possible. It would have been easier if it were more of a "slop in the system", meaning that a large range of values for the constants would still lead to the world we see. As the years progressed, the fine-tuning dilemma only got worse.

Most important was that throughout the standard model's development, gravity remained stubbornly outside its purview. Physicists adamantly believed a theory of quantum gravity must exist, with a particle (which they took to calling a "graviton") to mediate its force, but that prospect remained far over the horizon. By the end of the 1970s that horizon began to be obscured by gathering storm clouds.

TROUBLE AT THE DAWN OF TIME

The Big Bang had become *the* dominant model for the birth of time and the universe. But as the decade of disco and punk rock raced along, new problems surfaced. The questions facing the Big Bang came from a variety of directions. Some had emerged from the domains of observational astronomy, while others sprouted from the growing interface between cosmology and particle physics. By the end of the 1970s, at least three of these questions turned out to be especially pressing.

– The Causality Problem –

Using the cosmic microwave background, astronomers could make detailed measurements of the properties of the early universe. In particular, they could extract the temperature of the early cosmic plasma with great precision. Looking at the sky in different directions, they could compare conditions such as temperature in widely separated regions of the early universe. To their astonishment, they found temperatures unchanging regardless of which direction they looked. Even when their instruments were pointed in opposite directions of the sky they still found the cosmic plasma's temperature to be the same, down to one part in ten thousand.[20] Why would this temperature be exactly the same everywhere? Since the universe began as a chaotic, expanding fireball, it would be reasonable to expect that some parts of the fireball ended up with slightly different conditions than others.

The fundamental problem astronomers faced with a universe with conditions of almost perfect uniformity can be posed simply as one of cause and effect. Points on opposite sides of the sky were too far apart in the early universe to ever have been connected. Regions of the universe that are now a hemisphere apart in the sky were so far apart when the CMB photons decoupled that they never, ever could have been in contact. A light signal could not bridge their separation in the time from the Big Bang to the moment when CMB light waves were freed. This is the root of the cause-and-effect dilemma.

Astronomers know how fast the universe is expanding now. They also thought they had a pretty good idea of how fast that expansion progressed from the first few seconds after the Big Bang onward by using the Friedmann-Lemaître solutions. With this understanding, it was easy to see that regions of space now on opposite sides of the sky could never have been so close together to be causally connected. That means the perfectly constant temperatures implied by the CMB were an amazing cosmic coincidence. For astronomers, it was like opening the morning paper to find that every city on the planet had the exact same temperature down to four decimal places.

But astronomers do not like profound coincidences. Coincidences are the kind of thing that keep them up at night wondering what else is going on. If light did not even have time to make it from one cosmic neighbourhood to the next, then the different neighbourhoods could not have known about one another. They did not have time to pass information—in the form of light waves—back and forth to allow for such perfect synchronization in temperature. This causal conundrum was a monster problem for astronomers and physicists. If they could not solve it, their hopes for a rational cosmology would collapse.

– The Flatness Problem –

Recall that the Friedmann-Lemaître solutions for cosmic evolution contain a critical number determining the shape of the universe as well as its fate. As we saw in Chapter 6, omega (Ω) is the matter-energy density of the universe expressed in terms of a special or "critical" value. Finding omega's true value had become a holy grail of Big Bang cosmology. By the 1970s, the best answer pointed to omega being slightly less than 1. (The values seemed to hover around $\Omega = 0.05$.) That was close enough to the magic number 1 to cause a real problem for Big Bang theorists.

According to the theory, omega will, in general, change its value as the universe evolves. Only if omega is exactly 1 will it stay exactly 1 for all cosmic time. If omega equalled exactly 1, it also meant the geometry of cosmic space was, and always had been, exactly flat.

If the universe began with omega greater or less than 1 by even a tiny amount, then cosmic expansion dramatically alters its value over time, sweeping it to extremely small values like 10^{-40} or extremely large values like 10^{40}.[21] While $\Omega = 0.05$ might not seeem so close to 1 for most of us, for a cosmologist expecting a number like a million trillion billion, its close enough to the critical value of $\Omega = 1$ to make it seem that something fishy was going on. Either cosmic density really had the critical value of 1 and our measurements were missing the rest of the matter-energy, or omega at the Big Bang had a specific value that over time changed such that it was close to 1 at this particular moment in cosmic history. If omega was less than 1 at the start of cosmic evolution, it would have to be set with a special value out to sixty decimal places to end up with what astronomers were seeing in the 1970s. This demand was certainly a fine-tuning of cosmic initial conditions and astronomers were stuck with yet another coincidence. Everything astronomers knew told them an omega of almost 1 was bad news for the theory, and it demanded an explanation.

– The Magnetic Monopole Problem –

The third problem for Big Bang cosmology arose from the domain of particle physics. The universe is full of electric charges. Some are negative, like those of electrons, and some are positive, like those of protons. Physicists call these different, separate polarities electric monopoles. One of the strange things about our world is that there are no magnetic monopoles—every magnetic field comes with north and south poles connected. There are no particles that carry only a north or only a south magnetic "charge". Physicists have always wondered why no separate magnetic charges were ever seen. By the 1970s, theorists attempting to dig deeper into the structure of the world were trying to develop what they called grand unified theories, or GUTs, of particle physics. GUTs were designed to go beyond the standard model. They were the grand attempt by theoretical physicists to understand the structure of matter and its interactions at a deeper level. Particle physicists had long suspected that the four known forces shaping the universe were simply different facets of a single, as yet undiscovered "superforce". They were passionate about searching for a unification of these forces. Their best grand unified theories demanded the creation of magnetic monopoles in the soup of particles that developed after the Big Bang. In spite of years of searching, however, no monopole has ever been seen.

For particle physicists the lack of monopoles was an embarrassment. Along with the horizon and flatness problems, the monopole problem stood as an unmovable challenge to Big Bang orthodoxy.

INFLATION: COSMOLOGY AND PARTICLE PHYSICS RENEW THEIR VOWS

Inflation theory began as a bold attempt to solve the panoply of paradoxes that plagued cosmologists at the end of the 1970s. The basic idea of inflationary cosmology was simultaneously radical and elegant. With inflation, cosmologists imagined that the part of the universe we can see underwent a brief period of rapid expansion in its very early history. Here, "early" is an understatement. The era of inflation began when the universe was a mere 10^{-33} second old. That is less than a million billion billion billionth of a second after the Big Bang.[22]

During inflation, the universe increased in size by a factor of approximately 10^{40} (that is, 1 followed by forty zeros), enlarging its scale from a fraction of the size of a subatomic particle all the way up to the diameter of a softball. This expansion on steroids takes place in just 10^{-33} of a second. After inflation shuts down, the universe resumes the more leisurely expansion we see today. For comparison, in the last half of the universe's life (the last seven billion years) the cosmic scale has increased by less than a factor of ten.

While it may seem like a crazy and even unnecessary addition to the Big Bang, inflation's brief period of hyperexpansion solved all the problems with standard cosmology. With inflation, the causality problem disappears because every part of the universe we see today *was* in causal contact before space was stretched to the extreme. The flatness problem is also solved: inflation's rapid expansion naturally pulls space out flat, so omega is forced to a value of 1 no matter what its original value was. Physicists were happy to have a theory requiring no fine-tuning of the universe's initial conditions (it was assumed that future measurements would eventually find "missing mass-energy" and yield a flat-space value of $\Omega = 1$). Finally, inflation also did away with the magnetic monopole problem. Space is so diluted from its brief period of inflation that the density of monopoles is diluted as well. Monopoles are pulled so far away from one another as space expands wildly that the odds of us observing one now becomes essentially zero.

Thus, with a single change to the Big Bang (an early, brief period of hyperexpansion) all the problems were resolved. It was hard for physicists and astronomers not to take notice.

Every good idea needs a champion and inflation found its hero in Alan Guth, a physicist at MIT. In 1981, Guth wrote a paper describing the inflationary model for cosmology.[23] While some of his ideas had been proposed before, Guth brought them together in a coherent, accessible way and added the catchy brand name of inflation. What mattered most however was that Guth was not an astronomer. He was, instead, a particle physicist and he built his inflation theory using tools from the empire of grand unification theories. Guth would use ideas from the same GUTs domains that demanded monopoles to articulate reasons why they could never be observed.

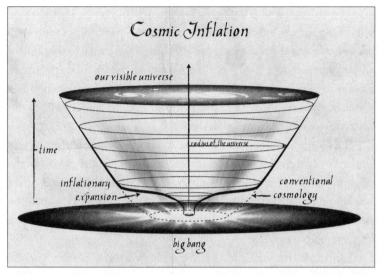

FIGURE 8.3. Inflation saves cosmology. Inflation imagines taking a small, causally connected patch of the universe and blowing it up into everything we can see, resolving all the classic Big Bang paradoxes in the process.

GUTs predicted that if one could "heat up" the universe to ever higher energies, the different forces we see today would sequentially "melt" into the superforce as ice crystals would melt into liquid water. Heating up a tiny speck of the universe was exactly what particle physicists try to do in their giant accelerators, smashing subatomic particles together with tremendous energy.

Of course the entire universe had already been through a stage when it was as hot as any particle physicist could ever desire. Cosmology naturally gave physicists the GUT laboratory they wanted. The critical link between inflationary cosmology and grand unification was the energy source that drove inflation, something inflation cosmologists called the "false vacuum".

Inflation may seem like the ultimate free lunch—taking a speck of the natal universe and blowing it up into everything we can see. But there has to be some mechanism, some source of energy to make it work. Drawing on grand unification theories, Guth and others imagined the early universe to be pervaded by a field of energy, a background mimicking empty space, but with energy to burn. This energy-rich background was the "false vacuum". Physicists were used to this kind of all-pervading energy field from their quantum explorations of matter. Extending uniformly throughout space, the false vacuum energy field was unstable, and sooner or later it would have to "decay" into a real vacuum, releasing the energy in the field.

Inflation was the result of that decay. Kicking in just as the universe cooled to the right temperature, the energy given up in the decay of the false vacuum acted as a kind of antigravity. It ripped space-time apart, accelerating its expansion by providing just the right force at just the right time to inflate a speck of the cosmos into the entire observable universe—a big rip right after the Big Bang.

Uniting astronomy and the hairy edge of theories from particle physics, inflation became a new "standard model" for the birth of the universe—an inescapable prologue to element cooking, the CMB and all the rest. But inflation wasn't the only addition to standard Big Bang cosmology in the last decades of the twentieth century. Throughout the 1980s and 1990s, physicists and astronomers would be forced to struggle with new and unexpected discoveries, and each of these new actors would have to be added into the cosmic story of creation.

THE DARK UNIVERSE, PART I: MATTER

The last decades of the twentieth century would mean a profound step into the dark for cosmology. To understand this movement, however, we have to briefly step back fifty years. Though few astronomers recognized it, their first introduction to the dark universe came long before the triumph of the Big Bang. In 1933, an iconoclastic Swiss expatriate astronomer stumbled upon the discovery of what would eventually come to be called *dark matter*.

Fritz Zwicky has been described as the "most unrecognized genius of twentieth-century astronomy".[24] A man of both terrible anger and uncompromising humanity, Zwicky was prone to heated arguments

FIGURE 8.4. Astronomer Fritz Zwicky, the man who first "saw" dark matter.

with colleagues. A famous galaxy catalogue compiled by Zwicky opens with a rant against other astronomers (names provided), calling them "fawners" and "thieves" who stole his ideas and hid their own errors. A fellow astronomer at Mount Wilson once feared Zwicky was going to kill him after a particularly nasty exchange. But Zwicky had a profoundly human and charitable side as well. He was devoted to those he counted as friends and after World War II he led efforts to ship books to the ravaged libraries of Europe. For all his personal contradictions, however, it was his science that made Zwicky stand out. He was the essence of creativity in science, seeing astonishing solutions to problems others had yet to even recognize. In no domain of science was Zwicky's leadership more apparent than in the recognition of dark matter.

Galaxies took on new importance in the wake of Hubble's 1925 discovery that they were separate systems of stars. By 1933, it was recognized that galaxies were not uniformly distributed in space but were grouped together in large-scale structures of various size.[25] At the time, Zwicky was studying what appeared to be the largest of these galaxy clusters. As part of his work he first added up the mass in all the cluster's component galaxies. Then he examined the velocities of the galaxies within the cluster. Comparing the two, he found a paradox: the galaxies were all moving so fast that they could not be bound by the cluster's total gravitational force. If the cluster was simply a collection mass (the galaxies) bound by gravity like a planet, then the individual galaxies were like rocket ships that were moving so fast they had reached escaped velocity.

According to Zwicky's calculations, the cluster should have flown apart billions of years ago. Yet there it was in the sky, an obviously dense collection of galaxies in the midst of a cosmic void. Zwicky concluded that if the "luminating matter" he could see (the galaxies) was not enough to keep the cluster together, then there must be more matter in the cluster he couldn't see. "The average density of the Coma system must be at least four hundred times greater than what is derived from observations of the luminating matter", he wrote. "Should this be confirmed, the surprising result thus follows that *dark* matter is present in very much larger density than luminating matter" (emphasis added). It was a stunning result. It was also, for the most part, forgotten, or at least put on the shelf of as yet unexplained results. Astronomers were not ready to admit that what they saw was not all that was.

By the 1970s, astronomers such as Vera Rubin had begun to rediscover dark matter in their studies of individual galaxies and their rotation. Compiling data for one galaxy after another, Rubin found each one was spinning too fast to be explained by the matter emitting light. Rubin's data suggested that the luminous parts of a galaxy are pulled into rapid rotation by vast halos of dark matter surrounding them. Her results implied that the galactic whirlpools that had so enchanted us in image after image were not the whole story. The bright stars and luminous clouds of galactic gas were nothing but bright Christmas lights hung on a vast, invisible tree. In truth, galaxies were mostly dark.

The term *dark matter* was more a description of ignorance than anything else. All astronomers knew about dark matter was that it produced a gravitational force driving luminous matter to orbit their galactic centres at high speeds. As evidence piled up that dark matter outweighed normal matter by a factor of ten to one hundred, the race was on to understand both its nature and its place in physics. By the mid-1990s, scientists had concluded that the omnipresent dark matter was essentially different from the material you and I are made of. "Our kind of stuff" is called baryonic matter and it is composed of protons and neutrons. Using a variety of techniques, the possibility of a universe dominated by "dark baryons"—dead stars, or even huge quantities of rocks floating in deep space—had been ruled out. Whatever dark matter was, it clearly did not respond to electromagnetism or the strong nuclear force. It was not "our kind of stuff".

As computers reworked culture in the 1980s, they also reworked cosmology. Computer simulations rapidly became an important tool in all branches of astronomy, including the study of the universe as a whole. Using the fastest supercomputers of the time, astronomers began running simulations of cosmic history after the era when CMB photons would have decoupled. At this point, galaxy clusters would have begun to form across vast regions of space. The simulations, paired with new observational projects to map the cosmic distribution of galaxies, provided critical clues to the nature of dark matter.

Outweighing baryonic matter by a factor of as much as a hundred, the distribution of dark matter becomes the key player in the era of galaxy formation. Small lumps and bumps in the dark matter left over from the early universe become the seeds for galaxy clusters. Just as the computer simulations of cosmic-scale galaxy cluster formation were ramping up, observational astronomers began compiling their new large-scale surveys of the sky to hunt for traces of the original lumpiness. These maps revealed how galaxies and galaxy clusters were distributed across space at scales of hundreds of millions of light-years or more. The goal of the new computer simulations was to reproduce the observed distribution of galaxy clusters by tracking their formation and evolution in the first few billion years of cosmic history. The computer simulation project was a stunning success. Simulations *could* recover the

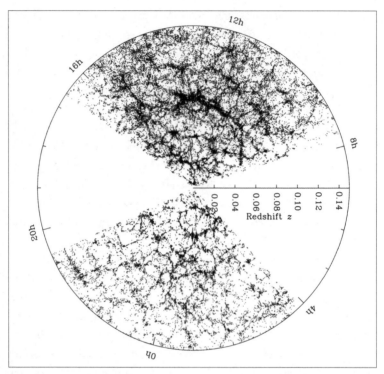

FIGURE 8.5. Large-scale structure of the universe. This image shows the distribution of galaxies on a scale of 2 billion light-years (the outer circle). The distribution of galaxies has been mapped in two thin slices, one in the north galactic cap and one in the south. The slice in the south is thinner, which is why it looks sparser.

data but only if certain kinds of dark matter were used. If the dark matter moved too fast, the gravitational collapse producing galaxies and galaxy clusters never happened. Whatever the dark matter was, it had to be "cold", meaning it moved slowly relative to the speed of light. Hot (fast) dark matter was out. Cold (slow) dark matter was in.

By the turn of the new millennium, cosmologists had to fit a new player, cold dark matter, into their models. Particle physicists were ready to supply many (some said too many) theoretical candidates for cold dark matter. Almost all were some form of weakly interacting massive particle (WIMP for short). These particles felt only the weak nuclear force and gravity. Physicists were heading into new territory, as the WIMPs were not part of the standard model. While that prospect of finding new physics beyond the standard model was exciting, the reality was that no form of dark matter particle had shown up in any laboratory.

The galaxy cluster observations and cluster formation simulations of the 1980s were the beginning of the heroic effort to map the cosmos across all visible space and time. It was an effort that required the ever faster and ever more sophisticated computers that appeared throughout the decade. A new tool for scientists, these supercomputers created unimagined possibilities for mapping cosmic evolution. The next, wholly unexpected step in cosmology would find map making and a new concept—cosmic acceleration—closely connected. And just at this moment a similar pairing of maps and acceleration would find its place in the construction of human time as well.

READING, ENGLAND • MARCH 8, 2004

"Keep the blue dot on the red line and everything will be OK. Blue dot on red line. That's all."

He was late already and that wasn't good. The meeting started at two-thirty and here he was, miles outside Reading. The flight had been late into Gatwick, the paperwork at the rental car counter had taken too long and his directions had been marginal at best. The rental clerk asked if he wanted a SatNav with his car and in his panic he said yes. He hoped it was going to help him. The deal would fall through if he couldn't make this presentation.

The bloke with the sleeve tattoos at the rental pickup gave him a three-minute tutorial on the GPS thingy mounted on the dashboard. Together they typed in the address, and bang! There it was—his current position was a blinking blue dot on the map, his route a red line snaking from the airport to the company's local office. "That's all you need," Tattoo Guy said. "Just keep the blue dot on the red line and you're going to be fine."

And it was. Every turn was called out to him long before he needed to make it. "Left turn in one mile," announced the SatNav's pleasant-sounding female voice. "Yes, yes, yes," he called back. "Cheers."

He looked up at the sky, thinking about the web of satellites sailing some-where overhead, beaming their signals every which way, tracking his posi-tion down to a metre or so. It was amazing. At his dentist's office just a few months ago, he'd read about GPS in a popular science magazine. Now here it was, saving the day.

"Right turn in one hundred metres," said the GPS. The dashboard clock read 2:41. He had already called ahead on his mobile. They said they could wait if he wasn't too late. "Left turn in fifty metres." He was going to make it.

"I love you," he told the dashboard box, its blinking blue dot hovering on the red line.[26]

MAPPING TIME TO SPACE: THE GLOBAL POSITIONING SYSTEM FINDS ITS WAY

Reconstructing time has always been a process inseparable from new encounters with space. Time and space are paired in human life by the simple concept of travel: the time it takes for an edict from the Roman emperor to travel across Europe, the time it takes for a gold-laden ship to travel from the Americas to Spain, the time your family takes to reach Cornwall when travelling by car from London, the time it takes for that critical e-mail message to travel across the network.

Our encounters with space through travel, however, are always mediated by our ability to determine location. Travel requires know-ing where you are going. And because the human experience of time is intimately connected with our experience of space, we must pay at-tention to the way space has always been mediated by maps. The long saga of longitude—and its connection with the determination of time— was a very public, culture-wide battle over the ability to make accurate maps and know location relative to them. There is also a personal and more closely experienced encounter with maps as well. As individuals, we deal with maps on a variety of levels, from the internal topologies of neighbourhoods we carry in our heads to the country-spanning maps we began using once automobile-powered travel became the norm in the 1920s and 1930s. The advent of silicon-based material engagement would change our encounter with maps on all levels. Just as e-mail and personal information management reshaped our experience of the day by changing our expectations of time, the radical electronic technologies associated with GPS would change our encounters with space leading to its own changes in time. And, as with e-mail and the Outlook universe, GPS would mean a profound and ubiquitous acceleration of human culture.

It began, as these things often do, with the military. Just as the Royal Navy's disastrous fogbound loss of warships in 1707 led to accurate methods for longitude determination, the U.S. military's need for hyperaccurate navigation led to the Global Positioning System.

Just days after Sputnik achieved orbit in 1957, scientists at MIT's Lincoln Laboratory discovered that they could use frequency changes in the satellite's pulsing radio beacons—à la the Doppler shift—to accurately pin down Sputnik's position.[27] Recognizing that the process could be reversed by finding a position on the ground from a satellite in a known orbit, the U.S. Navy conceived its TRANSIT satellite-based navigation system. Launched in 1964 TRANSIT used signals in a network of six orbiting satellites to give nuclear submarines accurate positioning at sea.[28] The TRANSIT concept was a success, but its coverage was too sparse; sometimes submarines had to wait hours before the satellites passed overhead and could be used to nail down a position.

Looking to build a more reliable and continuously available system, the Pentagon initiated the Navstar (Navigation System with Timing and Ranging) program in 1973. The theory behind Navstar was to ring the planet with a web of satellites and use their signals to allow position determination at any time and any place.[29]

Each satellite would broadcast its own location in orbit and the exact time, down to a billionth of a second. Ground-based receivers would pick up this information and use it to calculate their position on the planet to within a hundred metres or so. Four satellites were needed for each position determination: three to triangulate the three-dimensional position in space (the range) and one satellite to make any needed corrections for time differences.[30] Range is found by comparing signal travel times. The receiver on the ground compares its own clock with the time measured by satellite. The difference between the two times is then used to calculate its distance from the satellite. Since light travels at 186,000 miles (299,790 kilometres) per second, a time difference of one-thousandth of a second implies the ground-based receiver must be 186 miles from the satellite.[31] By comparing the receiver's time and the time signal sent by the three satellites, scientists could get hyperaccurate determinations of the receiver's position in space—its longitude, latitude and altitude.[32]

The story of GPS is a story of material engagement involving both of the twentieth century's great scientific revolutions: relativity and quantum physics. Hyperprecise position measurement requires hyper-

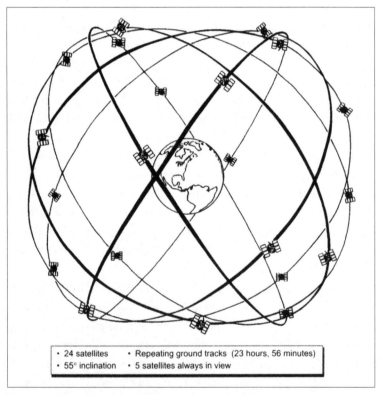

- 24 satellites   · Repeating ground tracks (23 hours, 56 minutes)
- 55° inclination · 5 satellites always in view

FIGURE 8.6. Schematic of the GPS satellite network in Earth orbit taken from an early RAND Corporation study.

precise clocks. This is where quantum mechanics enters the story. Each GPS satellite carries an atomic clock on board. In the 1950s, scientists had started to use quantum-based understanding to control atoms with exquisite accuracy. Rapid-fire transitions of electrons within atoms (the fabled quantum jumps) were manipulated to yield a steady atomic pulse accurate to one second every thirty thousand years. By the 1970s, engineers devising the GPS system could make good use of this quantum time technology. Each satellite would carry four onboard atomic clocks, maintaining time with a precision of a few parts in ten billion over a few hours, or better than ten nanoseconds. Expressed as the distance travelled by the satellite's signal, each nanosecond corresponds to about thirty centimetres.[33]

Relativity made its appearance in the technology because the precision required for distance determination was also GPS's own worst enemy. Each satellite moves at high speed in its orbit (3.8 kilometres/ hour). Each satellite is also moving through the gravitational distortion of space-time from the Earth's mass. Thus, Einstein's theory of relativity (both the special and general versions) could not be ignored. Relativistic time dilation from the satellite's motion (in the form of special relativity) causes the onboard satellite clocks to fall behind ground clocks by seven microseconds per day.[34] Because of their distance from the Earth and the gravitational bending of space-time, the satellite's clocks will also tick *faster* than identical clocks on the ground. General relativity predicts that GPS clocks will run ahead of ground-based clocks by about forty-five microseconds per day.[35] Thus, the same theory of relativity that mapped out cosmic structure could not be ignored in mapping out terrestrial positions with GPS. Without correcting for relativistic effects, GPS would quickly become useless. As Richard Pogge of Ohio State University put it, ignoring relativity leads to an error "akin to measuring my location while standing on my front porch in Columbus, Ohio, one day, and then making the same measurement a week later and having my GPS receiver tell me that my porch and I are currently about 5,000 metres in the air somewhere over Detroit."[36]

By combining quantum mechanics, relativity and space-age rocket science, the Navstar/GPS system promised a new age of position determination for the military. Throughout the 1980s the network of orbiting GPS satellites was slowly put into place. The success of GPS for military applications got a very human face in 1993 when a U.S. Navy pilot, Captain Scott O'Grady, was shot down over Serbia during the Bosnian conflict. After four days of hiding from Serbian forces and living on grubs, O'Grady was dramatically rescued by the 24th Marine Expeditionary Unit. Behind the heroism of Grady and his rescuers was the GPS device in his life vest allowing the Marine unit to precisely determine O'Grady's position and swiftly sweep him up and out of harm's way.[37]

Even before the full network of satellites was put in place, however, events determined that GPS would enter the public domain. In 1983, Korean Air Flight 007 was shot down by Soviet fighters after it got lost over Soviet territory. The attack killed all 263 passengers on board, including a U.S. senator.[38] In response, President Reagan declassified GPS, moving it from a purely military effort to a public project. In 1993, President Clinton authorized the military to reduce the amount of "fuzz" it added to the signal made available to the public. This fuzz was called "selective availability" because it meant that the various clients using GPS got different levels of position accuracy. The military, of course, had access to the highest accuracy. In 2000, selective availability was stripped away and GPS began its encroachment into public life and time.[39]

The first consumer GPS devices became available in the late 1980s and early 1990s from companies such as Magellan and TomTom. These products made inroads quickly, sprouting applications for everything from the obvious car-mounted units to devices for meter reading and barcode ID positioning.[40] As the 1990s progressed, GPS devices rode the wave of increasing power and sophistication that was advancing all forms of electronic material engagement. Once selective availability was dropped, GPS was freed to work its way into culture on new and deeper levels. Of particular importance was the integration of GPS with mobile phone technologies.

Looking at a film from the 1980s, it is hard to believe that people managed to make their way around modern civilization without mobile phones. The shift from a telephone connected to wire to a telephone in your pocket occurred so quickly and completely, it is as if a Grand Canyon of experience opened between our former lives and the lives we lead now. Watching people staring into their little plastic rectangles as they stand on line in any airport or on a train platform, it is hard not to ask, "What did we do with ourselves before?" Mobile phones, like e-mail and the Internet, formed the substrate of new material engagement and transformed human time in the digital age. But mobile technology was always as much about space as it was about time. When you make a call, the mobile network is tracking your position through electromagnetic signals that travel from the phone to the local network towers. With this emphasis on position determination it was only a matter of time before mobile phones and GPS found common ground.

It was not a stretch to adapt the clock synchronization so vital to GPS to support mobile position determination. The killer app for this hybrid of GPS and mobile phones was locating emergency callers. In the aftermath of 9/11, the U.S. Federal Communications Commission required that emergency GPS location capacity be included in all U.S. mobile phones.[41] From there it was just a short step to including GPS as a feature in the rapidly advancing smart phone market. In 2006, several major American carriers picked up on mobile phone GPS, touting its functionality on their new high-end smart phones. Then Apple released the iPhone and the contours of human-machine interface shifted. The iPhone was wildly successful in redefining what was possible from a handheld device. Because the iPhone was, essentially, a fully Web-connected (and eventually GPS-connected) handheld computer, our interaction with time and space shifted yet again. As competitors took up the iPhone's challenge, silicon-chip-based material engagement entered the era of "cloud computing". All human knowledge on all subjects—from the nearest Indian restaurant to the history of Indian cooking—was always instantly available in the wireless aether, "the cloud". Not having instant access to information, including information about where you were, came to seem like a relic of a different age.

With the advent of GPS-enabled mobile phones, hyperaccurate space was woven together with hyperaccurate time as the new standard for a cultural life that had accelerated to the speed of the cloud. In this process, we traded the internal arena of personal, inhabited time for a more crowded, more public domain in which we were always available, always locatable and always part of a social network. We were always at work, even on weekends or family trips. We were always visible, even if in a distant country, through our status updates and microblog posts. For good or for ill, acceleration did not just mean going faster; it also meant a seemingly endless expansion of an ever-connected public space that became steadily more difficult to avoid.

THE DARK UNIVERSE, PART II: ENERGY

Dark matter did not fit into anyone's scheme for cosmology but that did not mean it constituted a fundamental problem. In attempting to move beyond their standard model, particle physicists had always expected to discover deeper laws and new forms of matter. The fact that most of the universe's mass appeared in this dark form was fascinating but could easily fit into the inflationary Big Bang models without angst. Dark energy, the second great discovery of the post–Big Bang era, was another story entirely. While dark matter was hinted at as far back as Zwicky's work in the 1930s, dark energy appeared suddenly and without warning.

In 1998, two highly competitive groups of astronomers were each completing a multiyear project with the same goal: they hoped to extend Hubble's distance-velocity law farther into cosmic space than anyone had before by using a special type of exploding star as a beacon. Recall that Hubble found a straight-line relationship between galaxy recession speeds and galaxy distance. The farther a galaxy was from us, the faster it was moving away—exactly the expectation for an expanding universe. General relativity, however, predicts that Hubble's straight-line relation should be a local phenomenon only on cosmological scales. Peer far enough out into space and the Hubble law should change due to gravity. Since everything is pulling on everything else, the expansion of the universe should be slowing down. At great distances, cosmologists expected to see Hubble's straight-line relation bend over—effects of the universe's gravitational braking.

The two research groups, one based at Berkeley and the other at Harvard, were racing against each other to find the magnitude of the universe's deceleration. It was a critical project since the rate of cosmic braking is directly related to the all-important cosmic density parameter, omega. The determination of the cosmic deceleration meant the determination of the total mass-energy content of the universe. It would be a result worthy of a Nobel Prize.

Each group used the same method. First they looked for special types of supernovae in distant galaxies. Supernovae are the apocalyptic explosions of geriatric stars. Only a few years earlier, a special kind of supernova, called Type Ia, had been found to make an excellent probe for cosmic distance. Type Ia supernovae were standard candles in the same way as Henrietta Leavitt's Cepheid Variables, which had allowed Edwin Hubble to nab the distance to the Andromeda galaxy. Unlike Cepheid Variable stars, however, supernovae are so bright they can be seen halfway across the known universe. This made the perfect tool for pushing Hubble's law deeper into cosmic space.

Finding a distant Type Ia supernova demanded use of giant instruments such as the ten-metre Keck telescope on top of the Mauna Kea volcano in Hawaii (almost four times larger than Hubble's massive Hooker scope). Each team pushed hard, using these powerful instruments, to find the bend in Hubble's law and announce a value for cosmic deceleration. But things didn't go quite as planned.

"I was just quite frankly denying this was happening", recalled Brian Schmidt of the Harvard team.[42] As data were gathered and analysed, Schmidt's group was stunned to find no evidence for deceleration. Instead, everything pointed in the opposite direction. According to the supernova observations, the expansion of the universe was actually *accelerating*. The universe was expanding faster now than it had been billions of years ago. Over in Berkeley, team leader Saul Perlmutter was having the same problem and the same reaction. The Berkeley group's data (using different supernovae) also showed acceleration. "Is there something wrong with this?" Perlmutter asked his team. After exhaustively checking and rechecking their data, both groups bit the bullet and announced their results. Cosmic acceleration became worldwide news.

Overnight, cosmology was turned on its head. Rocky Kolb, a leading cosmologist who literally wrote the book on the early universe, was stunned: "I still have a hard time believing it."[43] In spite of the community's incredulity, further studies rendered the cosmic acceleration a remarkable fact that would force cosmologists to rethink their business.

As Newton had showed four hundred years earlier, acceleration needs a force. And, as the physicists of the eighteenth and nineteenth centuries demonstrated, forces need energy. The discovery of cosmic acceleration meant that space was being forced apart and must, therefore, be pervaded by a new form of energy acting like antigravity. While Newtonian gravity only produces attractions, Einstein's famed cosmological constant had shown physicists that repulsion and gravity could go hand in hand. The supernova data made it clear that some form of antigravitational energy had to exist. Since nothing was known about this newly discovered energy other than its space-stretching ability, it too was given the "dark" moniker.

Theoretical cosmologists were quick to build models for what was now being called dark energy. The most obvious candidate was Einstein's long-dismissed cosmological constant. Recall that this constant implied a repulsive force that pushed space-time apart. Einstein had originally introduced it as a way of keeping his preferred static universe from collapsing. Perhaps what the great scientist considered his biggest blunder had in fact been an act of fantastic prescience. For many researchers cosmic accleration implied that the cosmological constant was back with a vengeance. The value for the cosmological constant extracted from the supernova data was, however, puzzling. After Einstein gave up on the idea, physicists assumed the cosmological constant did not exist, or thinking mathematically, its value was exactly zero. But physicists soon recognized that quantum mechanics and its all-important concept of uncertainty implied that even the vacuum of empty space was seething with energy. Because quantum mechanics demands that no system, including the vacuum, can be specified without some uncertainty, "virtual" particles can appear and disappear against the vacuum's background state of zero energy. When physicists calculated this "vacuum energy", as it was called, the number they found was huge and would lead to a huge cosmological constant. Such a cosmological constant would have torn space apart billions of years ago which clearly did not happen. Thus physicists had always assumed that some mechanism led to a perfect cancellation of the quantum vacuum energy, resulting in a cosmological constant of zero.

The observations of cosmic acceleration pointed to a cosmological constant that was far smaller than anyone would have predicted from vacuum fluctuations but still far larger than zero. If dark energy was a cosmological constant, someone needed to explain how it took such an unexpected value. Of course, the link between the newly discovered cosmic acceleration and the brief early period of acceleration that defined cosmic inflation was not lost on scientists. Some researchers proposed that, like the inflation field of the early universe, the cosmos was now pervaded by a similar energy that would also eventually fade away. Drawing on terminology from Aristotle's 2,500-year-old cosmological texts, the theoretical energy field was given the name *quintessence*, and a host of new studies explored its potential as a source of dark energy.[44]

Alongside the cosmological constant and quintessence, other options for dark energy sprouted like wildflowers in spring. None of them, however, was an obvious or perfect solution. Dark energy had stormed onto the stage and physicists were left to hurriedly pick up the pieces.

RIPPLES OF CREATION: THE TRIUMPH AND TRAGEDIES OF INFLATIONARY COSMOLOGY

The cameras were on and the scientists were arranged at a table at the front of the room. It was June 5, 1992, and the new results from the Cosmic Background Explorer (COBE) satellite were set be announced. COBE's mission was to chart the cosmic microwave background in the sky at a higher resolution than ever before. Scientists were about to get their best, most detailed map of the universe's all-important fossil radiation. As a first step, the COBE team confirmed, once again, that every corner of the sky glowed with the relic photons. These light particles emerged from the seething cosmic plasma that was 13.7 billion light-years away and 13.7 billion years in the past. Then the COBE scientists had gone further, looking for small variations in the CMB from one point in the sky to the other. These variations translated into telltale lumps and bumps of slightly hotter or slightly cooler primordial gas. For years cosmologists had predicted such tiny variations, small knots in the otherwise smooth cosmic gas emerging from the Big Bang.

Cosmologists believed the knots had to be there. They were the seeds that had grown into the rest of cosmic history. According to Big Bang theory, all the structures we see in the universe today—galaxy clusters, galaxies, stars, planets and people—must have started as those tiny perturbations. Begin with those tiny lumps, the story goes, and gravity does the rest. Through gravity the dense lumps pull the surrounding gas towards them. The lumps get bigger and denser, pulling even more gas inward from ever-larger distances. In time, the lumps grow into galaxies and galaxy clusters. COBE's job had been to find those first steps in the ladder of cosmic structure building. The scientists at the press conference were there to affirm that COBE had done its job.

On the wall behind the scientists was an all-sky map of the inhomogeneities in the CMB. The map was a mottled oval showing amorphous blobs of red for slightly cooler (and slightly denser) gas and blue for slightly warmer (and slightly more rarefied gas)—the earliest seeds of all cosmic structure. The project scientists were beside themselves with joy. George Smoot, the project scientist for COBE, leaned into the microphone and described his reaction on first seeing the completed map: "It was like looking at the face of God."[45] Though he would catch hell from some quarters for his poetic slip of the tongue, Smoot's characterization was not altogether incorrect. COBE had caught echoes from just after the dawn of time, and in so doing, had retrieved one of the holy grails of Big Bang cosmology and its modern incarnation, inflation.

Years before COBE, in the early 1970s, two cosmologists, an American named Edward Harrison and a Russian named Yakov Zel'dovich, had used the observed distribution of galaxies in the sky to push back-

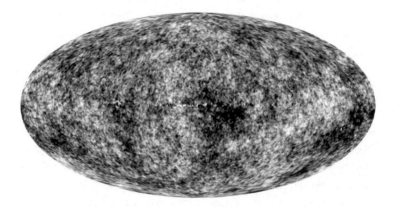

FIGURE 8.7. The ripples of creation. All-sky map of fluctuations in the cosmic microwave background, produced by the Wilson Microwave Anisotropy Probe (WMAP), an orbiting telescope that followed COBE. White represents denser regions, and dark represents under-dense regions of the early universe a mere 380,000 years after the Big Bang.

wards and calculate the distribution of perturbations that must have existed in the cosmic plasma after the Big Bang. Their calculations showed how gravity took hold of lumps in the gas after radiation and matter decoupled (when the CMB formed). Working backwards, Harrison and Zel'dovich calculated what the distribution of lumps, or the "spectrum of perturbations", must have been to produce what we see today.[46]

A decade later, Alan Guth introduced inflation, and physicists began probing its consequences. In 1982, researchers discovered that inflation theory naturally recovered the Harrison and Zel'dovich spectrum of cosmic perturbation. The inflation field powering hyperexpansion in the very early universe was inherently a quantum mechanical entity—it was a quantum field. Since the very essence of quantum physics is its inherent randomness, the decay of the inflation field must have imposed random perturbations on the rapidly stretching universe. These quantum fluctuations then grew with space-time, transmitting their lumps, bumps and wiggles all the way to the creation of CMB photons three hundred thousand years after the Big Bang. The matter/radiation decoupling of the CMB was the critical turning point in the story. Before the CMB photons were released, matter and radiation were so closely connected that gravity's attempts to build structure were lost to radiation's ability to smooth it all away. Once the CMB radiation was decoupled from matter, the cosmic gas was ready to be structured by gravity.[47]

It was a significant early triumph for the new inflationary theory. The recovery of the Harrison-Zel'dovich perturbations by quantum fluctuations showed that inflation was good for something. A decade later, the COBE data showed that the same distribution of lumps and bumps in the Harrison-Zel'dovich perturbations was there, clear as day, in the CMB. In 2005, a decade after that, a second satellite, called WMAP probed the CMB at ten times the resolving power of COBE. WMAP's portrait of the CMB sky allowed scientists to go beyond COBE and make exacting explicit comparisons between different versions of Big Bang theory, including the inflationary Big Bang. Once again inflation was the winner. The inflationary scenario was picking up at least some forms of experimental verification.

The precision testing of model and data extended beyond inflation's prediction of the CMB's minute bumps and wiggles. Mapping the distribution of galaxies across billions of light-years had become a small industry in the 1980s and 1990s. Using dedicated telescopes hooked up to networked computers and huge databases, projects such as the Sloan Digital Sky Survey automated the observation of hundreds of thousands of galaxies. Generating terabytes of data, the 3-D distribution of galaxies was slowly teased out of these observations. Just as structure in the CMB maps could be linked backwards to inflation, structure in the galaxy distribution maps across the entire universe could be linked back to the CMB. Together a consistent story of cosmic evolution emerged all the way from 10^{-33} of a second out to our current place in time 13.7 billion years later.

In this way, by 2005, the WMAP data could be combined with other observations to nail down parameters cosmologists had spent decades arguing over. The exact amount of normal baryonic matter: 5 percent. The exact amount of the now accepted dark matter: 25 percent. Most astonishing was that the newly discovered dark energy was also indirectly locked in by observations at 70 percent. Add it all up and you got

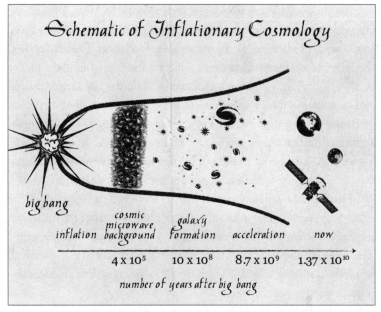

FIGURE 8.8. The ages of inflationary cosmology. Schematic diagram of cosmic history flowing from the Big Bang (left) to today (right). The width of the figure represents the radius of the universe as time progresses.

an omega of 1. Big Bang cosmology, with inflation included, had now become an exact science.

That was inflation's triumph. But hiding offstage was also a story of its failures. Inflation was not really one theory; it was many. While the WMAP data supported inflation's prediction for the cosmic density perturbation spectrum, many versions of inflation included quantum fluctuations. The WMAP data did nothing to sort through competing versions. In addition, enough unknowns remained in even the generic versions of inflation to keep many scientists sceptical. No one, for instance, knew what constituted the inflation field or from what physical principle it originated. Worse still, the inflation theory required a good deal of its own fine-tuning. Descriptions of inflation's evolution have to be tweaked in just the right way to keep the universe from breaking apart into empty, disconnected regions or not working at all.

Even more damning for some scientists is the fact that dark energy, the greatest discovery in cosmology since the CMB, came as a surprise. It was never predicted as an inevitable part of inflation. The acceleration the universe is experiencing now seems, on the face of it, like a milder version of the hyperexpansion that occurred during the early universe, and yet inflation theory offers no link between the two. The current era of cosmic acceleration must be grafted onto inflation like a Jonagold apple tree branch onto a Granny Smith tree trunk. For many scientists, a true, complete theory would not have to be grafted on this way. Inflation, in a word, should not have been taken by surprise.

Finally, and most important, inflation was still a theory of *after*. The orthodox versions of inflation still started their story after the initial and impossible singularity. They still began with a mysterious beginning. The universe and time still started without explanation, and we are all still left wondering why. Inflation, for all its promise, left the most important question unanswered.

By the first decade of the twenty-first century, the entire field of cosmology and particle physics seemed ready, one way or another, to push against the temporal frontier of the beginning. A chorus of scientists began demanding an answer to the question: "What came before the Big Bang?" As the new millennium began, radical new visions of the universe and time would be offered as science and the culture it was embedded within would find itself standing, once again, at the edge.

Chapter 9

WHEELS WITHIN WHEELS: CYCLIC UNIVERSES AND THE CHALLENGE OF QUANTUM GRAVITY

Eternal Time Through Repeating Time

ATHENS · THIRD CENTURY BCE

The handful of students was leaving the Stoa of Zeus, just north of the Athenian agora. Some were more advanced in their studies; others, like the young man from the noble family, were new to philosophy.

He had travelled here against his father's wishes. "You should concern yourself with our estates and the affairs of state," his father told him, shaking his head. "I can't see what these Stoics have to offer you other than wild ideas."

"Questions," the young man reflected as he walked. "That is what they offer me."

The young man had many questions about life, about the world and about their tangled origins. The questions had plagued him since his first childhood encounters with philosophy (during the very studies his family encouraged), and his heart leapt at the possibility of answers.

He had heard Chrysippus speak before. Chrysippus was a Stoic and, like others of that school, he taught that the world was composed primarily of fire. All we see, all we encounter, is nothing more than fire and its transformations. Most important, Stoics such as Chrysippus claimed an understanding of the universe's origins.[1] This was the young man's most burning question and his reason for coming to the see the philosopher today. The old man had

251

not disappointed him. In Chrysippus' answer the youth felt like the sky itself had opened up before his mind's eye.

There was no origin. Time and the universe repeated in endless cycles. Beginning in fire, ekpyrosis, each cycle ended in fire—creation followed by destruction followed by creation, from eternity and to eternity. It was so perfect, and so . . . elegant.

As he crossed the market by the piers the young man took no notice of the merchants with their large clay urns of goods and the clamour of ships being unloaded. Chrysippus' words still echoed in his ears: After the conflagration of the cosmos everything will again come to be in numerical order, until every specific quality, too, will return to its original state, just as it was before and came to be in that *kosmos.*

"So simple," murmured the young man, unaware of anything but the endless blue sky in his mind's eye. "So simple . . . so beautiful."

CATASTROPHIC FAILURES AND ETERNAL HOPES: THE CYCLES OF CYCLIC COSMOLOGY

Across more than twenty centuries of human cosmological investigation, the possibility that time simply repeats itself has been powerfully seductive. Forcing the cosmos to just begin ex nihilo holds so many paradoxes that the logical alternative—an eternally existing cosmos—seems too clean a solution to easily give up on. But the great triumph of Big Bang cosmology in the 1960s showed that the universe had evolved. It had changed.

The universe was *not* at rest, as Einstein and so many others had hoped. Static models were doomed by Hubble's discovery of cosmic expansion. Steady-state models, in which the universe moves but always looks the same, offered another kind of eternity. But the discovery of the CMB doomed Fred Hoyle's popular version of that idea too. In seeking a way out from the seemingly absurd notion of time "just beginning," scientists felt they had to explore repeating cycles within the context of a relativistic space-time framework. The theories *were* explored in great detail but each attempt ended in catastrophic failure.

Cyclic cosmological solutions to Einstein's equations were recognized early on. Friedmann's seminal papers in the 1920s revealed three possibilities for a universe that seemed to be expanding. In two of Friedmann's solutions, expansion continued forever. In the third, expansion was followed by contraction. If the cosmological constant was assumed to be zero in these models, then it was just the matter density of the universe—the all-important omega parameter—that determined which fate awaited the universe. If the store of mass-energy was above the critical value ($\Omega > 1$), then gravity would eventually halt cosmic expansion and the universe would turn back in on itself like a deflating balloon.

If expansion and contraction happened once, there was no reason to believe it could not happen many times. Thus Friedmann called his third solution a "periodic world." He imagined that time "could vary from minus infinity to plus infinity, and then we come to a real [endless] periodicity."[2] Einstein, in his own distaste for a universe that "just" begins, explored Friedmann's periodic-world solutions for a time in his own cosmological studies.[3]

Friedmann was, however, a mathematical physicist and did not attempt to think about his solutions in a physical context. Richard Tolman, a Caltech theorist with a background in physical chemistry, took up a study of Friedmann's oscillating universe models. Tolman's research ended in a devastating critique of cyclic cosmology. His objection to the cyclic models arose from the same science that explained cycles in steam engines: thermodynamics. Filling Friedmann's oscillating universe with matter and radiation, Tolman showed how the accumulation of cosmic entropy had to sound a death knell for repeating cosmic cycles.

Recall that the second law of thermodynamics states that transformations of energy doing work (like expanding a universe) always create entropy. While entropy can be thought of as waste heat, it is more helpful to picture it as a measure of disorder. The second law demands that energy transformations in an isolated system increase its entropy. Thus the disorder within the system must also increase. Since the universe as a whole is the very definition of an isolated system, Tolman could apply the second law to it and track the long-term evolution of the periodic universe cycle after cycle. He found that every cycle generated more entropy. With nowhere to go, the entropy would be carried through to the next cycle.

If we let a bag of Lego stand in for the universe, we can metaphorically see Tolman's entropy problem. Imagine beginning with a block of Lego all neatly snapped together. The disorder and hence the entropy of the bag is small. If you break all the blocks apart, their disorder—their entropy—increases. If you throw them back into the bag in this state, they now take up more room. This is, essentially, the situation Tolman found in tracking the entropy of an oscillating universe. After each bounce, increasing entropy inflated the cosmos to a larger size, delaying the eventual turnaround to the next contraction. Each cycle got longer, with a trend pointing towards a final cycle that would last forever. Tracking backwards, Tolman also showed that time between cycles got progressively smaller. Go back far enough and the cycle duration heads to zero. Thus, entropy forces a beginning for even a cyclic universe. An eternal past was not an option.[4]

Though Tolman's entropy crisis prevented most scientists from exploring cyclic models further, hope remained that the physics of the bounce itself might somehow eat the entropy and save the oscillating model. But in the 1950s and 1960s, a group of Russian theorists studying the problem killed that hope too.[5] Exploring the fate of small space-time ripples, or irregularities, during the cosmic contraction phase, the Russian theorists showed that these perturbations were driven into catastrophic amplification. As the universe closes in on itself, any tiny wiggles in the space-time fabric grow wildly in one direction and then the next. Like a ball of dough kneaded by a strong baker, space-time is stretched and pulled this way and that until it becomes so chaotic that what emerges from the Big Crunch looks nothing like our universe. Reaching to the very real-world experience of electrical appliances for a metaphor, American physicist Charles Misner called this fate the "Mixmaster universe", and it formed the second blow to oscillatory models.

By the 1990s, the entropy problem, the Mixmaster disaster and the

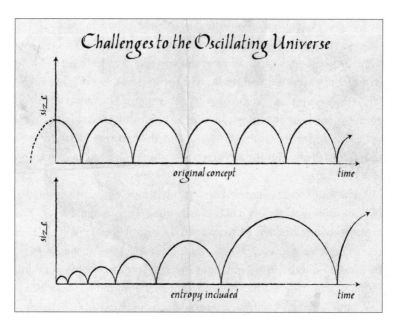

FIGURE 9.1. Oscillations in relativistic cosmology. Originally cosmologists hoped to use Friedmann's closed-universe solutions to create repeating cycles of Big Bangs and Big Crunches. By considering entropy from one cycle to the next, Tolman showed that such cycles could not continue forever and still required an original Big Bang.

eventual discovery that the density of the universe was too low to imply a future Big Crunch had doomed the relativistic version of the oscillatory universe. For the idea to be revived it would have to come in a radically different form compared with a standard version explored for seven decades by general relativistic cosmologists.

THE HOLY OF HOLIES: A THEORY
OF QUANTUM GRAVITY

The singularity. It always came down to the singularity. Ever since Friedmann and Lemaître discovered their expanding universes hidden in Einstein's equations, the damnable singular behaviour at $t = 0$ has plagued cosmology. Run the film of cosmic expansion back all the way to the beginning and the radius of the universe shrinks to zero. There is, literally, no room for all the stuff in the universe. Every point in the fabric of space is piled on top of every other point. The universe's mass-energy is crushed into a single geometric point with zero volume. At $t = 0$ the cosmic density, temperature and even the curvature of space reach to infinity. That point for physicists is nothing other the abomination of a singularity.

The infinities encountered in a singularity are not like the infinities philosopher-cosmologists had been wrestling with for millennia. They are not the infinities of the five great cosmological questions, as in "Will the universe exist forever?" or "Is space infinite?" Instead, the singularity with its infinite density and infinite temperature speaks of something far more mundane. Rather than philosophical profundity, these infinities imply a failure of physics. The singularity's impossibly high temperatures and densities near $t = 0$ simply mean Einstein's equations are no longer working and are no longer an accurate description of reality. This kind of problem can happen even in terrestrial physics. Infinities sometimes appear in equations describing phenomena as ordinary as fluid flow or the conduction of heat. When they do appear, physicists know the equations have been pushed beyond usefulness. The prescription for moving forward always involves finding a new set of equations with a deeper description of the underlying physics. Thus, the cosmic singularity at the Big Bang's beginning was a stop sign.

A similar kind of blow-up occurs at the centre of a black hole. Singularities appear in the general relativistic description of a massive star collapsing in on itself. The density at the centre of the collapsing star is driven to infinity along with the curvature of space-time. In the end, the collapse produces a black hole with a singularity in the middle—an apparent tear in the fabric of space-time.[6]

There had always been hope among general relativists that singularities and their seemingly violent end of space-time might be avoided for both black holes and Big Bang cosmology. In the 1960s and 1970s, however, Stephen Hawking and Roger Penrose probed deeply into the structure of general relativity and discovered that the singularities were unavoidable. Collapsing stars always ended in a singularity and Big Bang cosmologies always began with them. The main message of these singularity theorems was to show that the singular beasts lurk within general relativity as limits, places where the equations can go no further. To make any additional progress meant moving beyond the equations of general relativity.

While general relativity's problem with singularities has been known for years, its solution has eluded physicists for just as long. General relativity is a classical theory. It treats space-time as a smooth fabric—an infinitely divisible continuum. But quantum physics had shown scientists that on its deepest levels nature never appeared as a continuum. The principal lesson of quantum mechanics had been to teach physicists that everything in nature came in discrete bundles. Nature, at its root, is granular. Energy, momentum, rotation, spin—at the smallest levels, nature was not continuous but was quantized, discrete, bundled the way a beach resolves itself upon close inspection into a multitude of tiny flecks of sand.

Thus, physicists understood that at some point Einstein's classical continuum-based equations must cease to hold true and space-time itself must become quantum mechanical. They could even calculate the physical scales where this occurs. Below the Planck length, an impossibly small 10^{-35} of a metre, smooth space-time had to break down.[7] The limit can also be expressed as a time scale, which is particularly useful for cosmology. When the universe was younger than the Planck time of 10^{-44} of a second, space-time must have taken on its true quantum mechanical guise. How are we to imagine a quantum background for all existence? In thinking about these domains, physicists will sometimes speak of a space-time "foam", with bubbles of space-time separated by, literally, nothing. The bubbles of space-time are reality and between them is nonexistence.

If you don't know how to picture foamy, or quantized, space-time, don't worry. No one else does either, at least not in a fully consistent way. In spite of more than five decades of effort the best minds in physics have yet to develop a complete theory of quantum gravity. More important than grand unified theories, quantum gravity has been the one true holy grail of physics for decades.

In the last years of the twentieth century, the search for a grand unified theory that united the strong, weak and electromagnetic forces lost momentum. The pathways first explored in the 1960s and 1970s were not panning out. In the wake of stalled efforts, quantum gravity took on new urgency as a way of vaulting over the problems of GUTs and going straight to superunification. A true account of quantum gravity would, most physicists believed, become a theory of everything (TOE), a grand overarching theory that could explain all particles and all forces. Physicists were convinced that achieving this lofty goal would vault them through the Big Bang's singularity to see what lay beyond and before.

THE MANY REVOLUTIONS
OF STRING THEORY

Though there are many routes to developing a theory of quantum gravity, as of this writing all of them remain tentative. The most obvious approach would be to take Einstein's equations for space-time and simply quantize them. Physicists have done this before in other domains of physics. You start with the classical continuum equation and cast it in a form that behaves quantum mechanically, with all the associated uncertainties, probabilities and discrete jumps.[8] But what works for something like an electron in a silicon wafer could not be made to work with general relativity. The straightforward path to quantizing the equations of general relativity has been explored many times, always leading to a dead end. The straight-line route to quantum gravity appeared closed and physicists were forced to search through stranger theoretical terrain. It was in one of these unexpected corners that string theory appeared. In an astonishing twist of fate, what some scientists see as the best hope for quantum gravity began as a possible route to describe just the strong force.[9]

String theory began in 1968 when Gabriele Veneziano at CERN, the European Organization for Nuclear Research, proposed a new equation describing the behaviour of the strong nuclear force. A few years later, his equations were reinterpreted and shown to describe particles as vibrating strings. The move from thinking of a quark as a zero-dimensional "point particle" to thinking of it as a very small but still extended one-dimensional string gave the new theory attractive characteristics. Point particles pose their own singularities since they must be treated as if their size (radius) is zero. Divide by zero and, of course, you end up with infinity. The problems point particles forced on physicists must therefore be cleaned up, or avoided entirely, through mathematical gymnastics. In the string picture these kinds of difficulties did not appear, making it an appealing new route of research.

After getting stuck in a briar patch of technical problems, string theory rebounded in the early and mid-1980s. In what is sometimes called the first revolution of string theory, physicists were able to show how many of the known particles and forces in their standard model could be accounted for as vibrations of the elemental strings. Things got really exciting as scientists recognized that the string picture led naturally to a consistent description of the quantum force carrier of gravity—the graviton. By substituting vibrating strings for point particles, a theory that began in the realm of GUTs had expanded to include gravity and, seemingly, hinted at a theory of everything. For theoretical physicists, the early explorations into string theory were a heady time. As physicists Neil Turok and Paul Steinhardt put it, "The string picture is beautiful in that one basic entity—string—can potentially account for the myriad elementary particles observed in nature. . . . Almost overnight, it seemed, the focus of research shifted from particles to strings."[10] While the bits of string are much too small to be seen directly, if string theory is correct, their properties will describe all known and knowable particles, including those that correspond to the four forces.

There was, however, a price to be paid for this triumph. String theory is highly mathematical in a way that would make Pythagoras smile. Its levels of abstraction reach far higher than those of general relativity or ordinary quantum physics and require a serious leap of theoretical faith. String theory can recover known particles and forces only if the universe has more space then we perceive, implying a cosmos with so-called extra dimensions. In addition to the three dimensions we are familiar with, height, breadth and width, string theory requires the universe to possess seven additional dimensions of space.[11] The equations fail with anything less.

Why can't we move into these seven additional directions? String theory answers that question by curling the extra dimensions up on themselves (physicists say the extra dimensions are "compactified"). To picture this, take a sheet of paper and roll it up into a superthin straw. Viewed from far enough away, the 2-D sheet of paper appears to have become a 1-D line. This is how seven extra dimensions of string theory are "hidden". Each extra dimension loops around on itself on such tiny scales that we can't experience them (though physicists hope measurable effects of this dimensional looping may exist).

For some physicists this multiplication of dimensions was too strange. The invisible, extra dimensions became reason enough to re-

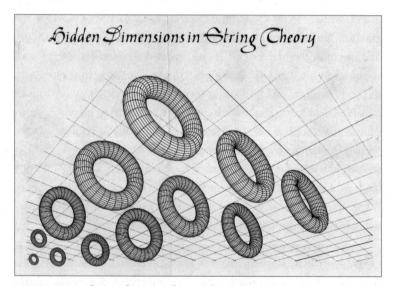

FIGURE 9.2. String theory and extra dimensions. String theory demands that the universe have seven extra, "hidden" dimensions of space. We would not experience these extra dimensions, even though they exist at every point in our 3-D space, because they are curled up on themselves.

ject string theory. For others, it was a small price to pay for such a large step towards a quantum theory of gravity.

Things got muddied again after the first string revolution. Theorists had realized that there was not one but five distinct ways to construct a string theory. Then in the mid-1990s a second revolution occurred as physicist Ed Witten showed that all five versions of the theory were just images of a deeper, unified construct he called M-theory. (He never said what M stood for, though others took it to mean "mystery" or "mother of all theories".) A critical feature of this second revolution was the appearance of a new set of actors on the string theory stage—the multidimensional branes (or membranes).

In the exploration of string theory's remarkable mathematical terrain, physicists found that in addition to vibrating one-dimensional strings, multidimensional vibrating membranes were also possible. A 2-D example of a membrane would be the tightly stretched skin of a drumhead. Physicists call this a 2-brane. Higher-dimensional branes (called p-branes, where p is the number of dimensions) could exist as well. These branes became fundamental players in string theory's eleven-dimensional universe (ten space dimensions + time dimension). Physicists explored a cornucopia of hyperdimensional geometric possibilities with wiggling strings and branes colliding to produce sprays of oscillations that would, if the theory is correct, appear to us as particles both known and unknown.

In 1999 Brian Greene published his wildly popular book *The Elegant Universe* and string theory became a household word popping up in everything from TV programmes to pop songs. Articles, books and websites on the theory multiplied, all dedicated to articulating, elucidating and/or obfuscating its meaning for the general public. But by this time, the string tide had already swept across the world of physics. Though many detractors remained, string theory came to dominate the landscape of theoretical fundamental physics. Cosmology could not help entering the string theory revolution too. For two veteran theoretical physicists, string theory's core ideas would serve as inspiration for making a complete break with the past, severing all ties with both Big Bang cosmology and inflation.

WHEN WORLDS COLLIDE: A STEP
TOWARDS A NEW CYCLIC UNIVERSE

For Paul Steinhardt and Neil Turok, the Big Bang ended on a summer day in 1999. The two scientists were sitting together in Cambridge, at a conference they had organized on string theory and cosmology. String theory was well into its next revolution and hopes were riding high that a true theory of quantum gravity was at hand. Given the field's progress, Steinhardt, Turok and others felt the time was right to explore string theory's overlap with cosmology.[12]

Were strings a way to power inflation in the natal universe? Could string theory explain how the twenty constants in the standard model were fixed and the universe fine-tuned for life? Most important, could strings and branes offer a way to get inside the singularity and explain the origin of the universe? These and other questions formed the raw material of the conference.

Steinhardt and Turok were both seasoned veterans of the particle physics/cosmology overlap. Steinhardt, a professor at Princeton, was responsible for important early work on inflation.[13] Turok, at Cambridge, had built a career exploring particle physics and its consequences for cosmic evolution. Both came to the conference with doubts about cosmology's dominant theory for the early universe. By 1999, Steinhardt was beginning to harbour serious questions about inflation even though he had helped establish the theory. It had grown unwieldy, and he was bothered by its continuing inability to explain the beginning. Turok had never fully embraced inflation. There were too many unanswered questions about the fundamental mechanism powering hyperexpansion to make him comfortable. It was with this mind-set that Steinhardt and Turok came to be sitting at opposite ends of a lecture hall when the same alternative to time's origin appeared to them both, separately and suddenly.

Steinhardt and Turok were listening to a lecture on brane-worlds by Burt Ovrut, a physicist from the University of Pennsylvania. Ovrut's goal was to explain why gravity appeared so weak compared to the other three fundamental forces. His work began with string theory's concept of extra dimensions but went a step further. For Ovrut, one of these extra dimensions might not be curled up at all. If the extra dimensions were uncurled and extended, they would allow more space and more possibilities for his physics.

Ovrut was exploring a model in which our entire three-dimensional world—including all the protons, electrons, galaxies, planets and people—constitutes one brane (a 3-brane) in a space with more than three dimensions. Other 3-branes can exist in this hyperdimensional universe and they might have their own versions of protons, electrons, galaxies, planets and (perhaps) people. Thus our 3-D universe is just one 3-brane in a higher-dimensional space (called "the Bulk"). Ovrut concentrated on a model with two of these brane-world separated from each other in the "extra" dimension, like two sheets of newspaper hanging parallel to each other. The point of this theoretical exercise was to show that three of the four known fundamental forces could be made to "live" only *on* the 3-brane. Only gravity, the fourth force, would extend throughout the entire Bulk. Forcing electromagnetism and the strong and weak nuclear forces to operate only within the three-dimensional membrane was like forcing ants to move only along a 2-D sheet of paper, without being allowed to jump off the sheet. Ovrut's theory of the universe conspires to force our experience on (actually, in) the 3-brane even though there is more space in the extra dimensions. Gravity, in Ovrut's theory, was the only force that filled all of this hyperdimensional space, including the space between the world branes. By diluting gravity throughout the Bulk, the concept could offer an explanation for its weakness as a force.

Listening to Ovrut's lecture from opposite ends of the lecture hall, Steinhardt and Turok were both hit with the same thought: what if the branes could move? Rushing to the podium after the lecture, Steinhardt and Turok peppered Ovrut with questions and quickly recognized that each had the same idea. Could a collision between the brane-worlds mimic a Big Bang? It was one of those flashes of insight scientists live for. This simple idea would launch a collaboration that would produce one of the more thoroughly explored alternatives to the Big Bang: a hyperdimensional cyclic model.

The heart of Steinhardt and Turok's insight lay in the fact that each universe—that is, each 3-brane—exerts forces on its neighbour. As they

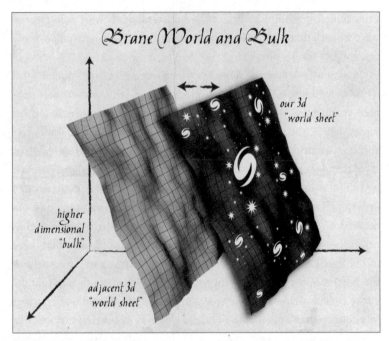

FIGURE 9.3. Branes and cosmology. Our 3-D universe as a sheet or "membrane" in a higher-dimensional space (called "the Bulk"). Other 3-D universes might exist as well and be separated from us by a short distance in one of the extra dimensions of the Bulk.

listened to Ovrut's talk, both physicists imagined what would happen if the branes could move towards each other along the direction defined by the extra dimension. As the two branes drew closer, the "interbrane" forces would become stronger, releasing apocalyptic reserves of energy. Could the eventual collision between the separate 3-D universes—defined by branes—act just like a Big Bang? Could it produce all the cosmic evolution we now associate with that Big Bang? In essence, Steinhardt and Turok hoped to substitute the collision between the branes for the mysterious singularity of classic Big Bang cosmology.

The picture we had in mind was of two widely separated, parallel branes stretching to infinity in three directions. A tiny force existed between the two branes, causing them to attract and move very slowly toward each other along the fourth dimension over a long, perhaps infinite period of time. The force grew ever stronger as they approached, speeding their motion toward the collision. At the bang, the kinetic energy of the branes would be converted into hot radiation.[14]

If the idea worked, there would be no need for the singularity. More important, there would be no need for a beginning. They would have a cosmological theory of "before".

The devil would lie in the details. Luckily, the mathematical machinery for working with the brane collisions was exact enough to allow for fairly explicit calculations. The deeper Steinhardt and Turok plunged into the nuts and bolts of their model, the greater the number of happy surprises that awaited them. One of the most important results to fall out of their calculations was the production of a cosmic microwave background with exactly the right spectrum of bumps and wiggles.

As the world-branes pulled towards each other, the force driving the collision created space-time ripples "like bedsheets waving in the wind". The origins of the ripples were fundamentally quantum mechanical. As the sheets drew closer, quantum mechanics and its inherent probabilities demanded that some parts of the sheet should feel a slightly stronger force while others would feel a slightly weaker force. The consequences were startling—some parts of the branes (the peaks of the ripples) would collide before other parts. As energy was released in the collision, the rippling branes naturally created regions of hotter, lower densities and cooler, higher densities. Thus, the seed for hot and cool regions seen in the CMB was laid down in Steinhardt and Turok's model not by inflation after the Big Bang but by forces occurring before their brane-world collisions.

Steinhardt and Turok then showed how the story of cosmic evolution on our 3-D slice of the universe played out exactly like the classic Big Bang story. The 3-branes rebound from the collision, each one moving apart from the others along the extra dimension. But the collision releases so much energy that the space *within* each brane is set into intrinsic expansion, just as in classic Big Bang theory. The ripples imprinted by the brane collision are carried along with the expansion, becoming perturbations in density and temperature. Thus, the physics within our 3-D universe go on just as they would in the standard Big Bang theory. Nucleosynthesis builds up the light elements. The CMB photons decouple from matter. And just like in the standard Big Bang model, the gravity working within our 3-D space eventually grabs hold of the ripples and turns them into galaxies and clusters of galaxies.

With a nod to the Stoics, the two scientists named this first version of their model the ekpyrotic universe—meaning "born in fire". There were no cycles yet. They just imagined that the branes slowly approached each other, perhaps over an infinite past. Everything focused on the one collision. The remarkable conclusion of their calculations was that this brane-world collision could account for all the features of the post–Big Bang universe just as well as the Big Bang.

BANGS WITHOUT BEGINNING: DARK ENERGY AND THE CYCLIC MODEL

When the ekpyrotic model was launched into the world of physics, it met with a mixed reaction. Many scientists were heavily entrenched in inflationary cosmology and saw no need to look elsewhere. Others thought it was too exotic an alternative to be of much use. In 2001, one particular criticism stung Steinhardt and Turok hard. The ekpyrotic model began with perfectly flat, perfectly smooth brane-worlds. This was necessary if forces between the branes were to imprint space-time ripples of the right form to recover the observed CMB perturbations. Preexisting bumps and lumps would ruin the calculation. Critics rightly argued that this was an unnaturally stringent initial condition. In the wake of the critique, Steinhardt and Turok were left looking for a mechanism that would naturally smooth out the brane-worlds as they approached their collision.

There was a critical difference between 1999, when the two scientists first thought of their brane-world collision idea, and 2001, when they confronted the problem of brane smoothness. In those three years, scientists had absorbed the shock of cosmic acceleration and the dark energy it implied. But dark energy did not fall naturally out of inflationary cosmological models—it remained a new fact that everyone had to deal with. As they mulled over their brane-world smoothing dilemma, Steinhardt and Turok realized that dark energy might not be an unwanted intruder into cosmology. It might be a gift.

In the wake of dark energy's discovery, inflation suddenly needed two forms of invisible unknown energy: the first to drive inflation and the second to drive the cosmic acceleration we see now. That was a problem for inflation. Steinhardt and Turok, however, soon recognized that dark energy could do double duty in their own theory if they made a simple addition to the story: allowing ekpyrosis to happen more than once.

The two physicists already knew that the space within each 3-D universe would be set into expansion after the brane-world collision. Once they added dark energy, that expansion was accelerated, stretching space so dramatically that it became, over eons, very empty and very smooth. Thus, dark energy gave Steinhardt and Turok the kind of space they needed to make ekpyrosis work, but only if they could force the branes back together. Here, dark energy enters the picture in a second way. Inside the 3-D universes, dark energy makes space accelerate. In the space between the 3-branes, however, Steinhardt and Turok saw how to make dark energy an attractive force that could draw the 3-D "sheets" back together. Using dark energy in this way forced the second brane collision, which then would be followed by another rebound, as in the original ekpyrotic model. Like a spring strung between two plates, dark energy keeps the brane-worlds colliding over and over again, even as it keeps the space inside the branes expanding forever. For Steinhardt and Turok, dark energy not only became the key to solving their smoothness problem but also pulled the entire vision of the model together, creating an entirely new version of the cyclic universe.

Steinhardt and Turok's new ekpyrotic cyclic model avoided all the pitfalls of the previous incarnations of the idea. The Mixmaster disaster never occurs because dark energy produces high pressures, keeping the space inside the branes and between them from undergoing any wild gyrations. The entropy dilemma, which killed relativistic cyclic models, is also avoided due to the extra dimensions of the new model. Closer examination of Tolman's calculations show that it is not entropy itself but the amount of entropy in every cubic centimetre of space (that is, the density of entropy) that dooms eternally recurring cycles. If 3-D space were to contract, then all the entropy would indeed be squeezed into a tiny space, and its density would drive the next bounce to ever larger sizes and ever longer durations. But in the new cyclic model, the 3-D space inside the branes where matter, radiation and entropy live never contracts. It is the space between the branes, the extra dimension only known to the Bulk, that contracts. Since the 3-D space inside the branes is continually and eternally expanding, the entropy generated in each bounce is continually diluted.

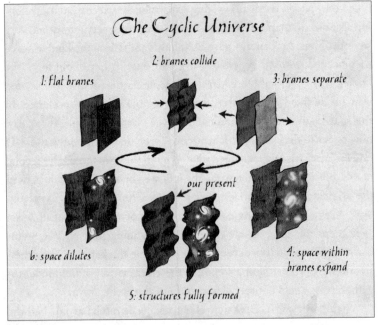

FIGURE 9.4. A cyclic brane world cosmology. Steinhardt and Turok's cyclic model in which two branes (one of which is our universe) collide, draw apart and collide again. Each cycle takes trillions of years.

In Steinhardt and Turok's views, all the problems that killed previous oscillating universe models were solved naturally within the framework of the model. No theoretical bells and whistles were needed. Whatever questions arose, the new cyclic model was always poised to answer them. As the two scientists described it:

> The model did not need even one iota of change. It was only necessary to discover what was already there all along, waiting to be noticed. This eerie experience is typical. . . . Whenever a problem has arisen, it's turned out that model already contains the ingredients necessary to address it. Not a single new element has been added to the picture since it was first envisioned.[15]

This kind of "eerie experience" is what makes a good theory in a scientist's eyes. The new cyclic model's prescience in dealing with problems convinced Steinhardt and Turok they'd stumbled upon a valid theory for the universe that had nothing to do with bangs or beginnings.

CONFRONTING REALITY: INFLATION, THE CYCLIC MODEL AND DATA

Attempts to imagine alternatives to the Big Bang did not end with the discovery of the cosmic microwave background. A small contingent of astronomers and physicists had remained unconvinced and determinedly worked on alternative cosmologies. As the decades wore on, the problem they faced was the growing mountain of evidence supporting the Big Bang's basic image of a universe expanding from a hot condensed state. To challenge the Big Bang, all those separate lines of evidence needed to be addressed: the expansion of space, the abundance of light elements, the existence of the CMB, the pattern of temperature and density fluctuations in the CMB, the distribution of galaxy clusters in space. Each line of evidence points to a universe that is evolving from a hot dense state, and each line of evidence is interwoven with the others. The distribution of galaxy clusters, for example, can be directly tied to the pattern of temperature and density fluctuations in the CMB, as lumps seen in the CMB are what grew into the clusters we see today. Any plausible alternative theory of cosmology must recover these connections in a natural way. But most models haven't been up to the task.

The brane-world, cyclic model of Steinhardt and Turok has, to an impressive extent, been an exception to this rule. It makes a series of concrete, testable predictions that stand up well to the data cosmologists already have in hand. Most of these predictions are the same as those for inflation. Like inflation, the ekpyrotic cyclic model predicts a universe with flat space.[16] And, like inflation, it anticipates a tight relationship between CMB fluctuations and the distribution of galaxy clusters. Thus, inflation and the ekpyrotic cyclic model appear on equal footing. As Steinhardt and Turok put it, "Despite the basic differences between the inflationary and the cyclic pictures, there are surprising mathematical symmetries that ensure the two agree very closely."[17]

This ability to address the majority of experimental tests in a natural way with no special fine-tuning makes the cyclic model unusual in terms of alternative cosmologies. In a sense, this is more than inflationary models yield. They must add dark energy and the inflationary quantum field as separate, unrelated entities rather than as natural parts of a unified model. There is, however, a future test using gravity waves that should be able to distinguish between these two very different stories of cosmic history. Gravity waves are moving ripples in the fabric of space-time. Like the ripples expanding away from a stone dropped in a pond, gravity waves are generated whenever mass-energy moves through the space-time background. While they have never been directly observed, exotic astronomical objects such as binary neutron stars provide indirect evidence for their existence.[18] Scientists are currently building sensitive detectors to pick up these waves from gigantic events nearby, like the merger of black holes. Space-time ripples are also expected to have been generated en-masse during the chaotic first moments of cosmic history. Predicting the form and strength of these waves constitutes a critical future test for all cosmological models. Inflation makes very different predictions for these fossil gravity waves than the cyclic model does. The detection of these waves is likely decades away, as it will require extended space-based instruments. But if and when the gravity waves are found, they will determine which vision of time and cosmic evolution must be abandoned and which can lay some claim to the truth.

RAJAGAHA, INDIA • SIXTH CENTURY BCE

The old Brahmin adjusts his tunic and waits for the class to settle down. The students are noisy and distracted, chatting among themselves amidst the wide-leaved plants of the garden. It was going to be a hot day in the city but the old man is willing to wait. He wants today's lesson to sink in before he sends them back to their families and their studies.

"Today," he begins, "I will tell you the story of Indra and the ants so that you might understand where and who you are in this great world."

Indra was the king of the gods. Brave, noble, possessed of a compassionate heart, he looked after both the divine and human worlds with the steady hand of a wise father. After defeating a great dragon that had destroyed the city of the gods, Indra called on Vishvakarman, master of the arts, to rebuild the great metropolis. Vishvakarman worked tirelessly and created shining palaces with marvellous gardens, lakes and towers. But Indra was not satisfied. "Give me bigger ponds, trees, towers and golden palaces!" he demanded. Whenever Vishvakarman was done with one thing Indra wanted another. The divine craftsman fell into a deep despair. In desperation he complained to the Brahma, the Universal Spirit, who abides far above the gods. Brahma comforted him: "Go home; you will soon be relieved of your burden."

Early the next morning a Brahmin boy appeared at the gate of the palace asking to see the great Indra. "O king of the gods, I have heard of this palace you are building. How many years will it take to finish this rich and extensive residence? Surely no Indra before you has ever succeeded in completing such a task."

Indra was amused by the boy. How could this child have known any Indras other than himself? "Tell me, child," he said, "how many other Indras have you have seen or heard of?"

The boy replied in a voice as warm and sweet as milk, but with words that sent a chill through Indra's veins. "My dear child," said the boy, "I knew your father. And I knew your grandfather. Also I know Brahma, brought forth by Vishnu from a lotus growing from Vishnu's navel. And Vishnu, too, the Supreme Being, I know.

"O king of the gods, I have seen the dreadful dissolution of the universe. I have seen it all perish again and again, at the end of each cycle. At that time every single atom dissolves into the primal pure waters of eternity, whence originally all arose. Who will count the universes that have passed away, or the creations that have risen afresh, again and again, from the formless abyss of the vast waters? Who will search through the wide infinities of space to count the universes existing side by side, each containing its own Brahma, its Vishnu and its Shiva? Who will count the Indras in them all?"

As they were talking a procession of ants had made its appearance in the hall. In military precision, the tribe of ants paraded across the floor. The boy noticed them and suddenly laughed. "See these ants in their long parade? Each of them was once an Indra. Like you, each by virtue of his deeds ascended to the rank of king of the gods. But now through many rebirths each has become again an ant. This army of ants is an army of former Indras."[19]

The king of the gods was speechless. The boy turned and left. After many days alone, Indra called his architect and thanked him for his work. "You have done enough," Indra said. "You may rest now."

The old Brahmin finished his story. He closed his eyes, took a deep breath and exhaled in meditative repose. His students, normally so boisterous and noisy, sat in stunned silence. A small smile appeared at the corners of his mouth. He had opened their imaginations, if only for a moment.

TIME CYCLES AND CREATION

A universe of eternal cycles, endlessly resurrected from the ashes of its own destruction, is a seductive vision that has been offered as a narrative in cosmic myth and cosmic science many times. Such a cosmos has no beginning and no end, just endless rebirths stretching from the infinite past to the infinite future. The difference between myth and science, however, lies in science's dual imperatives of theoretical consistency and empirical evidence. As attractive as Lemaître might have found the "phoenix universe" implied in his and Friedmann's general relativistic solutions, deeper theoretical explorations of the idea always led to disaster. The buildup of dense entropy from one cycle to another doomed cyclic models, as did the chaotic space-time flailing of a Big Crunch's Mixmaster phase. Classic general relativity combined with the quantum physics of matter was simply not a framework that allowed for the existence of a cyclic universe. Only by going outside classic relativity could Steinhardt and Turok's new cyclic model succeed. Only by embracing a string-theory-inspired hyperdimensional brane-world vision of the cosmos could they move past its pitfalls. To embrace the elegance of an eternally regenerating universe, the new cyclic model needed more space. But was this too high a price to be paid?

By the end of the twenty-first century's first decade, enthusiasm for string theory and its revolutions had waned for many physicists.[20] Earlier in the decade, string researchers had discovered that their still-emerging theory would not lead to a unique specification for cosmic evolution. Instead of a single all-embracing string theory prediction for what our universe should look like, there were many theorists who began to talk about a "landscape" of many universes.[21] In the new picture, our universe would not be uniquely determined by string theory but would become one instance of a vast range of possibilities. According to best estimates, the number of potential universes in the landscape would be around 10^{500}, which for all intents and purposes was about infinity. For some, string theory's apparent failure to explicitly predict the features of this one universe we inhabit was the last straw. Instead of a theory of everything, some critics began calling it a "theory of nothing".

Steinhardt and Turok's cyclic model never required the full machinery of string theory. But by beginning with the hyperdimensional brane-world idea, it accepted certain key assumptions about reality (the extra dimensions) that lay above and beyond what we directly experience.

Of course, string theory is not the only way to create a route to quantum gravity. Other ideas provide rich and fertile ground for theorists to explore. These other paths have not received as much attention or effort as string theory but progress has been made. A theory called loop quantum gravity is perhaps the best-known of these.[22] Loop quantum gravity seeks to work out the details of quantized space-time in the form of elemental atoms of existence. Originally developed by researchers such as Abhay Ashtekar, Ted Jacobson, Carlos Rovelli and Lee Smolin, the field has grown more slowly than string theory. In the wake of the general disappointment over string theory's perceived failings—that is, the string landscape—more attention is now being paid both to loop quantum gravity and to other routes to quantum gravity. Loop quantum gravity in particular has even matured to the point where researchers have begun looking at its cosmological alternatives. Loop quantum cosmologists such as Martin Bojowald of Pennsylvania State University have even found evidence for their own version of cyclic models, as atomized space-time may yield routes through the dreaded singularity of a Big Crunch. Bojowald's popular book *Once Before Time*, published in 2010, offered a layman's description of loop quantum cosmology.[23]

There are other versions of cyclic cosmologies as well. In 2009, University of North Carolina physicist Paul Frampton published a popular book, *Did Time Begin?*, on his own version of a cyclic model. In the same year as Bojowald's book came out, none other than Roger Penrose announced what he claimed was evidence for previous cycles existing within the CMB. His ideas would also appear in a popular book called *Cycles of Time*.[24] Clearly a tide of new theories reaching back to before the Big Bang had begun to wash into the popular imagination from the domains of academic cosmology.[25]

Thus the ancient dream of endless cycles is very much alive as a possibility for tomorrow's cosmology. The Vedic and Stoic visions of creation followed by destruction, followed by creation again and forever, remains a very viable alternative. Whether it's Steinhardt and Turok's colliding brane-worlds or another form of quantum gravitational theory providing the framework, the ancient mythic call to cycles remains and cannot yet be discounted.

Chapter 10

EVER-CHANGING ETERNITIES:
THE PROMISE AND PERILS
OF A MULTIVERSE

Eternal Inflation, Arrows of Time
and the Anthropic Principle

LONDON · FEBRUARY 4, 1950, 8:50 P.M.

It was a cold night in the middle of a long cold winter. But the terrible tem-
perature outside made staying home and listening to the radio that much
more delicious. And it wasn't just any radio programme she was giving up
a Saturday night for. It was that astronomer Fred Hoyle and his wonder-
ful lectures on space and the stars and the whole universe. Fred Hoyle was
worth so much more than that skinny boy, Tommy McEwan, her friend Lilly
wanted to set her up with tonight.

    She was fifteen years old and, in spite of all the things everyone told her
to the contrary, she longed to be an astronomer. She devoured everything she
could find on physics and maths at her school's library and she never stopped
pleading with her parents for a telescope. When her teacher, the only person
who encouraged her scientific ambitions, told her about Professor Hoyle's
Saturday night BBC lectures, the radio became her new religion. Tonight
was one of Hoyle's last programmes. Her head was already spinning. He was
talking about the science of the entire universe, a field he called cosmology.
He obviously was not happy with a theory other scientists seemed to believe.
She loved the way he explained it all in his crooked Yorkshire accent:

The assumption [is] that the universe started its life a finite time ago in a single huge explosion. . . . This big bang idea seemed to me to be unsatisfactory even before detailed examination showed that it leads to serious difficulties.

After that Hoyle began talking about his own theory. He called it a steady-state model. The universe was expanding, just as that American Edwin Hubble had discovered, but the expansion had no beginning and no end. Even though space expanded, Hoyle said of his theory, the universe would always look the same. He asked her to imagine a film of the universe taken frame after frame over billions of years.

What would the film look like? Galaxies would be observed to be continually condensing out of the background material. The general expansion of the whole system would be clear, but . . . there would be a curious sameness about the film . . . The overall picture would stay the same because of the compensation whereby the galaxies that were constantly disappearing through the expansion of the universe were replaced by newly forming galaxies. The casual

FIGURE 10.1. Voice of the Cosmos. Fred Hoyle's BBC radio lectures on astronomy in 1950 captivated listeners. It was during his lecture on cosmology that Hoyle invented the name "Big Bang" as a term of derision. Hoyle favoured his own "steady state" version of cosmology.

observer who went to sleep during the showing of the film would find it difficult to see much change when he awoke. How long would a film show go on? It would go on for ever.[1]

Forever. The sound of it rang in her head. She threw a coat over her night-gown and ran outside into the frigid night air, her slippers crunching on the week-old snow. There weren't many stars to be seen from the back garden of their London council flat, but it didn't matter. The bright city lights could not drown out the beauty of the ones she could see. She stood looking at the black and the stars for a long time, superimposing her own vision of Hoyle's steady-state universe on them: galaxies expanding, new galaxies forming forever.

"Forever," she said out loud, watching the words turn to mist in the cold air. "How long is forever?"

THE LURE OF MULTIPLE UNIVERSES

Eternity can wear many masks. A universe running through endless cycles of creation and destruction offers one route beyond, and before, the Big Bang. It is an elegant, aesthetically pleasing solution to the dilemma of time's origin. But it is not the only solution.

The desire for a timeless universe that changes locally but is framed within a changeless cosmic architecture also holds an aesthetic appeal. From Aristotle to Newton to Einstein, philosophers and scientists have all been drawn to visions of a universe in eternal repose. But Hubble's 1930 discovery of an expanding universe, now coupled with the 1999 discovery of accelerating expansion, makes the old vision of stasis impossible. For those seeking to replace time's origin with eternity running forward and backwards, stasis must be replaced by a steady state—a universe that is always changing yet always looks the same.

Fred Hoyle's tireless attempts to create a steady-state cosmology in the 1950s were driven by this scientific aesthetic. But Hoyle and his collaborators' efforts were killed when the world spoke so clearly for itself. Material engagement in the form of astronomically tuned microwave antennas provided an unambiguous signal, selecting between competing cosmic histories. Once the cosmic microwave background was discovered, it became clear that the universe now looks nothing like the universe 13.7 billion years ago. The universe is not in a steady state. It is always changing and does not always look the same. Thus Hoyle's steady-state model could not provide a scientifically supportable theory for a universe free of time's beginning.

The siren call of eternity did not, however, disappear—it was waiting to find a new expression. It is no small irony that new possibilities for eternity and a steady-state universe appeared from within the very theory intended to save the Big Bang. Just a few years after the original formulation of inflationary cosmology, some of its most creative advocates found a way out of the Big Bang's paradox of before and after. Inflation became eternal inflation and the one universe became a multiverse.

As the last decade of the twentieth century ended, ideas once considered the domain of science fiction were put on sure mathematical footing. Multiple universes became the stuff of foundational physics. Stranger still, some scientists began to ask if the very existence of life might somehow play a role in selecting the nature of the cosmos we see. Another solution to the question of "before" the Big Bang, very different from cyclic models, appeared in the form of a theory called eternal inflation. Looking over the consequences of the new theory, however, many scientists saw it demanding a redefinition of scientific cosmology's fundamental purpose. Was it a solution with too high a price?

To understand eternal inflation and its universe of universes, the story must begin with inflation's beginnings—its mechanisms for making cosmic mountains out of infinitesimal molehills.

DEVILS AND DETAILS: MAKING INFLATION

Beginning with Alan Guth, cosmologists invoked inflation as a mechanism for taking a small piece of the infant universe and blowing it up to become the entire universe we can see. By stretching a sliver of reality by the gigantic factor of 10^{40}, all the paradoxes of classic Big Bang theory were resolved: space-time was stretched to flatness, regions that appear unconnected today shared an initial harmony at the beginning of time, unwanted magnetic monopoles were so diluted as to be unobservable.[2]

Inflation accomplished all this paradox-busting by filling the early universe with what cosmologist Sean Carroll calls "dark superenergy".[3] Inflation's titanic push, stretching space apart, worked in any region of the universe where this dark superenergy was active, hyperexpanding a speck of creation into everything we can see today. The trick to making inflation theory work, however, was getting the superenergy to start and stop at just the right moment.

Originally, Guth and others hoped to power inflation by using energy locked up in a quantum field associated with grand unification. In physics, a field is a physical entity that extends across space. This is the opposite of a particle, like an electron, that is localized in just one place at just one time. Fields became actors in physics as far back as the 1850s. The electromagnetic field reaching out across space from magnets and electrical charges is the archetype of a physical field. Learning how to describe fields in terms of quantum mechanics took some time, but by the 1940s and 1950s the main principles had been worked out. When the smoke cleared, scientists realized that quantum field theory even blurred the distinction between localized particles and extended fields. In this theory, all the particles we know of—the electron, the neutrino, and so on—are simply localized oscillations of an extended quantum field.

Guth and others hoped to tap the vast energies in the primordial GUTs quantum fields that must have filled the early cosmos and use them for inflation. One of these fields, they argued, could power inflation if it extended across post–Big Bang space in just the right way. This inflation quantum field (associated with some form of quantum particle) would fill space by creating an all-pervading energetic background. If the energy of this background—the false vacuum—could be transformed at just the right moment and in just the right way, then a small region of space could be driven into a fantastically rapid expansion. In the language of physics, a phase transition would occur. Like cold water turning into ice, the false vacuum of the inflation quantum field would transform into the true vacuum of empty space, releasing tremendous energies in the process. After the phase transition was complete, the residual energy of inflation would fill our newly expanded piece of the universe with matter and light particles, continuing on with the usual history of the hot Big Bang.

Making inflation work, however, demanded a description of the potential energy that was locked up in the false vacuum quantum field. We all have experience with potential energy in the context of gravity. A heavy ball, like a bowling ball, placed on a high shelf has more potential energy than the same ball lying on the floor. Knocking the ball off the shelf and letting it plunge to the floor converts the gravitational field's potential energy into the bowling ball's kinetic energy.[4] The urgency you'd muster to pull your foot away from a ball speeding down towards you is stark testimony to the reality of potential energy. Getting cosmic inflation to work meant finding a quantum field with the right form of potential energy. The field would have to drop from its false-vacuum latent-energy state to its true-vacuum low-energy state in just the right way, and at just the right time, to hyperexpand a small speck of post–Big Bang space-time.

The details depend on what physics calls the quantum field's potential energy curve, which describes how the potential energy changes when the field itself changes. Any bumps and wiggles in the potential energy curve will change the evolution of the field, in the same way that a ball rolling down a lumpy hill will speed up or slow down as it traverses small depressions. For inflation, the "depressions", called local minima, were critical, as they shaped the transition from the false vacuum to the true vacuum.

The success of inflation depends greatly on the shape of the quantum field's potential energy curve. Guth's original proposal for inflation was almost stillborn because the potential energy curve he used was taken from grand unified field theory and seemed to stop short of making the universe we see today. In Guth's first proposal, now called "old" inflation, the dark superenergy field began resting in a local depression

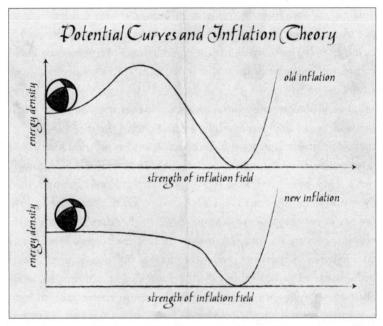

FIGURE 10.2. The potential energy curves for inflationary cosmology. These plots show different models for how the energy in the quantum field generating inflation change as the field itself changes. The potential curves determine how an inflating universe evolves. Some curves lead to a cosmos like ours and others don't.

somewhere high up on the potential energy curve. This was the false vacuum state. Any piece of the universe caught in this state would be inflating, growing and stretching.

Guth then imagined that small regions of space, whose fields were trapped in a local minimum, would spontaneously decay, like a radioactive nucleus, to a lower position on the curve, corresponding to the true vacuum of the universe. This would be a phase transition similar to what happens when bubbles of steam pop into existence within a pot of boiling water. Different parts of inflating space would make the transition at different times, leading to bubbles of true vacuum. Guth hoped that eventually all the bubbles of true vacuum would collide and merge to form one spatially expanded true-vacuum universe. But the timing didn't work out. Space in the false-vacuum state was still inflating so fast that new bubbles of true vacuum would be swept apart too quickly to interact. Trying to tweak the theory so bubbles of true vacuum could form faster and collide only forced inflation to stop before it could work its full magic.[5]

The problems with Guth's idea were fixed a few years later when theorists began abandoning explicit grand unified theories and realized that other kinds of quantum fields could be used as inflation fuel. These quantum fields were not associated with any known GUT's field (or particle). They were inventions, obeying the basic rules of quantum field theory. Relinquishing the GUT's framework was not considered a loss. No one knew what kinds of quantum fields might exist in the early, early universe because no complete theory for that era existed. Most physicists felt free to construct models of possible inflation quantum fields with associated particles they called "inflatons". These alternative models did not begin with the inflaton field stuck in a false-vacuum local minimum; instead they allowed it to begin at the top of a potential-energy hill and slowly roll down to the true-vacuum state. A ball rolling down from the crest of a hill after a random nudge was the analogy physicists were using to construct their new versions of inflation. By assuming that the initial high point in the curve was unstable, physicists realized, a small nudge would be all that was needed to start the inflaton rolling down its potential curve. The extended roll was important for the theory because it kept space in the false vacuum state long enough to form a local bubble of true vacuum.

This "new" inflation overcame the timing issue plaguing Guth's original model and allowed it to yield the observable universe with all the properties we now see. New inflation was a spectacular theoretical success and soon won over many converts. Different versions of the basic theory rapidly multiplied as scientists explored various potential energy curves for the inflaton and their consequences for cosmic evolution. But it was not long before people started thinking about the rest of the universe—the part that did not make the transition from false vacuum to true vacuum. What about these other regions of cosmic spacetime that remained in the false vacuum state?

In the late 1980s physicists began exploring inflation's consequences for the entire post–Big Bang universe, and not just the little speck that grew into our cosmic home. Their conclusions expanded the meaning of inflationary cosmology and the very definition of the universe it was meant to explain.

INFLATION BECOMES ETERNAL: MAKING THE MULTIVERSE

Inflation, in its way, is an elegant solution to the paradoxes haunting Big Bang cosmology and particle physics. Fill the early universe with dark superenergy to hyperexpand space just when needed, then sit back and watch all the problems with monopoles, causality and so on wash away. But the cost of inflation's solution to the paradox was a recognition that the Big Bang created more universe than what we see. Just as important was the realization that the other regions of creation existed in an entirely different state than our slice of the pie did. Once that genie—the rest of the universe—was let out of the bottle, physicists found it difficult to force it back in.

Alexander Vilenkin and Andrei Linde were two of the first physicists to begin thinking about the rest of the universe under the auspices of inflation.[6] Both scientists began exploring the relationship between the parts of the post–Big Bang universe that had decayed to true vacuum and the parts that continued to inflate in the dark-superenergy-sodden false-vacuum state.[7]

If a ball is resting exactly at the top of a hill, all that is needed is a gentle breeze or the tremor from a passing lorry to send it rolling down. With inflation, it is quantum mechanics that determines when a transition up to false vacuum occurs and when the fall back to true vacuum follows. Inflationary cosmology uses the random fluctuations associated with quantum physics to kick a small region of space up to its false-vacuum perch and then back down to the true-vacuum state lower on the potential energy curve. The random quantum kicks come about once every 10^{-33} of a second. In some regions of space, the random kicks will build up enough to push the inflation field out of its false-vacuum state and send it "rolling" downhill to the true vacuum. Exploring this process in more detail, Vilenkin realized that inflating space always won out over noninflating space in terms of sheer volume. Once inflation began, there would always be some parts of the universe left in the false vacuum. Inflation, in other words, just kept going. It was eternal.

Together Vilenkin, Linde and others began working out the consequences of this eternal inflation idea. Through their work, a new vision of cosmic history and cosmic cartography was formed. If inflation continued even after our observable universe had formed, then what was to stop other small regions of space that were still inflating from making their own transitions to the true vacuum in their own sweet time? The answer was nothing. Under eternal inflation, different regions of space make their own phase transition from false vacuum to true vacuum, each creating a "pocket universe" like our own in the process. Each pocket universe is, in general, cut off from the others because the inflating space in between carries them apart at such tremendous speeds. It is worth noting that while nothing travelling through space can exceed the speed of light, this speed limit does not apply to how fast space itself can expand. Thus, the different pocket universes are carried out of causal contact with each other even though they are all part of the same grand, infinite space-time fabric.

By allowing different regions to drop into the true vacuum on their own, eternal inflation makes a radical prediction about the universe— that there is no longer just one.

If inflation continues forever, then true-vacuum pocket universes are constantly forming from the background of false-vacuum hyper-expanding space. All these universes, together with the inflating space between them, form a new entity: a multiverse. The multiverse is a universe of universes. Once a pocket universe like ours forms, then its history is no different from what occurs in the standard post–Big Bang model. Space expands, matter cools, nuclei form and the CMB eventually fills the observable sky. As Vilenkin stated:

Because of inflation, the space between these islands rapidly expands, making room for more island universes to form. Thus, inflation is a runaway process, which stopped in our neighborhood, but still continues in other parts of the universe, causing them to expand at a furious rate and constantly spawning new island universes like ours.[8]

But there is more to the story. Not only does eternal inflation predict other universes forming in the multiverse, most of them won't look anything like the one we live in.

There are many ways for a small shard of space to make a transition from the false vacuum. Once inflation ends, the energy of the false

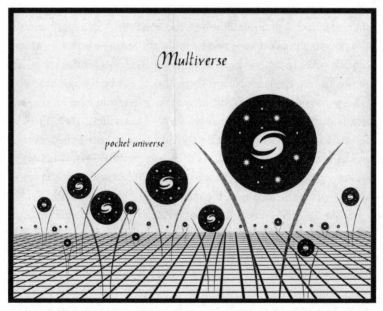

FIGURE 10.3. Eternal inflation and the multiverse. Eternal inflation assumes that many different regions of space-time as a whole can make the transition from false vacuum to true vacuum. Thus a multiverse of "pocket universes" arises. Each pocket universe can have its own version of physics.

vacuum is freed up to fill each pocket universe with particles, allowing normal cosmic evolution to proceed. But the exact form of those particles, and the forces they obey, will vary from one pocket universe to the next. This is similar to ice forming in freezing water. Each ice crystal gets its own orientation from the random motions of water molecules in that region. Before freezing, the water looks the same from one region to the other, but in the phase transition that constitutes ice formation that symmetry is broken. Ice crystals forming on one region of a pane of glass will not be aligned with crystals forming on another region.

Drawing on the analogy of freezing water, eternal inflation theory predicts that each pocket universe might form with different physics. The initial symmetry and unified physics of the false vacuum are broken during the transition to the true vacuum state inside a pocket universe. In some pockets there might be no electromagnetism, while in others electromagnetism might be a thousand times stronger than it is in ours. Some pockets have two forms of the strong nuclear force and no weak force. In others it will be reversed. The possibilities are nearly infinite. In fact, the landscape of string theory can be brought into play at this point, providing 10^{500} different ways to go from false to true vacuum. If string theory is the correct theory of everything, then it will be the foundation for inflation. The 10^{500} dimples in string theory's own "energy landscape" of possibilities then becomes 10^{500} possible laws of physics for the pocket universes. Every possibility is a pocket universe existing within the totality of the multiverse.

From the 1990s onward, eternal inflation gained recognition as an almost inevitable consequence of plain old inflation. While there were ways to construct inflation theory so that *all* space-time made the transition from false vacuum at once, these types of inflation were not its most general form. Left to its own devices, an inflating cosmos appeared likely to become an eternally inflating cosmos complete with separate pocket universes and different laws of physics. In this way, *multiverse* was added to the lexicon of cosmology. Entire conferences attended by sober physicists were dedicated to the theoretical exploration of the multiverse's properties. And while scientists spoke of pocket universes with different laws of physics, the goal was always to find the meta-laws (such as string theory) that would govern the multiverse as a whole. Through it all, recognition of a shift away from the Big Bang was becoming apparent. The eternity implied in eternal inflation seemed to nullify time's beginning.

Infinity in the past *and* the future had found its way back into cosmology.

PIERCING THE MULTIVERSE'S ARROW OF TIME

"The real problem is not the beginning of time but the arrow of time,"[9] said Sean Carroll, a theoretical physicist at Caltech in Pasadena. Carroll is a prodigious thinker on cosmological issues both inside and outside academia, and from his point of view, the problem has never been what came before the Big Bang; it's the very notion of "before" and "after." "The thing we really have to understand," he says, "is why cosmic time has a direction at all."[10] For Carroll, the multiverse of eternal inflation offers a new form of the steady-state model writ large.

The arrow of time is one of deepest problems in physics. Big Bang cosmology cannot escape its conundrums. The problem is simple and lives at the interface between the second law of thermodynamics and the rest of physics. Ever since Newton, the equations describing physics have been time-reversible. The laws governing individual objects do not care about time's direction. Imagine a film of two billiard balls colliding in space. There is really no way to say if the film is being run forward or backwards. The same is true of two atoms colliding. The equations of quantum physics, like Newton's equations, do not have a direction of time built into them.

If, however, you mix a zillion atoms together into something like a dozen eggs, then everything changes. Suddenly past and future look very different. As everyone knows, you can't unscramble an omelette made with those eggs. With large collections of atoms, thermodynamics enters the picture and with it comes that all-important quantity, *entropy*.

We have already encountered different ways of thinking about entropy, including as a measure of a system's disorder. The flip side of this definition is to see how entropy is directly related to the number of ways a system's parts can be arranged. Together these definitions link the important second law of thermodynamics to a direction of time.

Pop a balloon in a big empty room and the atoms will fill the entire space. Heat up a cup of coffee and it will cool down to the temperature of the room. In both cases, the system slides through changes until an equilibrium is reached. But in physics, equilibrium and maximum entropy mean the same thing, and both are related to the system achieving a state of maximum disorder. What does maximum disorder mean? It's just another way of saying the maximum number of ways for arranging the system's parts. There are, for example, only a few ways to get atoms squeezed into a corner of the room. The atoms have to be packed in close and there is not much room between them. There are, on the other hand, a lot of ways to distribute the same atoms across the entire room. So the path from low entropy (all atoms in one corner of the room) to maximum entropy (atoms filling the room) defines the arrow of time. The route to equilibrium discriminates what is *before* from what is *after*. Once equilibrium is reached, nothing changes, and the difference between past and future has no meaning.

With this definition in mind, the Big Bang's entropy problem can be simply stated: our universe has been evolving for thirteen billion years, so it could not have begun in equilibrium. Somehow, all the matter—energy, space and time—in the universe began in a state of low entropy. That is the only way change and evolution could have occurred. That is the only way we could start with a Big Bang and end up with the wonderfully diverse cosmos of stars, planets and people we find today.

The idea that the Big Bang had low entropy might seem strange at first. It was a hot, chaotic mess, right? How could the empty, cold universe we find today, a universe full of stars, planets and mosquitos, have higher entropy? The answer is gravity. The universe is not a box of atoms; it is an expanding space-time full of gravitational effects.[11] By including gravity in calculations, the movement from a smooth primordial soup of particles to the clumpy universe of galaxies we see today can be understood for what it is—an increase in entropy. But you don't have to work through these details to make the connection between evolution, entropy and equilibrium. If the universe evolved (as it did and still does), then it must have started with lower entropy than it has now.

Once you recognize that point, as mathematical physicist Roger Penrose did in the 1970s, you have a problem. If entropy is equivalent to the number of ways to arrange a system, then the low-entropy beginning suggested by the Big Bang is remarkably unlikely.

If you had a bag filled with a million black marbles and only a few white marbles, you would be pretty surprised to reach into the bag and pull out a white marble. That is the situation facing cosmologists trying to explain the low-entropy Big Bang. "There are, literally, an infinite number of ways to set up the initial universe", says Sean Carroll, "and pretty much all of them have high entropy." But these numerous high-entropy universes would begin in equilibrium, making evolution and change impossible.

The probability that our universe began in a low-entropy state is so astonishingly unlikely that it's almost embarrassing. Without a compelling explanation, physicists have to admit our Big Bang must somehow have been fine-tuned. It began improbably far out of equilibrium and can, therefore, point the arrow of time in one direction. We have encountered fine-tuning before, in the standard model of particle physics. Recall that the twenty or so constants needed in the standard model are amazingly sensitive. If nature had "chosen" even a slightly different value for these numbers, the universe would look so different that life as we know it could never form. The low-entropy Big Bang requires a similar kind of fine-tuning. With so many ways to make a universe begin with high entropy, how did we end up in one that began with low entropy? For physicists, fine-tuning is akin to saying that a miracle occurred—which doesn't sit well with most scientists. The beauty of science's approach to the world is its emphasis on plausible and purely physical explanations, even in a field like cosmology. So for Carroll and others the real problem is not just explaining what happened before the Big Bang but also explaining our universe's unlikely low-entropy origin, which tipped time's arrow forward.

"Our universe" is the key phrase that Carroll and others lean on for their solution. If our universe is one of many, then a new kind of steady-state model becomes possible in which even the arrow of time is explained. Eternal inflation's multitude of universes gave Carroll the raw material he needed to create a time-symmetric cosmology. As Carroll explains it:

> People immediately asked if eternal inflation could work in both directions. That means there would be no need for a single Big Bang. Pocket universes would always sprout from the uninflated background without beginning or end. The trick needed to make eternal inflation work is to find a generic "starting point". This would be an easy-to-achieve condition that would occur infinitely many times and allow eternal inflation to flow in both directions.

It's important to see exactly what Carroll means by time symmetry. In any particular pocket universe, the local arrow of time will simply run from low entropy to high entropy. For someone living in that universe, the past points to the time of low disorder and the future points towards the direction of higher disorder. What Carroll is looking for is time symmetry for an entire multiverse. From a god's-eye view, the entire web of evolving pocket universes should look the same if the film of its global evolution ran backwards or forward.

In 2004, Carroll and his student Jennifer Chen found exactly that— a version of eternal inflation without any global arrow of time. "All you need", said Carroll, "is to start with some empty space, a shard of dark energy and some patience." As counterintuitive as it sounds, detailed calculations show that an empty space-time driven to expansion via dark energy is the one with the highest possible entropy. Empty, flat space-time is, therefore, the condition a generic universe will evolve towards. It is the most generic form of cosmic equilibrium—the one you'd expect to occur for a randomly chosen universe with a randomly chosen starting point. It's also the perfect generic starting point for Carroll.

A background of dark energy in an empty, flat space-time is also crucial for Carroll and Chen because quantum physics states that any energy field produces random functions. These rapid, quantum spikes in the dark energy can serve as the trigger for the next step in their story. Wait long enough and strong fluctuations can momentarily push some tiny region of the empty background up inflation's potential energy curve into a false vacuum state. The result is a new region of inflation creating new crops of baby universes from the empty space. Note the term *baby universe* is used rather than *pocket universe*. Even if the details of quantum gravity remain unknown, physicists expect that quantum fluctuations can give rise to separate, disconnected domains of space-time. Carroll used this possibility to go further with his multiverse than the usual ideas of eternal inflation. Baby universes literally tear themselves away from the space-time of their parents and, in doing so, begin in a low-entropy state. By the nature of their creation, the baby universes begin with low entropy and can go on to inflate and evolve on their own.

"Some of these baby universes will collapse into black holes and evaporate, taking themselves out of the picture", says Carroll. "But other universes will expand forever. The ones that expand eventually thin out. They become the new empty space from which more inflation can start." Since the ever-expanding universes evolve towards the generic empty-space condition, the whole process is renewed again and again.

More important, the direction of time does not matter in the process. "That is the funny part", says Carroll. "You can evolve the little inflating universes in either direction away from your generic starting point." When Carroll speaks about different directions in time, he is taking the god's-eye view, looking at the evolution of different individual baby universes from the point of the multiverse as a whole. "You can do [inflation] going backwards in time from the initial state; the generic evolution is the same. The universe will empty out and eventually begin to spontaneously inflate. So in the super-far past of our universe, before our Big Bang (which is nothing special), we will find other Big Bangs for which the arrow of time is running in the opposite direction." On the largest scales, the entire cosmos is like a connected foam of baby universes, which are completely symmetric with respect to the multiverse's time. There is no direction of time for the multiverse as a whole.

There is a compelling vertigo in Carroll and Chen's solution. Their multiverse always exists and always will exist. It is dynamic and evolving but is, in a statistical sense, always the same. The Big Bang is just *our* Big Bang and not unique. We define it as our past because that is the direction where entropy was lower. But limits on the entropy of the whole multiverse are never reached because more baby universes can always be created and inflated. New universes flow continually into the multiverse's past and future. The distinction loses any absolute cosmic meaning. The multiverse stands outside of time because there is no universal flow of its time in any particular direction. The question of "before" has not just been answered, it has been overwhelmed.

From Carroll's perspective, eternal inflation's multiplication of universes may offer a way around both time's beginning and its arrow. But for many scientists, eternal inflation of any form has not been a welcome development. The multiplication of universes (all unobserved and per-

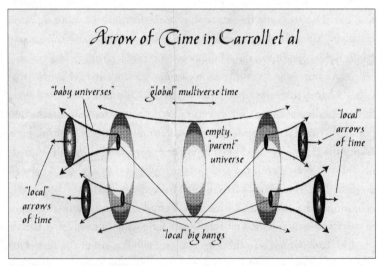

FIGURE 10.4. A steady-state time symmetric multiverse. Beginning from the generic condition of flat space, separate "baby universes" can be created. Each baby universe will have its own thermodynamic arrow of time. The multiverse as a whole, however, need not have a global direction of time.

haps forever unobservable) seems more akin to a *Star Trek* episode than to serious science. But for others the universe's radical redefinition of the cosmos not only answers cosmology's oldest questions but pushes the definition of science in new directions.

OUR MEDIOCRE UNIVERSE: THE ANTHROPIC PRINCIPLE MAKES NEW CONVERTS

Fine-tuning is a real problem for physicists and cosmologists. From the twenty constants needed for the standard model to the Big Bang's demand for a special, low-entropy initial conditions, the imperative for the universe to be built "just so" (down to many decimal places) for life to form has haunted the project of fundamental physics for decades. The goal of physics had always been to find timeless laws that specify the exact form of nature and its evolution. But the more scientists probed the universe, the more they saw happy accidents picking out special conditions and values of constants. All these accidents appeared in just the right form to create a cosmos where life could grow. It was a dilemma sure to please deity-happy advocates of intelligent design. With fine-tuning, they could claim physics itself provided evidence for a superintelligence turning knobs on his creation, dialing in exactly the right conditions (out of an infinity of possibilities) to make this implausible cosmos. Such an easy way out would be an anathema to the project Thales, the creator of the Greek rational tradition, began 2,500 years ago. For many scientists the multiverse, with its infinity of universes, seemed to provide a clear explanation for the dilemma. But to fully invoke its power, scientists would have to confront the dreaded *A*-word of cosmology: the Anthropic Principle.[12]

The Anthropic Principle has been hovering in the background of cosmological thinking for decades.[13] In its simplest form, it states that the universe and its laws must take a form consistent with our existence within it. This may seem like a tautology at first—so obvious it's not worth stating—but over the years, some physicists and cosmologists have been keen to show how the Anthropic Principle could be used to make cosmological predictions. In the 1950s, none other than Fred Hoyle stumbled upon an early example of anthropic reasoning when he predicted that specific nuclear reactions had to exist for carbon (essential to our existence) to be built up within stars. Hoyle's nuclear physics prediction was confirmed just a few years later in experiments. A more recent "success" of the Anthropic Principle came in 1995 when Nobel Prize–winning physicist Steven Weinberg wondered why the cosmological constant was so much smaller than particle physics calculations would suggest.

Recall that physicists have long known that quantum mechanics predicts that the vacuum is anything but empty space; rather, it is a state seething with virtual particles that pop into and out of existence while never violating the rules of uncertainty. Particle physicists saw that this quantum vacuum could be expressed as an energy permeating all space, exactly like a cosmological constant. But their predictions for the size of the cosmological constant based on vacuum energy was so large that the universe as we know it never could have formed. Somehow, they reasoned, the quantum vacuum fluctuations must cancel each other out. Most researchers thought the cancellation would be complete, leading to a cosmological constant that was exactly zero. Using anthropic reasoning relating the necessity of forming galaxies as a precursor to the formation of life, Weinberg went further and was able to derive an upper value to the cosmological constant. Any value larger than his prediction, Weinberg argued, tore protogalaxies apart before they could fully form. For galaxies and hence life to exist, Weinberg concluded, the constant had to be just so big, but no bigger. The discovery of dark energy in 1999, if interpreted as a cosmological constant, gave a value exactly in the range Weinberg predicted.[14]

The Anthropic Principle takes many forms. Some forms are so weak as to be useless and exist as a kind of parody of the idea—"the existence of life tells us the universe has to allow life to exist". Some forms are so strong as to make most materialist scientists balk—"the laws of physics must take a form that makes life a necessary feature of cosmic evolution". After its introduction in the late 1960s and 1970s, most scientists rejected anthropic thinking, seeing it as so obvious as to be useless or so constrictive as to be an exercise in mysticism. Hoyle's often-cited success, for example, was dismissed as not being really anthropic. Carbon is just as important for limestone as it is for life, after all. The problem was always seeing how the laws for this one universe could be linked to our existence. But if more than one universe existed, then anthropic reasoning would take on an entirely new meaning.

A multiverse makes anthropic reasoning a question of statistics. Eternal inflation gives us a vast ensemble of universes, each of which will have different physics and distinct constants guiding that physics. With so many possibilities, our appearance in this particular pocket universe with its apparent fine-tuning of constants and initial conditions becomes something less than a mystery. Even if life is exquisitely sensitive to this universe's constants of nature, the multiverse's infinite, or nearly infinite, sample of pocket universes means constants favourable to life must have happened somewhere. We, of course, are in one of those universes that won the life lottery. But there are other universes with other physics, and many of those will be sterile. In some of those other universes, dark energy pushed space apart so fast that galaxies never formed and no structure or stars exist. Matter and energy spread out like wisps of wind blowing across an endless void. In other universes, the nuclear physics of carbon and other elements is just different enough to keep stars from cooking heavy elements in their cores. Stars burn, but life and even planets might be barred from existence. There are many possibilities. In a multiverse, some scientists argue, it will be anthropic reasoning that provides a way to sort through them all.

Some researchers have taken this logic further. It has long been a tenet of science that explaining phenomena based on the observer's perspective was a form of special pleading. Scientists often invoke what they call the Copernican principle to eliminate claims that an observer is in a special time or a special place that led to special results. Using the multiverse and anthropic arguments, some cosmologists argue that we shouldn't expect to end up in a "special" universe either. That would just be more fine-tuning.

Reach into the multiverse's bag of pocket universes and choose one at random. On average, you would choose a universe that possessed the average kind of conditions for the multiverse as a whole. The same rule applies for the universe we find ourselves occupying. If the Copernican principle holds, then our universe should be close to the average universe. A scientist should, therefore, be able to use our existence to say something about the actual statistics of pocket universes across the multiverse. We should not expect anything less than to find ourselves in an average, mediocre universe. Depending on how you look at it, this prospect is either humbling or insulting.

There is, however, enormous and contentious debate within the cosmological community over what "statistics across universes" means. Some scientists argue that it's impossible to define an average universe, while others offer what they claim are working definitions. But below the technical debate lies a deeper issue about the direction and aim of cosmological science in a post–Big Bang era.

MANY UNIVERSES, ONE SCIENCE

For many researchers in the field, eternal inflation is now seen as a natural consequence of inflation. With that step, the radical possibility of a multiverse becomes a real possibility. But it remains unclear if, in lieu of observational evidence one way or the other, the costs of accepting the multiverse as a research paradigm outweigh its benefits.

Whatever their personal religious affiliations, many physicists share the vision of a nonreligious God of sorts, and this is reflected in their holy grail—the search for the ultimate description of ultimate reality.[15] For many the multiverse, with its implied anthropic logic, is nothing less than an affront to that quest. Following the bright line of reasoning from Pythagoras to Kepler to Newton and onward, the modern enterprise of physics has been a search for a single, unified description of the world. That description, so sparse it could fit on a T-shirt, would unequivocally determine the origin, form and evolution of this one universe. That describes the search for the holy of holies in physics. The effort put into the standard model, the quest for grand unified field theories and the hopes pinned on quantum gravitational theories of everything were all driven by this ancient desire to find this one true description of reality. Once the final equation is determined, everything we see, everything we experience, the whole of cosmic history—be it eternal or springing into time—should stand revealed. Unless, of course, the faithful fear, we retreat to an anthropic logic of the multiverse.

Leonard Susskind is not a member of the faithful. Susskind, the string theorist who coined the term "string landscape", has argued strongly that the Anthropic Principle's time has come.[16] He advocates its embrace as a means of making further progress in articulating why our universe has the structure it does. By thinking in terms of pocket universes and their statistics, Susskind and others suggest that it is time to change the goal of cosmological theorizing. Instead of finding the one eternal law of nature that ordained this universe, we must look for an eternal law governing all universes. The physics of our one universe then becomes an accident of statistics. It is a radical idea, one that many scientists rail against.

If inflation is true (and there is a slim but growing body of data to support it), does eternal inflation necessarily follow? And if eternal inflation is a consequence of inflation, does a multiverse without beginning or end have to exist? If the multiverse is real, does the project of physics and cosmology with its grand dreams of a final theory become a relic of a more naive era? The questions of multiverses and eternal inflation should, ultimately, be answered by data. But how long will it take to compile that data? In the meantime, what price will we pay for pushing beyond the Big Bang? These are questions facing the entire community of physicists, astronomers and cosmologists. For some the answer lies not in multiple universes or string-theory-based cyclic models but in something altogether new and radically different.

Chapter 11

GIVING UP THE GHOST: THE END OF BEGINNINGS AND THE END OF TIME

Cosmology's Radical Alternatives in Three Acts

THE BAVARIAN ALPS • OCTOBER 1963

It was supposed to be a simple weekend trip to the mountains. Now, as the train headed back to Munich, Julian Barbour wasn't sure where he was headed.

Barbour and a friend had just wanted to get away, recharge their mental batteries. Things were going well with his PhD programme and soon he would start a real research project in astrophysics. At least that had been the plan. Now an inescapable question gripped him, upending his tidy future.

What is time, really?

Two days before, he had been on a train heading up to the mountains. Reading the newspaper, he found a story about Paul Dirac, one of the founders of quantum physics. The article discussed Dirac's new thinking on time, space and relativity. It described how Dirac's new work had led him "to doubt how fundamental the 4-D requirement is in physics." Then Dirac asked the question that brought on this crisis: "Perhaps we should ask what is time itself."

The next day he woke with a terrible headache and his friend had to go hike up the mountain alone. All day as he lay in the dark room, Dirac's question haunted him. It was as if the great physicist's speculations had given him some kind of horrible form of mental indigestion.

"Perhaps we should ask what is time itself."

Now bleary-eyed and soul-wracked, he knew things would have to change. As the train headed back to Munich, he knew he could not spend a career writing acceptable "mainstream" scientific articles. He would have to leave academic physics and devote all his efforts to this one question. There was no other choice. He looked out of the window at the mountains receding in the distance. Now he had his own mountain to climb and he would need to husband all his time and effort for the ascent.[1]

THE REBELS

Julian Barbour's encounter with Paul Dirac's musings on time and space that day in 1963 set his life on a new path. "I knew it would take years to understand my question", says Barbour. "There was no way I could have a normal academic career, publishing paper after paper, and really get anywhere." With a bulldog's determination, Barbour left his PhD programme and settled in rural England, where he supported his family translating Russian scientific texts. Thirty-six years later, he published his hard-won answer in scientific monographs and a popular book, *The End of Time*.[2] Even though he had no academic affiliation, physicists across the world took notice of Barbour's creative theorizing.

Barbour is a rebel physicist, a researcher convinced that moving beyond and before the Big Bang will require more than just new theories of strings, branes or inflation. Instead, Barbour is ready to challenge the fundamental assumptions that physics has stood upon for centuries. He is not alone.

Across the globe a small but determined group of scientists has taken on the rebel's mantle as well. Each, in his or her own way, believes that progress at the frontiers of physics and cosmology has stalled. A renewed path forward will, in their eyes, require a radical departure from current methods and models. For some of these researchers, their path out of the mainstream was forced on them through work within it. Others have experienced a growing discord with the direction cosmology and fundamental physics have taken, leading them to ask deeper and more fundamental questions. For each of them, string theory and its

10^{500} possible solutions seem like a dead end, and the multiplication of unobservable universes in multiverse models appears more like science fiction than science. Something else, something better, something not yet imagined, must be waiting.

Each of the rebels sees a fundamental problem with physics and cosmology. In almost all cases the problem they see is time. Each is prepared to scrap the way physics describes time and begin again. Sometimes they are ready with radical solutions that, if correct, will profoundly alter the meaning of time on its own and in its role in cosmology. In other cases, the critiques lean towards metaphysics, asking what physics means, how it is carried forward and how it all begins with time. In every case, these rebels are willing to stand against the mainstream and re-examine how questions of time, physics and cosmology should be asked.

ACT I: THE END OF TIME

Julian Barbour's solution to the problem of time in physics and cosmology is as simply stated as it is radical: there is no such thing as time.[3]

"If you try to get your hands on time, it's always slipping through your fingers", says Barbour. "People are sure time is there, but they can't get hold of it. My feeling is that they can't get hold of it because it isn't there at all." Barbour speaks with a disarming English charm that belies an iron resolve and confidence in his science. His extreme perspective comes from years of looking into the heart of both classical and quantum physics. Isaac Newton thought of time as a river flowing at the same rate everywhere. Einstein changed this picture by unifying space and time into a single 4-D entity. But even Einstein failed to challenge the concept of time as a measure of change. In Barbour's view, the question must be turned on its head. It is change that provides the illusion of time. Channelling the ghost of Parmenides, Barbour sees each individual moment as a whole, complete and existing in its own right. He calls these moments "Nows".

"As we live, we seem to move through a succession of Nows", says

Barbour, "and the question is, what are they?" For Barbour each Now is an arrangement of everything in the universe. "We have the strong impression that things have definite positions relative to each other. I aim to abstract away everything we cannot see (directly or indirectly) and simply keep this idea of many different things coexisting at once. There are simply the Nows, nothing more, nothing less."

Barbour's Nows can be imagined as pages of a novel ripped from the book's spine and tossed randomly onto the floor. Each page is a separate entity existing without time, existing outside of time. Arranging the pages in some special order and moving through them in a step-by-step fashion makes a story unfold. Still, no matter how we arrange the sheets, each page is complete and independent. As Barbour says, "The cat that jumps is not the same cat that lands." The physics of reality for Barbour is the physics of these Nows taken together as a whole. There is no past moment that flows into a future moment. Instead all the different possible configurations of the universe, every possible location of every atom throughout all of creation, exist simultaneously. Barbour's Nows all exist at once in a vast Platonic realm that stands completely and absolutely without time.[4]

"What really intrigues me", says Barbour, "is that the totality of all possible Nows has a very special structure. You can think of it as a landscape or country. Each point in this country is a Now and I call the country Platonia, because it is timeless and created by perfect mathematical rules." The question of "before" the Big Bang never arises for Barbour because his cosmology has no time. All that exists is a landscape of configurations, the landscape of Nows. "Platonia is the true arena of the universe", he says, "and its structure has a deep influence on whatever physics, classical or quantum, is played out in it." For Barbour, the Big Bang is not an explosion in the distant past. It's just a special place in Platonia, his terrain of independent Nows.

Our illusion of the past arises because each Now in Platonia contains objects that appear as "records" in Barbour's language. "The only evidence you have of last week is your memory. But memory comes from a stable structure of neurons in your brain now. The only evidence we have of the Earth's past is rocks and fossils. But these are just stable

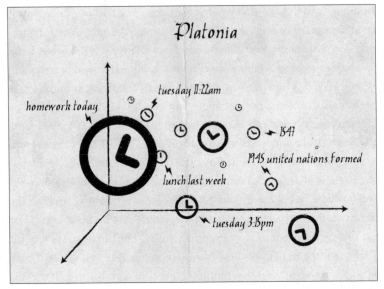

FIGURE 11.1. A world without time. In Julian Barbour's theory the true space of cosmological physics is "Platonia", which physicists call a "configuration space". Every instant is a distinct "Now" representing an arrangement of all the universe's matter and energy. Time does not flow from one Now to another. The Nows exist eternally but do have an arrangement determined by physics in the abstract space of Platonia. Thus some Nows are linked which gives the illusion of time's flow.

structures in the form of an arrangement of minerals we examine in the present. The point is, all we have are these records and you only have them in this Now." Barbour's theory explains the existence of these records through relationships between the Nows in Platonia. Some Nows are linked to others in Platonia's landscape even though they all exist simultaneously. Those links give the appearance of records lining up in sequence from past to future. In spite of that appearance, the actual flow of time from one Now to another is nowhere to be found.

"Think of the integers", he explains. "Every integer exists simultaneously. But some of the integers are linked in structures, like the set of all primes or the numbers you get from the Fibonacci series." The number 3 does not occur in the past of the number 5, just as the Now of the cat jumping off the table does not occur in the past of the Now wherein the cat lands on the floor.[5]

Past and future, beginning and end have simply disappeared in Barbour's physics. And make no mistake about it, Barbour is doing physics. "I know the idea is shocking", he says, "but we can use it to make predictions and describe the world." With his collaborators, Barbour has published a series of papers demonstrating how relativity and quantum mechanics naturally emerge from the physics of Platonia.

Barbour's perfect timeless arrangement of Nows into the landscape of Platonia is the most radical of all solutions to the conundrum of Before. But his audacity reveals an alternative route from this strange moment in science's history. In an era in which the search for quantum gravity has multiplied dimensions and the discovery of dark energy has sent cosmologists back to their blackboards, all the fundamentals seem up for grabs. Barbour is willing to step back even further and offer "no time" as a more basic answer to the question "What is time?"

Barbour's status as a radical was freely chosen, and it led him out of an academic career and straight into his small cottage in Oxfordshire. For other rebels, the path of apostasy was not chosen but thrust upon them.

ACT II: THE END OF CLOCKS

"I was not metaphysically predisposed to give up on the idea of immutable laws", says Andreas Albrecht, a physicist at the University of California at Davis. "I just stumbled on a problem and had to deal with it."[6] In fact, Albrecht, a highly respected cosmologist, was one of the first physicists to pick up on Guth's inflationary cosmology. It was Albrecht, working as a PhD student with Paul Steinhardt, who elaborated many of the key details of the theory.[7]

Andreas Albrecht's fascination with physics began early, especially his fire for its description of nature's laws. "I remember learning about atoms in my high school physics textbook", he says. "I got enthralled by the appendix on quantum physics. I just loved the idea that there were deeper laws behind what we see. After all these years it still keeps me going."

The deeper laws inspiring Albrecht are the same siren song heard by Pythagoras and Plato. From the Greeks to Kepler to Newton to Einstein, the idea of laws, eternal and immutable rules of physics cast in the eternal and immutable language of mathematics, propels young men and women to lives in physics. It's the Platonic laws of physics that catch their imaginations and fill them with awe. This was the inspiration that propelled Albrecht to a highly successful career as a quantum cosmologist. Ironically, it was also his work in this field that drove him to question the very fundamentals of his profession—the core idea of the eternal laws of physics.

In quantum physics, objects such as electrons have the strange property of not having any definite properties. In quantum mechanics, an electron can be in many places at the same time, existing in a kind of potential state until an observation nails it to a single location. In the language of the field, the electron is represented by a wave function, which is quantum physics' mathematical description of the electron and its properties. In classical physics, the description of the electron allowed it to exist at only one location at a time. The wave function is different and describes the electron as having many simultaneous and different locations. "Quantum physics inherently gives us multiple coexisting possibilities", explains Albrecht. "All those different coexisting possibilities are just an inescapable part of the theory."

Quantum cosmology is the attempt to explain the entire universe as a quantum object. Beginning in the 1960s, physicists such as Bryce DeWitt, Jim Hartle and Stephen Hawking began exploring quantum cosmological models designed to embrace the entire cosmos.[8] They were, essentially, looking for the wave function of the universe. Hawking's own pioneering studies of quantum cosmology (and the universe's wave function) had led him to propose "no-boundary" models of the universe, in which no origin of time ever appears (the subject of his famous book *A Brief History of Time*).[9]

Of course, without a full theory of quantum gravity, quantum cosmology can only sketch the possibilities. Researchers must explore the terrain in bits and pieces, and in hunting at the edges this way, they uncovered new and unexpected details that linked quantum physics and

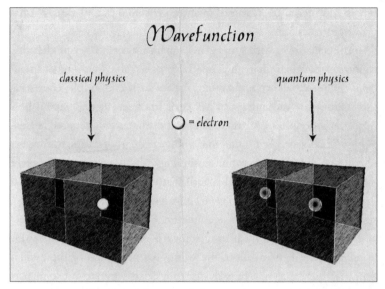

FIGURE 11.2. Quantum physics and the wave function. In classical physics every object has definite properties such as location. Thus a classical electron placed in a two-chambered box must exist in one side or the other. In quantum mechanics multiple possibilities exist at the same time. Thus before the box is opened, the electron exists in both sides of the box.

cosmology. It was in his exploration of those interstitial zones that Andy Albrecht stumbled on his "clock ambiguity" and the end, for him, of immutable laws.

"I started doing quantum cosmology in the 1980s", says Albrecht. "It was a very new, very hot topic then." Thanks to Stephen Hawking and other physicists, the broad outlines for thinking about quantum mechanics, general relativity and the universe as a whole had developed. This was possible even without a full theory of quantum gravity. Doing physics in this "half-sighted" mode is not new. As Albrecht explains, "When I was in high school we learned a lot about the atom even though we had not studied quantum physics yet. We could still do things using the incomplete tools we had learned already. When I study quantum cosmology I am doing the same thing." Inflationary cosmology is considered one form of quantum cosmology, and clearly it has proven its usefulness.

Albrecht works with simple "toy" quantum cosmology models that capture essentials from general relativity such as the expansion of space-time and essentials from quantum physics such as multiple, coexisting possibilities for the universe. "My goal has been to understand how quantum descriptions for the universe must behave in general", he explains. "That was how I found this strange property relating to the laws of physics." Deep in quantum cosmology's basic equations, Albrecht found an ambiguity that left his faith shattered.[10]

"The problem relates to time and what you decide to call a clock", says Albrecht. In normal life, we measure time by picking some object and letting it act as a measuring standard—a "clock". A clock can be water dripping from a faucet, the swing of a pendulum or the oscillations of a quartz crystal. In each case, physicists separate out some part of the world, some subsystem, and use it as a timekeeper. Albrecht found that in quantum cosmology, this separation was anything but straightforward.

"What does it mean to measure 'time'?" asks Albrecht. "You have to divide the world into the part you want to study and the part you call a clock. When I tried to implement this in my quantum cosmology equations I ran into a big problem."

The way time appears in the day-to-day way of doing physics (to a student in a laboratory, for instance) and the way it appears in quantum cosmology are profoundly different. If you try to calculate the motion of a billiard ball in normal physics, you just plug time into Newton's equations of motion and let the forces acting on the ball move it around. But when you try to explain the fundamental history of the universe, you can't just assume time—you have to figure out where it lives in the quantum cosmology equations. You have to figure out what part of your mathematical description represents time. The problem, Albrecht found, is that no unique formula tells physicists how to do this. There is no unique rule explaining how the universe evolved because there is no unique way to pull time apart from other pieces of the equation.

In a very real sense, Albrecht's issues arose when he tried to undo

Einstein's unification of time and space. "In quantum cosmology you have to decide which part of the equations describes time", explains Albrecht. "That choice turns out to be arbitrary." Space, time, matter—all the pieces of reality we are familiar with—do not appear in the equations as separate entities but are expected to "emerge" from a quantum cosmological model of the universe's birth. But Albrecht found that peeling time apart appeared to unwind the whole project of physics.

"I called it the clock ambiguity", says Albrecht. "Basically, different choices of a clock lead to different kinds of physics." It was as if the fundamental laws describing something as essential as a proton depend on whether you use a digital watch or a grandfather clock to tell time. The ambiguities in the laws of physics arise because everything physical is defined by time. "What is a ball?" asks Albrecht. "It's a thing that bounces, and that involves time. What is a ruler? It's an object that doesn't change its length. That involves time. Even elementary particles have their properties defined in terms of how they do or do not change in time."

In essence, the descriptions of "things" such as balls, rulers and elementary particles are exactly what scientists call the laws of nature. To his surprise and horror, Albrecht found that both time *and* the laws of physics rested on an ambiguous and arbitrary choice. "It sounds ridiculous", he explains, "but choosing which part of the equations represented 'time' turned the maths describing electrons into the maths describing photons." The physics changed. The laws changed. "I got entirely different kinds of universes depending on the clock I chose", he says.

This was not what Albrecht was expecting. "As physicists, we like to think of ourselves as hard-nosed people who stick to the facts. What I was finding seemed crazy." It was the implications that were really crazy. "It meant that the fundamental physical laws were not fundamental!"

Albrecht and his colleagues expected quantum cosmology to tell them exactly why the universe we live in looks the way it does. The clock ambiguity blocked their path. If Albrecht is right, then quantum cosmology can never fully predict the course of cosmic history because the laws determining that history can never be specified beforehand.

The idea that the laws of physics are somehow determined as the universe evolves is not entirely new. John Wheeler, a man many would consider one of the greatest physicists of the last half century, proposed something like it in a 1983 article called "Law Without Law".[11] Wheeler was looking for a way in which the laws of nature themselves are not eternal and independent of the universe but are somehow part of its evolution. While the idea has been considered intriguing, most researchers have not taken Wheeler seriously on this point. "So many times when Wheeler talked about 'Law Without Law' I thought, *That's nuts!*" explains Albrecht. "But now I am not so willing to dismiss it."

Reflecting on his dilemma, Albrecht realized that the problem he'd stumbled upon was special because he was trying to describe the universe as a whole. "This puzzle can only occur when you are thinking about cosmology", he says. "The choice of what part of the universe 'represents' time is not a problem for laboratory experiments because in that case it's easy to say, 'This is the clock' and 'That is everything else.'" But because cosmology describes all creation at once, there is a natural problem defining "inside" and "outside", including which subsystem represents time for everything else.

Realizing that the clock ambiguity occurred only when the universe was encountered at its most basic and all-encompassing level, Albrecht did not reject the strange direction of arbitrary law in which his research was leading him. "We could leave well enough alone and not go down this road", says Albrecht, "but as physicists, we seek a complete description of the universe. That means we have to go further than just what we expect from a laboratory perspective."

For Albrecht, the transition to this new way of thinking meant that what we call the universe may be just one patch of reality with one set of randomly determined laws. There can be other patches or other universes with other sets of laws, all equally random. In this way, Albrecht's view may connect to multiverse ideas now routinely discussed at cosmology conferences. There is, however, an important difference that puts Albrecht at odds with many of his colleagues. "When most people think about a multiverse and lots of different universes existing alongside each other, they believe they already have a foundation to build on", says Albrecht. "They believe they can, in principle, know the laws

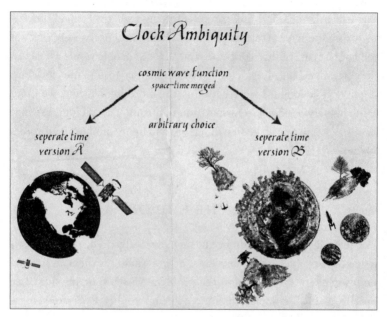

FIGURE 11.3. The clock ambiguity. When quantum cosmologists seek to split time from space in their "wave function of the universe", they confront a paradox. The arbitrary choice of time leads to radically different physics. Here choice A leads to our universe. Choice B leads to a very different set of physical laws and a universe with (for example) floating mountains.

that govern all these different universes at once." Eternal inflation, as practised by Vilenkin, Carroll and others, uses exactly this approach by developing the mathematical physics machinery that describes the multiverse as a whole. It's a path Albrecht believes may no longer be possible. "The clock ambiguity", he explains, "implies that the concrete set of physical laws we were all hoping for is just not there. You can't know what kind of laws will occur in any given universe until you sit in the middle of it and see what happens."

These are all startling and radical possibilities, but Albrecht has resolved to pursue them to their conclusion. Determined to keep an eye on the consequences that could appear in data from astronomy or particle physics, Albrecht is unapologetic about his research. "Maybe all this will fail, but I do think nature is forcing us to go beyond our immediate experience. Getting past my metaphysical preconceptions was the hardest part of this work", he says. "A lot of people would ask, 'How do we fix this?' meaning, 'How do we find a way around the clock ambiguity?' The longer I work with this, the more I don't think we can. I think it's very basic to *any* formulation of quantum cosmology. We may have to assume this is the way things are and figure out how it all comes together."

ACT III: THE END OF LAW

While Andy Albrecht might not be metaphysically predisposed to challenge the existence of timeless laws, others would say metaphysics and philosophy are exactly where the problem started. For physicist Lee Smolin of the Perimeter Institute and philosopher Roberto Mangabeira Unger of Harvard, the real challenge in moving forward is getting past an unspoken metaphysics that is leading the whole enterprise of fundamental physics and cosmology astray. In a move that is either courageous or foolhardy, Unger and Smolin are willing to challenge the philosophical limits of thinking that lead to multiverses and higher dimensions. In their eyes, these fantastical worlds are just that—fantasies—and progress forward may demand that we re-examine our most cherished beliefs in the eternal physical laws.

Smolin and Unger are an odd couple. Smolin is a highly regarded theoretical physicist who has worked in a wide range of fields, from particle physics to string theory to loop quantum gravity.[12] A popular author in his own right, Smolin has become a reluctant critic of string theory and the culture of present-day theoretical physics. He argues that a wider array of ideas and options must be explored. Unger is a well-known Brazilian philosopher and a professor at Harvard Law School.[13] He has written widely on the dynamic evolution of law in societies. Turning from the realm of philosophy to the pragmatics of politics, Unger has also served as Brazil's Minister for Strategic Affairs.

Both men, working from separate starting points, have found common ground in rejecting the idea of eternal, timeless laws as well as in the construction of a radically new starting point for physics.

Like most physicists, Lee Smolin began as a true believer in timeless law, but developments in cosmology forced him to rethink the unthinkable. "Thirty years ago, talk of other universes was not seen to be part of science", he says. "But there has been a gradual shift during which it became acceptable to work on theories that described possible other worlds, universes with a different number of dimensions than ours, or with different kinds of particles and forces."[14] According to Smolin, only over the last few years have these other worlds gone from being "just possible" to "hypothetically actual". "In certain circles", he says, "it's fashionable to talk of a cosmos containing a vast number of other universes with different dimensions, different fundamental particles and different laws. Some of the best young cosmologists now do most of their work on the details of such multiverse scenarios."

While Smolin has worked on his own version of a multiverse in the past, he now sees these theories as indicative of a wider problem in physics and is convinced that it needs to change on a fundamental philosophical level.[15] According to Smolin, hiding behind recent developments is an unspoken metaphysics that has two troubling features. First is the claim that our universe is just one of many. Second is the claim

FIGURE 11.4. Lee Smolin's *The Trouble with Physics*. Theoretical physicist Lee Smolin has become a critic of string theory and (with philosopher Roberto Unger) multiverse cosmologies, claiming they stray too far from concerns with the one universe we observe.

that time is not fundamental to reality but emerges from some deeper set of laws such as the quantum cosmological equations. Smolin argues that the equations themselves and the metaphysical approach they embody must now be considered suspect. In raising these deep objections to the way physics and cosmology are carried out, Smolin and Unger find common ground.

"The treasures of a scientist are his riddles", says Unger, speaking sadly of the modern history of physics, which led scientists to embrace theories with hidden dimensions and alternate universes. For Unger, these kinds of theoretical constructions are "allegories" because they are, essentially, fictions. As he points out, no one has observed another universe or an extra, hidden dimension of space. That fundamental fact constitutes the shaky ground on which entire fields of research in modern cosmology have been built. "The scientist should treasure the riddles he can't solve", continues Unger, "not explain them away at the outset."

According to Smolin and Unger, the invisible "fictions" of the new theories are created in response to some challenge posed by the world we directly observe. Physicists want a theory of quantum gravity, for example, and string theory looks like an attractive possibility. But string theory's project of exchanging point particles for microscopic strings could not be made to work if the world was only three-dimensional. In response, string theorists "added more space", allowing the world to include seven extra dimensions that are hidden from us. With this move, researchers built a working framework for quantum gravity in terms of vibrating strings, but the price they paid was the need to add seven invisible dimensions to reality.

In Unger's eyes, the multiverse counts as another fiction. Multiverse cosmologies do away with the dilemma posed by the twenty finely tuned constants of the standard model. In a universe of universes, the values of the constants we see in our cosmos are just an accident. Instead of being fundamental physics, the values of the constants become merely a matter of statistics occurring across many incarnations of the standard model in the many pocket universes that make up the multiverse.

In both these cases—the multiverse and string theory's hidden dimensions—Unger says the theories don't explain; rather, they explain away. For Unger, these unseen universes or dimensions drain the reality out of what we actually experience. "When we imagine our universe to be just one of a multitude of possible worlds, we devalue *this* world, the one we see, the one we should be trying to explain", he says. Both Unger and Smolin want to shift emphasis in physics away from these possible worlds back to the one real world. That world, they say, is saturated with time and free of timeless law.

In Smolin and Unger's view, it is critical that we recognize how the metaphysics of string theory and multiverse cosmologies continues the grip of an ancient idea whose roots were laid down with Plato. The job of theoretical physics, according to this venerable notion, is to discover timeless universal laws of nature. "By definition that goal is transcendental", says Smolin. "It requires grasping something true outside of time and beyond space." For both Smolin and Unger, the connections with religion are obvious. "Because of this implied transcendence", Smolin wrote in a recent article, "this goal just continues the still older metaphysical presuppositions of religion: that the world we experience is less real than a much greater and timeless reality of eternal truths." There is no need, say Smolin and Unger, to keep physics wedded to this kind of philosophy, especially if it forces so much fiction into the domain of science.

In his own work, Unger has already mapped out some of the territory Smolin now believes is the future of physics—a future without timeless truths.[16] Here the two men stand at the boundaries of science and philosophy. While Andreas Albrecht's clock ambiguity emerged from a direct exploration of quantum cosmology's equations, Unger and Smolin want to dig below any specific equations or physical formulations. In their view, developing equations for the new perspective will come later. For now, they want to argue broadly about the rationale for looking at those equations in the first place.

"We must try to imagine a new kind of law", Smolin says, "which applies only one time. Such a law need not and should not have any sense in which it exists outside of time. Nor could it be conceived of as apart from the universe it describes."[17] This distinction is crucial. If time is real, then everything in the universe is part of time and subject to it. The essence of time, as every human is painfully aware, is change. That simple fact gives Smolin and Unger one way of imagining what all their philosophy will mean for the future of cosmology and fundamental physics. What we call the "laws of physics" may have a story in time—they may evolve. History may matter even for the underlying rules of physics. "We may now have to think about laws where the distinction between a one-time narration of the one universe's history and the statement of principles governing that history weakens." In other words, the story and the laws unfold or "become" together.

Smolin and Unger do not yet have the concept of evolving laws worked out. They consider themselves at the beginning of a journey. While their ideas may seem radical, both men feel that cosmology's current, unspoken philosophy of invisible worlds is also a radical departure from traditional scientific practice. Given that state of affairs, their perspective must also warrant consideration. "What would it mean not to pull our punches?" says Unger, speaking of science's long tradition of dealing with the facts that this one world presents to us, and only those facts. "What would it mean to take the insistence on dealing only with this world to the extreme?" Unger asks. "What light would that cast?"

REBEL YELL: BEYOND TIME,
CLOCKS AND LAW

When confronting the questions of time and what came before the Big Bang, each of the rebel thinkers portrayed in this chapter offers a different perspective. Julian Barbour tells us that "before" the Big Bang is no problem because "before" and "after" never existed in the first place. There is no time and no change. Andreas Albrecht, one of the original leaders of inflation theory, follows his quantum cosmology equations to their logical conclusion and watches in wonder as the clock ambiguity swallows the possibility for eternal, timeless law. Lee Smolin and Roberto Unger look across the landscape of modern cosmology with its extra dimensions and other universes, declare them fictions and set off to deliberately forge another path. Each researcher is serious, sober and determined, but each is exploring a frontier that lies at the edge, or even past the edge, of what mainstream cosmological physics seems willing to consider.

In an era when mainstream cosmology offers so many strange-sounding alternatives to the standard model of the Big Bang, it may seem equally strange that any of these ideas can be considered as too much or too far. But science is, above all else, a community of thinkers. For hundreds of years, that community has developed its own consensus about what problems it thinks are essential and, more important, what methods it considers appropriate for attacking those problems. The options offered up by mainstream cosmology now may seem like a bewildering array of strange titles to the rest of culture: string theory, loop quantum gravity, holographic universes, brane-worlds and so forth. But these theories still live within the scientific community's shared sense of what is reasonable in terms of goals and methods. What puts something like the Anthropic Principle on the fringe is its implied demand to change the goals of science. The four rebels encountered in this chapter go even further: each one offers a much deeper critique of the assumptions underlying modern cosmological physics.

The presence of the rebels and their varied visions demonstrates the unsettled and thrilling character of this moment in cosmology building. We are clearly poised to push beyond the now sixty-year-old Big Bang model of a universe that began with a single moment of creation. Genesis is over. But what will take its place? New theories, many making strange demands of reality, are multiplying like rabbits. Observations that can settle the issue may be decades away at best. New discoveries, such as dark energy, might topple the current enterprise and send us off in a new direction at any time. With so much shifting and so much uncertainty, even the rebels must be given their due.

Chapter 12

IN THE FIELDS OF LEANING GRASS

Ending the Beginning in Human and Cosmic Time

NEW SEATTLE, THARSIS HILLS, MARS • 2256 CE

It was time.

With the mental equivalent of a blink she shut down the implant rest-ing behind her brain's frontal lobes. The DataStream she had been so deeply immersed in disappeared and it took a minute, as it always did, to get used to real input from her optic nerve. The wall screen told her it was 4:15 p.m. local Mars time. She had been In-World, immersed in the Network for just ten minutes, and yet her internal sense told her it had been hours. The DataStream was always like that. Direct connections to machine time meant her awareness was responding somewhere between femtosecond clock cycles and what her brain's biology allowed. That was a permanent state for a big chunk of the solar system's population these days, of course. They lived In-World most of the time, letting wetware and nanotech take care of their bod-ies. That was too much for her, though.

Still, she had got a lot done. Most of the homework set was complete. Now she had to get to the tram port to pick up her parents. They were so proud of their physicist-in-training daughter that they'd offered to take her shopping. Fine by her.

On the walk through the outer dome she was still mulling over the last problem. She had not been able to get the extra dimensions to curl up right. This was basic stuff and the foundation for the entire rest of the year. String

Theory 101. The huge extradimensional effects of modified string conformalism had been discovered in orbital colliders back in the 2100s. Now they had been relentlessly exploited in the microhole technologies that powered pretty much everything. She needed to get that problem worked out or she would look pretty stupid and Professor Yin-Paul would not be likely to consider her for his group.

Outside the dome the stars were becoming brilliant in the darkening, red-tinged Martian sky. In the distance she could make out the lights of the brane generators up on the Tharsis Bulge. That was where her hopes lay. If she made it into Yin-Paul's group, she'd be doing cosmology as she had always dreamed, probing the Bulk and the other brane that hovered like a ghost right here, just next to her and everyone else. Sometimes in the DataStream when her body seemed so far away she felt like she could almost move into that other, invisible dimension. God, she was so excited.

Someone bumped up against her and she was knocked out of her reverie. She shuffled along with the crowd to the tram. She didn't like being with all these people. They could have done all the shopping In-World together, but her parents were so old-school. Still, she was happy to have the time to spend with them.

STANDING ON EDGE

Books on cosmology do not usually come with a moral at the end. This one does. Actually, there are two morals braided together, just as the stories of human time and the human imagining of cosmic time have intertwined across fifty thousand years of cultural innovation. We have reached the end of the beginning in more ways than one.

From the creation of politically charged city-based empires five thousand years ago to the establishment of mercantile factory-based empires two centuries ago, human culture has re-invented itself again and again. The digital world we so quickly assembled in the last four decades has been yet another turning of the creative wheel. At every stage, the invention of a new time has been central to the cultural transformation. The lived day of a Babylonian merchant looked very different from the lived day of the English worker in a textile factory, which is itself so distant from the mobile-phoned, Facebooked, e-calendared days we move through today.

We have always been inventing new uses for time from our embodied encounters with the raw stuff of the world. Material engagement, the central driver of change, has become more abstract over the past fifty millennia—beginning with the heft of a stone held in a toolmaker's hands and arriving, most recently, at the computer-controlled, laser-guided etching of circuits onto silicon chips. But even as the distance between raw material and final technological product has increased, the time we live in continues to change.

Cosmology, our story of time and creation, has changed as well and it has done so in rough synchronization with cultural time. We began in a world where people's lives were inextricably woven into the fabric of sky, water and animal. Millennia later, humans were no longer part of nature but imagined themselves above the natural, as God supposedly had ordained. The monotheistic cultures, however, still placed human beings at the centre of a multilayered cosmos of angels, stars, planets and Earth. With the birth of modern science (and its Greek emphasis on reason), we became sequentially displaced from the cosmic centre. Now we are but one planet orbiting one star in one galaxy lost in a universe of galaxies so vast as to give new, visceral meanings to the infinite.

As the narratives changed, cosmic time and its origins changed as well. The human vision of the cosmos shifted back and forth between temporal trajectories first imagined in myths: a universe created, a universe eternal, a universe of cycles. While many of the imagined possibilities for cosmic time have been with us since the distant era of myth, each culture inevitably focused on one or another geometry of time. Neolithic farmers leaned heavily on stories of eternal return and cyclic universes. Christian theologians looked to scripture and its account of God's creation ex nihilo. With the advent of science, new vocabularies and new demands entered into the play of cosmology creation: gravity, entropy, relativity.

But always and forever, the creation of cosmological narratives remained married to the creation of human cultures. Human time and cosmic time remained closely linked, and those links were forged in the iron wheels of medieval clocks, the steel boilers of Victorian steam engines, and the silicon chips ticking off clock cycles in our modern-day computers.

Looking back across fifty thousand years and looking forward to our ever-accelerating future, we stand in a privileged position. We can, at long last, recognize the paired cycles of change in cosmic and human time and ask what they tell us about who we are, what we are and where might we be going. These questions are neither abstract nor inconsequential, for once we recognize the braiding of cosmic and human time we may also recognize the turning point both have reached. In this way, we come to the end of our own beginning as a species.

The first question we come to arises from the recognition that we have been creating time ever since we began shaping culture. Human time changes from one era to the next and there is nothing God-given or physics-given about the form of time each person is born into. When we teach our children how to read a clock, we are setting them into a specific framework of imagined time. When they enter school and find out that the sixth period is for maths but the seventh is for English they are, once again, being trained to exist within the context of a culturally defined time made real through specific forms of material engagement. By the time we go off to work, we are so deep in this cultural time we can barely see it for what it is—an invention.

How many of us are happy with the time we have? The acceleration of cultural life through electronically mediated technologies has brought us great boons, but in the rush of e-mails, mobile phone calls, texting and telecommunication we have been pushed to work more, produce more, consume more. And here comes our first moral. By recognizing that we have invented and are re-inventing time, we give ourselves the opportunity to change it yet again.

There is an urgency to this possibility that goes beyond mere personal choice. The acceleration of time over the last century was a byproduct of a culture fuelled by energy resources that are now clearly reaching their limits. It is ironic that the human story began as a narrative of climate, with our emergence out of the last ice age. Now our rapid exploitation of carbon-intensive petrochemical technologies makes the next step in our story climate-dependent once again. And there can be no doubt that whatever comes next will have to involve new inventions in time.

Choices are also the issue for the second question arising at the end of our beginning. Cosmology, our scientific narrative of the universe and time, is at its own precipice. The Big Bang as a sudden creation of cosmos and time without a "before" is poised to be swept aside. In its place, a bewildering array of alternatives has appeared: brane-world cosmologies, eternal inflation, multiverses, string theory landscapes, loop quantum cosmologies. Some of these ideas are more fully developed than others, and many require the addition of bizarre new players on the cosmic stage. The expectation is that ultimately it will be observations and data that decide among these alternatives, but what will count as a definitive test for such a new cosmology? How long will science, and the culture that supports it, wait for an answer?

More important, how will changes in cosmological conceptions of time parallel or interweave with culture's temporal innovations? These questions strike at the heart of our understanding of what science does, the nature of the reality it speaks to, and its interaction with human life. Thus emerges the second moral. By recognizing the way culture and its investigations of the cosmos have entwined across history, we see that simple narratives of discovery in the aether of pure thought are rarely accurate. We are poised to put an end to our naive conceptions of culture, science and cosmos, and develop a more nuanced perspective of our ongoing dialogue with the world.

Thus we end the book with an investigation of two questions that can serve to inform our next steps in both cosmic and human time: how does culture change its time in light of cosmology, and how does cosmology change its time in light of culture?

THE TYRANNY OF AN
EFFICIENT UNIVERSE

The time we live through today, the time that rules over our lives, rests on a single value: efficiency. Minimizing time spent completing tasks and producing products is the overarching temporal value of modern culture. It shapes the structure of our economics and, in the process, shapes everything from how we eat to how we educate our children. Efficiency is the principal "time invention" of modern Western culture. As Jeremy Rifkin wrote in his book *Time Wars*,

> With its [efficiency's] introduction, the modern temporal orientation is complete. Efficiency is both a value and a method. As a value, efficiency becomes the social norm for how all human time should be used. As a method, efficiency becomes the best way to use time to advance the goal of material progress.[1]

Efficiency is, according to Rifkin, the maturation of three different cultural innovations: the division of labour, mass production and the principles of scientific management. The division of labour and mass production were both essential facets of the industrial revolution. It was in Ambrose Crowley's era that the underlying principles for these facets was first articulated. The idea spread rapidly. Clockmakers, for example, recognized that they could produce more timepieces if they divided up workers so that "one man shall make the *Wheels*, another the *Spring*, another shall engrave the *Dial-plate*" and so on, until the entire process was "better and cheaper, than if the whole Work be put upon any one Man".[2] At the dawn of the 1800s, Eli Whitney took the process further with his new "American method" for the production of muskets. Whitney had standard interchangeable parts manufactured separately and then brought them together to be assembled by workers who might have no knowledge of how the parts were fashioned or how to build a musket from scratch.

Scientific management, the final cornerstone of the industrial pyramid and its new engagement with time, was the brainchild of Fredrick

Taylor. Taylor, born in 1856 to a wealthy Philadelphia Quaker family, began his career on the shop floor of a hydraulic works company fashioning water pumps. Surveying the activity around him with an engineer's eye, Taylor came to realize that production could be increased if each "work process" was broken down into its simplest elements, with systematic improvements made to each and every step. But systematic improvement required systematic study. For the rest of his life Taylor developed the principles of "scientifically" dissecting every aspect of work. His goal was to refine and streamline every step and every movement made by workers both on the shop floor and in the office. His efforts reshaped the world. As historian Daniel Bell wrote,

> If any social upheaval can ever be attributed to one man, the logic of efficiency as a mode of life is due to Taylor. . . . With scientific management, as formulated by Taylor in 1895, we pass far beyond the old, rough computations of the division of labor and move into the division of time itself.[3]

In slicing human activity into the abstraction of timed actions, Taylor's weapon was the stopwatch. Every movement was clocked down to fractions of a second and, in this way, was standardized. Opening a folder file drawer: 0.4 minutes. Opening a folder: 0.4 minutes. Closing a drawer: 0.26 minutes. By scheduling the most basic work tasks (in both blue- and white-collar jobs) people were, in effect, having their heads severed from their bodies, as decisions about their movement through time were made for them by "scientific" managers.

Taylor's ideas were adopted as the standard across the United States and Europe. Eventually his techniques spread across the world, as the industrial model for life and time became a global model. And Taylor was right. Production could be vastly increased through scientific study, management and analysis. There was a price to be paid for living with this new and more efficient time, however, and it has only been in the last few decades that the bill seems to be coming due for our new planetary culture as a whole.

The advent of digital technologies, of course, added a new dimension to the ever-expanding universe of efficiency. With machines that move in microseconds or less, the ability to co-ordinate, integrate and track human activity was both expanded and accelerated. Most important, the advent of digital communications technology meant that the separation between work and life was blurred beyond recognition. The BlackBerry quickly became the "CrackBerry" as workers moved into a state of being permanently at work. How many meals, kids' ballgames or walks in the park have been interrupted with "I just have to take this call" or "I'm just going to answer this text"?

From fast food for dinner to "quality time" with children, time management by the end of the twentieth century (with a healthy dose of personal information management) had become the aether through which all daily life moved. Everything had been sped up and nearly everyone was feeling the crunch. What we couldn't see until recently was that behind the acceleration of time was a background of energy and natural resources that could be assumed, or ignored, for only so long.

The acceleration in global industrial output in the last century was not simply a product of science and technology. What made that science and technology possible was a seemingly endless supply of cheap energy in the form of petrochemicals. So much useful energy can be extracted from a cubic centimetre of oil that virtually no other nonatomic resource can compare. The world's oil reserves are, essentially, millions of years' worth of stored solar energy. Oil is ancient vegetation slowly converted into the viscous ooze that now powers every aspect of our culture. Petrochemicals do more than simply heat homes, power cars and transport food. They *are* food in the sense that fertilizers derived from petroleum drive the world's prodigious agricultural output. We wear oil in fabrics and it's in the innumerable plastic products we hold in our hands. Our medicines would be impossible without oil and our digital world of computers could never have been assembled in its absence. In a single century, we have managed to run through a sizeable fraction of that stored energy. We have used it to build a culture that could literally move mountains, raise cities to the sky and give each of us the power of an army of servants to clean our dishes, wash our clothes and prepare our food.

But the energy we used to build this accelerated world has its limits and its consequences. By most accounts we are close to, or have already passed, the moment of peak oil, when we will be extracting as much of the easy-to-reach cheap oil from the ground per day as we ever will.[4] The spigots can be opened no further. Production will inevitably decline, while demand, with China and India already on line, will only increase. The hard-to-reach of oil is still out there to be mined, of course. BP's infamous *Deepwater Horizon* deep-sea platform was drilling one mile underwater off the coast of Louisiana in one of these hard-to-reach oilfields when, in the spring of 2010, the platform exploded. It took months to cap the well at the bottom of the sea and five million barrels of oil were inadvertently pumped into the Gulf of Mexico. From these kinds of deep-water fields to the tar sands of Alberta, Canada, it is becoming painfully clear that getting the hard-to-reach oil carries environmental costs we may not be able to pay.

But even if we could get at the oil, entropy and the second law of thermodynamics have once again risen to remind us that nothing comes for free. In using up all those million-year-old reservoirs of cheap energy to build our hyperefficient, hyperaccelerated culture, we have altered the Earth's atmospheric chemistry. Billions of tonnes of carbon dioxide as well as methane and other greenhouse gases have rapidly been added into the air, enhancing its capacity to trap the same solar energy that created the fossil fuels in the first place. Global warming, better described as climate change, is the ongoing consequence.

While it is impossible to predict the exact extent to which the climate will change, the evidence is overwhelming that it *will* change. With consequences that will lie between the unlikely poles of "no big deal" and "the end of the world", human culture will, in the next century, face an uncomfortable and perhaps dangerous journey between the rock of dwindling energy supplies and the hard place of climate change driven by our use of that energy. With the end of cheap energy sources, coupled with other forms of resource depletion such as the escalating global competition for fresh water supplies, we seem on the trajectory to another cultural shift.[5] Material engagement, this time in the form of material depletion, will once again be the engine driving the change.

Our ability to find new forms of material engagement or re-envision old ones will serve as bedrock for the new institutional facts that we'll imagine into being through choice or necessity. And, as has happened so many times in the past, we will imagine new forms of time to accompany our new institutions and new material engagements.

UNIVERSES IN THE MAKING: COSMOLOGY AND COSMIC TRUTHS

It stands twenty-five metres high and weighs seven thousand tonnes.[6] Three thousand kilometres of cable run into and out of its octagonal bulk.[7] The cables provide electric power and carry petabytes of data from its stacked panels of silicon wafers into banks of computers.[8] They call it Atlas. It is the primary detector for the Large Hadron Collider, a particle accelerator that represents the future of particle physics. Running through the centre of the Atlas detector is an evacuated metal tube that stretches off into the distance of the underground facility. Following the tube through its tunnel housing takes a visitor on a twenty-seven-kilometre (seventeen-mile) circuit.[9] Superconducting magnets filled with supercooled helium girdle the metal tube as it runs through the tunnel. Inside the tube protons are driven to speeds just a sliver below that of light. A beam of protons circulates in one direction on one side of the tube. Another beam, kept safely apart in the other side of the tube, circulates along the track in the other direction. In the middle of Atlas' great maw the opposing streams of matter are magnetically shunted towards each other. When the particles collide they shatter, sending a spray of debris in all directions. In the detector, traces of the collision are caught and measured. Data will stream into the computers to create a record of those brief instants when humans managed, ever so briefly and on ever so small a scale, to re-create conditions that have existed only once before in this universe's earliest history.

The Large Hadron Collider (LHC) took twenty years and more than $9 billion to build. Located outside of Geneva, Switzerland, it is an international facility operated by twenty countries through the efforts of thousands of scientists and technicians. It is the largest physics experiment ever created. The hopes of the entire field of fundamental physics ride with those circulating proton beams. From the first indirect evidence for string theory to direct detections of dark matter, physicists hope that the LHC will be the engine that finally takes them beyond the standard model. But in its sheer size, cost and organization the LHC represents something else as well. It is the culmination of a new way of doing science that emerged with, and from, the culture we have so recently constructed.

Physics and cosmological science were just as profoundly transformed by the new efficiency culture as was daily life. The years after World War II, in particular, saw the development of "Big Science", massive industrial-scale efforts dedicated to answering big questions. Particle physics led the way in the development of Big Science, building ever-larger particle accelerators of astronomical cost that required the staff of a fair-sized factory. Without these efforts, the standard model—with its leptons, quarks and force bosons—could never have been achieved. The space programme is another potent example of Big Science. The infrastructure required to design, build and launch a space-based research telescope of even modest ambitions extends across continents. The management of so much human effort, technical detail and funding contributed to, and relied upon, the rapidly changing capacities of material engagement and all the cultural behaviours that went along with it.

The hyperlinked Internet emerged, in part, as a research tool for particle physicists. The first Web browser emerged from a computational astrophysics lab. Just as these technologies exploded into the larger culture, rewiring the landscape of human behaviour, they were also reshaping scientific culture and possibilities. What would have taken scientists months just a few decades earlier now took seconds and could be repeated millions of times. The automated data collection used to produce cosmic maps of galaxies distributed across billions of light-years is just one example of the computer-facilitated efficiency that modern cosmology now relies on. The collaborations between hundreds of scientists, engineers and technicians making the WMAP satellite with its high-resolution view of the cosmic microwave background is another. The cosmological conclusions WMAP offered to thousands of scientists, who were linked together by electronic networks, would not have been possible without the same kinds of material engagement that powered Facebook, Wikipedia and Amazon.

It was against the background of these efforts that the bang in the Big Bang was challenged. Theorists followed their own imperatives in cosmology building as they contributed, and responded, to the large-scale interrogations of the world via experiments. Their efforts, expressed in the language of mathematical physics and theoretical models, now offer radical new narratives of the universe and time. The question is, what will happen next?

Wilson and Penzias' 1964 detection of the cosmic microwave background was an accident. That is what makes their story, and the story of Big Bang science, so compelling. They tripped over evidence for the true history of cosmic evolution as though they'd stumbled across the Rosetta Stone on a walk through the woods. There is a lesson for us in that discovery. The cosmic microwave background spoke so loudly for the Big Bang model that any dissent was immediately quashed. The evidence was that complete. Even a person staring at static on an ordinary TV screen could pick up a few CMB photons in the flickering chaos. In that sense, the material engagement of everyday life, in the form of television technologies, allowed everyone to "see" the Big Bang. Will the new alternative cosmologies find evidence as convincing?

String theory has provided deep mathematical insights into a diversity of fields. It may ultimately prove to be the foundation for an entirely new physics that will find its own confirmation in experiment. That would be a thrilling development. Experimental confirmation of unseen dimensions would radically expand the meaning of our concept of "cosmos". But string theory may also prove to be nothing more than mathematics, an intellectually rich but physically disembodied collection of interconnected ideas. The multiverse is also a thrilling possibility. If evidence of other universes were to manifest in our experiments and observations, it would throw open the doors of humanity's cosmological imagination and begin a new chapter in the human effort to place ourselves against the true cosmic background. But we may never find evidence for even one other universe in our maps of galaxies, the cosmic microwave background or anywhere else. Like the aether of a century ago, the multiverse may prove to be nothing more than scientists' deeply held wish.

In light of these questions, many scientists today criticize cosmology's reliance on currently unobservable entities, such as the extra dimensions in string theory or the multiple universes used in eternal inflation. String theory, in particular, has now existed in various forms for more than thirty years and has yet to produce any tangible connections with observation or experiment. Its proponents argue that we must wait and that it makes sense to do so given the extent of the theory's reach. But how long will the scientific community and the culture that supports it be willing to wait?

More important, what form will the evidence for a new and radical definition of universe take? The detection of the CMB announced a total victory for Big Bang models. Will the evidence for esoteric cosmological entities such as a multiverse be so clear as to muffle dissent the way the CMB did? Or will evidence for something like string cosmology lie so far in the realm of ninth-order effects in a perturbation spectrum that interpretations will differ for decades? What if no direct, compelling evidence arrives for hidden dimensions or other universes? Will cosmologists continue to infer their existence because adding those features to a theory is the only way to explain what is observed, such as the appearance of fine-tuning?

Finally, should the definition of science itself, and its fundamental goals, change in response to new theories and their confrontation with new data? The recognition that string theory can't uniquely predict our universe but instead provides an almost infinite landscape of possible universes was a disappointment for many. For others, including Leonard Susskind, it pointed towards a different approach, an anthropic approach, to cosmological science. How should science change as it stretches to embrace the oldest and deepest questions of existence and reality? Should it change at all?

These questions will obviously hinge on material engagement in the form of new technologies for physical experimentation and astronomical observation. Hyperprecise, space-based gravity wave detectors, may be launched within a few decades, and they represent one direction for new material engagement. Gravity wave detectors might provide evidence for alternative cosmologies. But it is not yet clear if the planned gravity wave detectors will be sensitive enough to see the signals predicted by the new cosmologies. Thus questions of alternatives to the Big Bang will also rest on the culture that supports science.

Strong and convincing evidence for string theory, brane cosmology and multiverse models may come in the next few decades. But if it does not come, then the entire approach now pursued so actively may fall by the wayside. If the predictions of the new cosmologies remain beyond the reach of experiments, then the efforts to pursue those cosmologies will wither. Like the aether of a century ago, thousands of scientific papers may become nothing more than historical relics of theoretical dead ends. Neither science nor culture will wait forever. Cosmos building will march on, especially when culture faces its own changes. In that case, the truly radical approaches to redefining cosmic time might begin to find support, encouragement and funding.

The principal point is that, as we have seen, culture always needs a cosmology to support its own institutional facts, its own organizing principles. From that perspective, it is not surprising that a culture that constructs its time in submillisecond clock cycles would find its image in a cosmology that focuses on events occurring 10^{-33} of a second after creation. We live in a scientific, technologically driven culture, and we look to science to provide our cosmological framework and orientation. A myth that ceases to be useful ceases to exist, said Karen Armstrong, speaking of human culture in millennia past. The same can be said of scientific narratives of the universe. Even today, human culture needs its dominant cosmology, and if culture changes, it appears that cosmos building will too. What we have learned about the time after the Big Bang is solid. The story of the universe expanding, of subatomic particles fusing into light nuclei, of cosmic microwave background photons being released to traverse space forever and of galaxies condensing out of a sea of hydrogen represents the best of what modern science has achieved. But the context into which we set that story is now become fluid. It is the bang at the beginning, the very meaning of beginning, that is up for grabs.

Seeing the braided evolution of cosmic time and human time points us to the deepest question of all: what is the nature of truth in cosmological science? How much of the cosmos do we have objective access to?

MASKS OF THE UNIVERSE

In 1996 physicist Alan Sokal put the finishing touches on a new manuscript and sent it off to an academic journal. The manuscript wasn't a description of a new experimental method or a new theoretical calculation, and the journal wasn't a scientific publication. Instead, Sokal sent his new work to *Social Text*, a journal focused on "postmodern culture studies", and his entire article was nonsense, a hoax.

The journal was dedicating an issue to the so-called science wars that erupted in the 1980s and 1990s when some scholars in the humanities began arguing that science was "socially constructed". In their view, there was no inherent truth to be found in scientific practice. Instead, the results of science were a kind of agreed-upon fiction, a game with made-up rules like those of bridge or chess. The language of social constructivist arguments drew heavily from postmodern studies of literature and could be highly obtuse and arcane. Sokal's article, "Transgressing the Boundaries: Toward a Transformative Hermeneutics of Quantum Gravity", appeared to favour the social constructivist argument claiming that quantum gravity was a cultural invention that depended solely and explicitly on linguistic conventions. This was not, of course, what Alan Sokal believed. Instead, he wanted to see if the journal would "publish an article liberally salted with nonsense if it (a) sounded good and (b) flattered the editors' ideological preconceptions".[10] The journal did publish the article and, once Sokal revealed his hoax, was humiliated. The firestorm of controversy that followed has yet to fully die down.

Social construction of science, the idea that science *does not* reveal aspects of the world's own structure, is surely a mistake. The world clearly pushes back even in the context of highly abstract, technologically dependent interrogations such as the study of the cosmic microwave background. But when dealing with science's encounter with all-embracing issues such as the five questions at the heart of cosmology, the polar extreme of social constructivism, what philosophers call naive realism, is just as much a mistake.

For centuries, philosophers and scientists have argued over realism and its opposite, anti-realism. As philosopher Samir Okasha put it,

> Realists hold that the aim of science is to provide a true description of the world. This may sound like a fairly innocuous doctrine. For surely no-one thinks science is aiming to produce a false description of the world. But that is not what anti-realists think. Rather, anti-realists hold that the aim of science is to provide a true description of a certain *part* of the world—the "observable" part. As far as the "unobservable" part of the world goes, it makes no odds whether what science says is true or not, according to anti-realists.[11]

Like all isms, these positions come with many, many variants. Naive realism is the idea that science gives us a perfect and complete description of reality, fully objective and fully independent. While naive realism makes sense in day-to-day life, it is a real problem in confronting the universe as a whole. The reasons for this are fairly obvious, and they are the ones Andy Albrecht cites in his exploration of the clock ambiguity: We remain locked within the system, the universe, we hope to describe. We have one and only one example of the universe we want to study, so its investigation is very different from exploring, say, the heat conduction of a metal bar in a lab. Most important, given the inherent possibilities of infinities in both space and time we cannot even be sure that our definition of "universe", the system we want to study, is correct as we begin the project.

The universe is not a bar of iron, a germinating seed, an atom in a magnetic trap or even a planet orbiting a distant star. It's the totality of existence, and because of this, we only "get" what we observe; we must build from that as best we can. And it is exactly with that recognition that richer and more interesting perspectives on interactions with culture come to the fore.

The argument traced across this book, the argument about the inseparable braiding of cosmic time and social time, could be interpreted in two very different ways. The first is not very interesting: technology changes and allows scientists greater capacities to study the world, and cosmology responds to that technology with new data and new models. This is the kind of easy, triumphalist narrative we get in secondary school. It's the "march of science" pablum that makes for easy reading and easier thinking. The second interpretation of cosmic and social time paints a richer picture. Material engagement shapes both the cultural and cosmological imagination. It creates horizons of possibility. The data of cosmology are the raw material. They are how the world pushes back in our interrogations. But the stories we imagine—the stories we can imagine—must always be framed by the cultural imaginations material engagement made possible.

Material engagement flowing both upwards and downwards creates the possibility of new imaginative landscapes for human culture, including science, to inhabit. The invention of those landscapes through culture building opens up a set of new possibilities for cosmology building that both enable and constrain the imagination. This shifting imaginative landscape focuses our cosmology-building efforts in new directions, creating new responses to questions that have been with us since the time of myth. Scientists, like everyone else, are born in the midst of culture's institutional facts. Our personal storehouses of metaphor, analogy and creative vision are shaped by the specific world we grow up within. From this cultural background come clockwork universes and Mixmaster universes as cosmologists seek to build their imaginative responses to the data that material engagement provides them. Thus, culture and cosmology flow back and forth through a paired process of invention, interrogation and response.

In 1948, the comparative mythologist Joseph Campbell looked at the panoply of world religions and myth systems and declared each to be a "mask of God". In 1985, cosmologist Edward Harrison looked at the history of cosmology and declared its shifting ideas to be "masks of the universe". Making a distinction between the one physical Universe and the many culturally imagined universes—between an ultimate reality and the realities we build through investigation and interpretation—Harrison wrote,

> Where there is a society of human beings, however primitive, there we find a universe; and where there is a universe, of whatever kind, there we find a society. Both go together, the one does not exist without the other. A universe unifies a society, enabling its members to communicate and share their thoughts and experiences. . . . Each determines what is perceived and what constitutes valid knowledge, and the members of a society believe what they perceive and perceive what they believe.[12]

His last point is the most radical: Each society perceives what it believes. Thus each culturally imposed framework acts as a filter constraining our conception of the universe while it enables our explorations.

Does this perspective rob science of its remarkable and inspiring power? Not at all. The Universe is there and it pushes back. The discovery of the scientific process, its methodologies and its all-important ethic of honesty in investigation (so critical to science) was a pivotal step in our evolution. When the practice of science became formalized beginning in the 1500s, it brought enormous new powers and energy to material engagement, greatly expanding our vision and capacities. Science has changed our ability to see the world's internal structure. But the recognition that we only see the universes and not the universe entire should move us past simple narratives of what science is and what it does. The recognition that cosmos and culture are so infinitely entwined can and should change our vision of both. What we need is a richer understanding of our proper relation to the Universe as a whole and the universes we create.

There is the old story of a group of blind philosophers studying an elephant. One feels the tail and declares an elephant is like a snake.

Another feels the ear and declares an elephant is like a palm frond. A third feels the foot and declares the elephant is like a tree. The relationship between the universe in and of itself and the universe each culture invents for itself is much like that between the philosophers and their elephant. Perhaps it is time to see the universe as an infinite elephant or, better yet, as a diamond with infinite facets. As culture changes it brings different facets into view. We gain a deeper understanding even as the universe in and of itself remains ultimately larger than all our accounts. In the end, it is our dialogue with the universe that matters most. Acknowledging the intertwined evolution of culture and cosmic vision does not diminish the power of science; it allows us to see more clearly our role as participants in the universe.

To put it bluntly, we can never be taken out of the narrative of creation. We are always, in some partial but essential way, its co-creators. In taking this perspective we make the most radical step of all. We begin to move away from a reflexive Copernicanism that made human beings irrelevant in the cosmos and instead recognize that there is a vital place for us—a life at the centre of the universes we manifest through the creative act of being human, creating culture and practising science.

THE FIELDS OF LEANING GRASS

The Slow Food movement was born in response to the fast-food industry. The 1986 brainchild of Italian writer Carlo Petrini, the movement was developed to protest the appearance of a McDonald's franchise in central Rome. Slow Food was explicit in its emphasis on time and material engagement. One of the things that mattered most about eating, in the Slow Food argument, was the satisfaction that came from time spent preparing a meal, the pleasure gained from sharing that meal with others, and the impact of that meal on ecosystems through the production of its raw materials. As the Slow Food mission statement puts it: "By reawakening and training their senses, Slow Food helps people rediscover the joys of eating and understand the importance of caring where their food comes from, who makes it and how it's made."[13]

In a culture built for speed, the idea of Slow Food hit a nerve. Spreading rapidly, the Slow Food movement now claims a hundred thousand members in 132 countries. In 2004, Carlo Petrini was named one of *Time* magazine's Heroes of the Year. The award, *Time* said, was given for "his willingness to consider that all the parameters—agricultural, industrial, commercial, ecological—constitute the real strength of Slow Food".[14]

Carlo Petrini knows the accelerated time we have been born into is a cultural invention and that invented time can be changed. There are many others who have made this identification and have their own vision for how to re-create time. Some see the terrible pressures of resource depletion bearing down on us and argue for a new economy based on new models for work, production and consumption. Others look at the personal impact of our existing time and argue for personal disengagement from demands for efficiency. "Lifestyle entrepreneurs" have appeared, seeking to create business models designed not to make them rich but to set them free. Each of these movements underscores that alongside its boons, the current form of culturally invented time has created a deep current of discontent.

Examining the way cultural time and cosmic time have emerged in tandem for fifty thousand years allows us to apply a broader perspective to movements that seek a "new time". We can first see that cosmology is not simply an abstract endeavour pursued in ivory towers. Rather, it is a vital part of finding meaning in human endeavour. We can also see that changes in cosmological visions of time accompany changes in cultural visions of time. The current precipice at which cosmological science now stands points to a coming shift in the basis of the human enterprise. The fact that new books appear almost monthly by scientists arguing for different cosmological theories beyond the Big Bang should tell us something. There is change in the air.

These changes may be subtractions just as easily as they may be additions. A culture stressed by climate change and resource depletion may find itself unable to fund the next iteration of Big Science needed to search for hidden dimensions or other universes. Alternatively, an energized pursuit of Big Science, along with new forms of material engagement, might reorder our concerns by providing new modes of energy and organization. Scientists such as Andy Albrecht see all the new cosmological visions of time as embracing a vision of temporality that is fundamentally more fluid and flexible. Perhaps one of these temporal visions will act as a seed for a human time that is also more fluid and flexible. Either way, recognizing the braiding of cultural time and cosmic time means recognizing their deep connections, and that means seeing the end of our own childhood as a species.

If, for the first time, we can recognize that our time has always been an invention, then we can ask more conscious questions about what is useful for us to invent next. If we can recognize the enigmatic entanglement between cultural time and cosmic time, we might stop looking for God in the form of "final theories" and find our rightful—and rightfully central—place in the narratives of creation. Either way we find ourselves back where we belong, at the centre of our universe—a universe suffused with meaning and potential. In recognizing the end of the beginning as we move beyond Big Bang cosmology, we might also recognize the end of our own beginning. From that vantage point we can gain a first glimpse at the beginning of our next evolutionary phase as a truly global culture whose material engagement embraces the whole planet.

In tracing the history of cosmology and culture, we have focused almost exclusively on the West, on European traditions and their impact. In looking for new metaphors that can embrace a new and proper understanding of cosmos, culture and time, we might turn to one of the cultures we have not followed. In viewing the tension between the relative world of human experience and an underlying, absolute reality—between the universe and the Universe—Buddhist philosophy emphasizes a doctrine called dependent arising.[15] From its perspective, everything in the universe—every object, idea, being, event and process—depends on everything else. Nothing ever exists entirely alone This would include the universe, the mask we put on the absolute.

The image called to mind is of a wheat field after the harvest. Tall columns of bundled wheat are set upright in pairs, each resting against the other. It makes no sense to ask which is the column that rests and which is the column that supports. Each rests against the other and each requires the other for support.

The claim is often made that the more we learn about the universe, the more meaningless it appears. Seeing the meaningful encounter between human beings and the universe for what it is can, perhaps, alter this oft-criticized nihilism: "Nothing exists entirely alone." Human beings build universes, and universes support human beings. Embedded in the world, we are the ones who look outwards, responding with effort and dedication to discover and shape what we find, honouring it all with study and inquiry. Seeing the interdependence of universes and human beings, we should not devalue our position in the cosmos even as we begin to comprehend the awe-inspiring vastness of that cosmos. Rather than make claims of final theories, perhaps we should focus on our ever-continuing dialogue with the universe. It is the dialogue that matters most, not its imagined end. It is the sacred act of inquiry wherein we gently trace the experienced outlines of an ever-greater whole. It is the dialogue that lets the brilliance of the diamond's infinite facets shine clearly. It is the dialogue that instils within us a power and capacity that is, and always has been, saturated with meaning.

Time after time, we have never been anything other than collaborators with the universe. Always and again we have been the co-creators of a time and a cosmos that exist together with us. That is what makes our story anything but insignificant and makes our universes anything but meaningless. We have always been weaving the fabric of our experience into a culturally shared time and, in the process, have become ever more intimate with a universe that has always invited our participation. With each step we gain a new perspective, even if it will never be the final perspective. That is just fine. With each step we gain a deeper sense of the awe and beauty that suffuse the universe's essential mystery. If we can trace our steps from the past and see our way clearly into the future, then certainly there is time enough for that great effort to continue with renewed clarity and purpose.

ACKNOWLEDGEMENTS

Writing this book has been a long journey of discovery that took me to many unexpected places and gave back much more than it asked (though it asked a lot). I am deeply indebted to the many friends and colleagues who helped me cover all that ground from the Palaeolithic hunter-gatherers to the Large Hadron Collider. Without them I would have gotten lost early and often and likely never found my way back.

Howard Yoon of Ross Yoon was the first to see that the book could be more than simply a presentation of new ideas in cosmology. I have been very lucky to have his guidance. Hilary Redmon at Free Press recognized far more than I did how much there was in the braided story of cosmos and culture, and without her the journey would not have even begun. It was a pleasure to work with an editor with an eye for detail and the biggest of big pictures. In writing the book I was very lucky to have Leonard Roberge act as editor for the first draft of the manuscript. Leonard not only helped me stay close to the narrative but also brought many excellent ideas and new perspectives on time and history to me. I am deeply grateful for his help. Sydney Tanigawa did a wonderful job honing the book's later draft. The original illustrations for the book were done by Sameer Zavery, and it was a great pleasure watching him take my crappy sketches and turn them into gracefully executed works of art. Finally, the gods of authorship decided to smile on the project when they sent me David Panzarella as a research assistant. David was a recently graduated English student at the University of Rochester when he joined the project and was tireless first in tracking down images for the book and then in taking on the enormous task of fact-checking and footnoting the manuscript. Any errors in the text are, however, my responsibility.

In writing a book like this I called upon the help of a number of scholars who were quite generous with their time. Anthony Aveni not only helped me understand key ideas but also read over a number of chapters. Discussions with Colin Renfrew provided me with important insights into the Palaeolithic and Neolithic eras. Discussions with Peter Galison helped expand ideas on the role of time and technology. Andy Albrecht was very generous giving time for a number of discussions and reading over key chapters. His insights into the development of alternative cosmologies were critical to the book. My good friend and co-blogger Marcelo Gleiser also provided important corrections in my discussion of cosmological issues and broader philosophical issues. Lee Smolin has always been enormously helpful in discussions of issues in the frontiers of physics, and I am grateful for his generosity. I greatly benefited from discussions at various times with Bruce Balick, Woody Sullivan, Julian Barbour, Paul Steinhardt and Roberto Mangabeira Unger. I also benefited from early conversations with Sean Carroll and Jennifer Ouellette.

Some of the cosmological ideas in this book were first explored in articles I wrote for *Discover* magazine. I have always been lucky to work with Corey Powell there, and it was a pleasure to collaborate with Fred Guterl again. I would also like to thank everyone associated with our National Public Radio blog 13.7 Cosmos and Culture. Many times early versions of ideas appeared in posts there, and I am thankful to Wright Bryan, Eyder Peralta, Ursula Goodenough, Stuart Kaufman and Alva Noe for the help and the community. And, as always, I am grateful to K. C. Cole for her vision and her friendship.

I am greatly in debt to my friends at the University of Rochester who read over chapters of the book and gave freely of their time for discussions of everything from the nature of language to the nature of space-time. Fellow astronomer Dan Watson acted as my scientific conscience, reading over the entire manuscript and offering sharp insights. Tom Dipero of the Department of Modern Languages and Cultures looked at the book with a cultural historian's eye and showed me aspects of my thesis that I had not seen. Alyssa Ney of the Department of Philosophy was invaluable in showing me underlying currents in the ideas I was exploring. I am also thankful to Nick Bigelow for his help and understanding. I am also indebted to Carol Latta for her help in so many forms.

Of course, all that time alone at the desk only made sense when there was something to do later on. Without the friendship and support of so many fine people I would have ended up with nothing to do on my off hours but mutter to myself and watch *Battlestar Galactica* reruns. If I didn't have June Avignone and Greg Van Maneen across the street to hang out with, I would still be writing. My fellow New Jersey expats were my anchor, and I will always love them for it. Tom and Mary Slothower remain my stalwart friends through thick and thin. Jill Pallum and Tim Mooris provided much needed letting off of steam on more than one occasion. An ongoing collaboration with artist Steve Carpenter has always been a source of inspiration. Conversations with Sara Silvo always kept me sane. I am grateful to Nancy Pignot for all her kindness. And, of course, without Paul Green and Robert Pincus the whole endeavour would be pointless. Margaret King, thank you forever. Adam Turner, that goes for you too. My sister, Elisabeth Frank, was always there to lend an ear, and after all these years how wonderful is that? Hendrick Helmer provided welcome comedy when I needed it. And Ingrid Frank and George Richardson—you are, of course, my original inspiration.

To my children, Sadie Ava and Harrison David, it's all good. Thank you for majoring in awesome. And Alana . . . I says what I mean and I mean what I says.

NOTES

PROLOGUE

1. I. Morris, *Why the West Rules—for Now* (New York: Farrar, Straus & Giroux, 2010). The developmental "cores" described in Morris' excellent book begin much earlier in prehistory.

CHAPTER 1: TALKING SKY, WORKING STONE AND LIVING FIELD

1. Bill Giles, "Katabatic Winds," BBC, http://www.bbc.co.uk/weather/features/az/alphabet31.shtml (accessed August 24, 2010).
2. We cannot be sure whether the shamans of Abri Blanchard were women. But, based on archaeological evidence, women are thought to have served as shamans in other Palaeolithic groups. Barbara Tedlock, *The Woman in the Shaman's Body: Reclaiming the Feminine in Religion and Medicine* (New York: Bantam, 2005), 1.
3. Paul G. Bahn and Jean Vertut, *Journey Through the Ice Age* (Berkeley: University of California Press, 1997), 17; Daniel Rosenberg, "Marking Time," *Cabinet* (Winter 2007).
4. Rosenberg, "Marking Time."
5. Ibid. The fragment was discovered in 1915; Marshack came across it in 1965.
6. Marshack describes the markings as "a serpentine figure composed of 69 marks, containing some 24 changes of point or stroke. . . . When the maker had finished this notation, the full serpentine figure represented two months or 'moons.'" Alexander Marshack, *The Roots of Civilization* (Mt. Kisco, NY: Moyer Bell, 1991), 45–48.
7. Other instances of prehistoric lunar time reckoning deserve consideration.

Marshack details several of these relics, including a chronology. Ibid., 96–97.

8. Anthony Aveni, *Empires of Time: Calendars, Clocks, and Cultures* (Boulder: University Press of Colorado, 2002), 58; Steven Mithen, *After the Ice: A Global Human History* (Cambridge, MA: Harvard University Press, 2003), 8.

9. P. Jeffrey Brantingham, Steven L. Kuhn and Kristopher W. Kerry, *The Early Upper Paleolithic Beyond Western Europe* (Berkeley: University of California Press, 2004), xiii.

10. Aveni, *Empires of Time*, 58.

11. Mithen, *After the Ice*, 3–4, 8–9.

12. Edward McNall Burns, *Western Civilizations: Their History and Their Culture* (New York: W. W. Norton, 1963), 9.

13. Anthony Aveni in *Empires of Time*, 33–69, examines the earliest attempts at codified time in a chapter on calendars from the prehistoric era to the Greek early classical period.

14. Mithen, *After the Ice*, 151.

15. Ian Tattersall, "Once We Were Not Alone," *Scientific American* (January 2000): 58.

16. Fred H. Smith, "Neandertal and Early Modern Human Interactions in Europe," *American Anthropologist* 110, no. 2 (June 2008): 257.

17. Neanderthal man had a considerably larger brain ($1,532$ cm^3) than anatomically modern humans ($1,355$ cm^3). But, taking into account the larger size of Neanderthal man, the ratio of brain size to bodyweight is roughly equal, at 3.08 in Neanderthal to 3.06 in modern humans. G. L. Dusseldorp, *A View to a Kill: Investigating Middle Paleolithic Subsistence Using an Optimal Foraging Perspective* (Leiden: Sidestone Press, 2009), 21.

18. The upper Palaeolithic toolkit would have included hand axes, scrapers and projectile points. Iain Davidson and April Nowell, eds., *Stone Tools and the Evolution of Human Cognition* (Boulder: University Press of Colorado, 2010), 4.

19. Research suggests that despite the million-year stasis in technological development, there are larger differences within the Middle Pleistocene (780,000 to 126,000 years ago) than popularly characterized. April Nowell and Mark White, "Growing Up in Middle Pleistocene," in *Stone Tools and the Evolution of Human Cognition*, ed. Iain Davidson and April Nowell (Boulder: University Press of Colorado, 2010), 71.

20. Steven Mithen, *The Prehistory of the Mind: The Cognitive Origins of Art, Religion and Science* (New York: Thames and Hudson, 1996), 43–45.

21. Renée Baillargeon, "How Do Infants Learn About the Physical World?" *Current Directions in Psychological Science* 3, no. 5 (1994); Elizabeth Spelke, "Nativism, Empiricism and the Origins of Knowledge," *Infant Behavior and Development* 21, no. 2 (1998).

22. Mithen, *Prehistory of the Mind*, 54–55.

23. Ibid., 71.

24. The circadian rhythm is entrained to the exposure to light and dark in an animal's environment. That is, it is governed by the cycle of daylight. In mammals, the circadian rhythm relies on two clusters of neurons above the optic chiasm. David Sadava et al., *Life: The Science of Biology*, 9th ed. (Sunderland, MA: Sinauer Associates, 2009), 1128.

25. The time it takes for the moon to complete one orbit is actually less than the synodic month. This period is called the sidereal month and it lasts just 27.3 days. The difference occurs because the Earth is moving in its orbit around the sun as the moon moves in its orbit around the Earth. Thus even if the moon completes one orbit in a sidereal month, it has to move a bit further to line up with the sun (the synodic month), which has shifted its position in the sky due to the Earth's motion.

26. Eliade notes that myths relate to "primordial Time, the fabled time of the 'beginnings.'" In this way myths correlate to the impulse of modern science to explain the beginning of earthly and cosmic time. Mircea Eliade, *Myth and Reality* (Prospect Heights, IL: Waveland Press, 1998).

27. Ibid., 9.

28. Denise Schmandt-Besserat, *How Writing Came About* (Austin: University of Texas Press, 1996), 122.

29. Mithen, *Prehistory of the Mind*, 47.

30. Ibid., 47.

31. Ibid., 47.

32. Ibid., 48.

33. Karen Armstrong, *A Short History of Myth* (New York: Canongate, 2005), 14–15.

34. Ibid., 13.

35. Ibid., 15.

36. Ibid., 15.

37. Ibid., 16.

38. Armstrong notes that often this sky god appeared to retreat and have no dealings with people. He is said to have "left" or "gone away." Ibid., 21.

39. Mithen, *After the Ice*, 12.

40. Peter Wilson demonstrates the radical nature of the shift from hunter-

gatherer to domestication and its impact on the evolution of culture in *The Domestication of the Human Species* (New Haven, CT: Yale University Press, 1988).

41. Colin Renfrew, *Prehistory: The Making of the Human Mind* (New York: Random House, 2007), 70.

42. An excellent exploration of the role physical embodiment may play in consciousness can be found in Francisco J. Varela, Evan Thompson and Eleanor Rosch, *The Embodied Mind: Cognitive Science and Human Experience* (Cambridge, MA: MIT Press, 1991).

43. Mithen, *After the Ice*, 178.

44. David Lewis-Williams and David Pearce, *Inside the Neolithic Mind: Consciousness, Cosmos, and the Realm of the Gods* (London: Thames and Hudson, 2005), 227.

45. Ian Shaw and Robert Jameson, *A Dictionary of Archaeology* (Padstow, Cornwall: Blackwell, 2002), 546; Rodney Castelden, *The Stonehenge People: An Exploration of Life in Neolithic Britain, 4700–2000 BC* (New York: Routledge, 2002), 101.

46. Renfrew, *Prehistory*, 133.

47. Colin Renfrew argues that the construction of Stonehenge brings about a "grander social reality" and simultaneously represents "the deliberate attempt to align the human society in question with the cosmos." Ibid., 155.

48. Armstrong, *A Short History of Myth*, 41.

49. Ibid., 43–44.

50. Ibid., 55–56.

51. Mircea Eliade, *Myth of the Eternal Return*, 54.

CHAPTER 2: THE CITY, THE CYCLE AND THE EPICYCLE

1. Tamsyn Barton, *Ancient Astrology* (New York: Routledge, 1994), 12. There is some debate about the exact date of Ammisaduqa's reign but some scholars estimate 1646 or 1581 BCE as a possible date for the beginning of his reign. Herman Hunger and David Pingree, *Astral Sciences in Mesopotamia* (Leiden: Brill, 1999), 38.

2. There are no records of Ammisaduqa enacting such a treaty but we know that astrological predictions played a role in arranging marriages. Barton, *Ancient Astrology*, 29.

3. Norman Yoffee, *Myths of the Archaic State: Evolution of the Earliest Cities,*

States and Civilizations (New York: Cambridge University Press, 2005), 210–11.

4. The Indus Valley society, called the Harappan culture after one of its major cities, flourished between about 2600 and 1900 BCE. Charles Gates, *Ancient Cities: The Archaeology of Urban Life in the Ancient Near East and Egypt, Greece, and Rome* (New York: Routledge, 2005), 67.

5. The city of Ur reached its apogee in the late third and early second millennia BCE. The First Dynasty of Babylon spanned 2000 to 1530 BCE, when Hammurabi's rule began. Ibid., 53, 56.

6. Barry J. Kemp, *Ancient Egypt: Anatomy of a Civilization* (New York: Routledge, 2006), 14.

7. Anthony Aveni, *Empires of Time: Calendars, Clocks, and Cultures* (Boulder: University Press of Colorado, 2002), 67.

8. Some research indicates that coin use may be traced back to early as 8000 BCE. These small cones, spheres, disks or cylinders served as counters for Neolithic people. Denise Schmandt-Besserat, *How Writing Came About* (Austin: University of Texas Press, 1996), 7–9.

9. Ibid., 117–18.

10. Henry George Fischer, "The Origin of Egyptian Hieroglyphs," in *The Origins of Writing*, ed. Wayne M. Senner (Lincoln: University of Nebraska Press, 1991), 67–70.

11. Gates, *Ancient Cities*, 71.

12. Ibid., 71.

13. Colin Renfrew, *Prehistory: The Making of the Human Mind* (New York: Random House, 2007), 99.

14. Aveni, *Empires of Time*, 114.

15. E. G. Richards, *Mapping Time: The Calendar and Its History* (New York: Oxford University Press, 1998).

16. Ibid.

17. Aveni, *Empires of Time*, 99–100.

18. Helge S. Kragh, *Conceptions of Cosmos: From Myths to the Accelerating Universe: A History of Cosmology* (New York: Oxford University Press, 2007), 7.

19. Ibid., 7.

20. Ibid., 8.

21. Stephanie Dalley, *Myths from Mesopotamia* (Oxford: Oxford University Press, 1998).

22. Otto Neugebauer, *The Exact Sciences in Antiquity* (New York: Dover, 1969).

23. Ian Johnston, "Device That Let the Greeks Decode the Solar System," *Scotsman*, November 20, 2006; Ian Sample, "Mysteries of Computer from 65 BC Are Solved," *Guardian*, November 30, 2006.

24. The Hellenistic period was an explosive moment in Western thought. It marks the advent of new methods of inquiry and interactions with the surrounding world. For more on the Tang Dynasty, see Charles D. Benn, *China's Golden Age: Everyday Life in the Tang Dynasty* (New York: Oxford University Press, 2004).

25. Aveni, *Empires of Time*, 36–46.

26. Ibid., 36.

27. Ibid., 37.

28. Ibid.

29. The following sections are quoted by line number from Hesiod, *Works and Days*, in *Hesiod*, trans. Richmond Lattimore (Ann Arbor: University of Michigan Press, 1972).

30. Aveni, *Empires of Time*. 43-44.

31. Ibid., 45.

32. Ibid.

33. *Kosmos* is a vast and complicated concept in ancient Greek thought, but the meaning most germane to our project is a universal, celestial order that links the Earth and the heavens. Paul Cartledge, Paul Millett and Sitta von Reden, *Kosmos: Essays in Order, Conflict and Community in Classical Athens* (New York: Cambridge University Press, 1998).

34. Edward Harrison, *Masks of the Universe* (New York: Macmillan, 1985), 50–51.

35. Ibid.

36. Ibid., 51.

37. Carl Huffman, "Pythagoras," in *The Stanford Encyclopedia of Philosophy*, ed. Edward N. Zalta, Winter 2009 ed., http://plato.stanford.edu/entries/pythagoras/#LifWor (accessed January 30, 2011).

38. Harrison, *Masks of the Universe*.

39. Kragh, *Conceptions of Cosmos*, 17.

40. Christopher Shields, "Aristotle," in *The Stanford Encyclopedia of Philosophy*, ed. Edward N. Zalta, Winter 2009 ed., http://plato.stanford.edu/archives/win2009/entries/aristotle/ (accessed September 27, 2010).

41. Kragh, *Conceptions of Cosmos*, 20.

42. Epicycles were used by a number of greek astronomers including Hipparchus sometime between 147 and 127 BCE. Ptolemy's theories on epicycles depend heavily on the frameworks Hipparchus put in place.

Hugh Thurston, *Early Astronomy* (New York: Springer-Verlag, 1993), 133, 143–170.

43. John Palmer, "Parmenides," in *The Stanford Encyclopedia of Philosophy*, ed. Edward N. Zalta, Fall 2008 ed., http://plato.stanford.edu/entries/parmenides/ (accessed October 21, 2010).

44. Daniel W. Graham, "Heraclitus," in *The Stanford Encyclopedia of Philosophy*, ed. Edward N. Zalta, Fall 2008 ed., http://plato.stanford.edu/archives/sum2011/entries/heraclitus/ (accessed November 10, 2010).

45. John Rist, *The Stoics* (Berkeley: University of California Press, 1998), 183.

46. Sylvia Berryman, "Leucippus," in *The Stanford Encyclopedia of Philosophy*, ed. Edward N. Zalta, Fall 2010 ed., http://plato.stanford.edu/entries/leucippus/ (accessed November 8, 2010).

47. Kragh, *Conceptions of Cosmos*, 18.

48. Marcelo Gleiser, *The Dancing Universe: From Creation Myths to the Big Bang* (Lebanon, NH: Dartmouth College Press, 2005), 3.

CHAPTER 3: THE CLOCK, THE BELL TOWER
AND THE SPHERES OF GOD

1. On occasions reformist Benedictine monks corresponded with Rome hoping for the appointment of similarly reform-minded bishops. Laura Swan, *The Benedictine Tradition: Spirituality in History* (Collegeville, MN: Order of St. Benedict, 2007), 35.

2. St. Benedict placed a heavy emphasis on his fellow monks honouring God through their work. Mayeul de Dreuille, *The Rule of St. Benedict: A Commentary in Light of World Ascetic Traditions* (Leominster, Herefordshire: Newman Press, 2000), 235.

3. The sacrist (less commonly, sacristan) was the monastic keeper of time as well as keeper of the ceremonial vestments. He directed the ringing of bells, though it's not clear whether he rang them himself. He may have had subordinates. David Knowles, *The Monastic Order in England: A History of Its Development from the Times of St. Dunstan to the Fourth Lateran Council, 940–1216* (New York: Cambridge University Press, 1963), 429–30; E. K. Milliken, *English Monasticism: Yesterday and Today* (London: George G. Harrap, 1967), 73.

4. The main offices of the Church are: matins (also called vigils) during the night, lauds at dawn, prime in the early morning, terce at midmorning, sext at midday, none in midafternoon, vespers in the evening, and compline before bed. Only matins, lauds and vespers were repeated in church. The rest

were said in private. David Knowles, *The Monastic Order in England: A History of Its Development from the Times of St. Dunstan to the Fourth Lateran Council, 940–1216* (New York: Cambridge University Press, 1963), 378, 449–51.

5. While this date is debatable, it marks the point after which medieval universities were founded and scientific thought regained a widespread foothold in European thought.

6. E. G. Richards, *Mapping Time: The Calendar and Its History* (New York: Oxford University Press, 1998), 219.

7. Ibid., 210; Anthony Aveni, *Empires of Time: Calendars, Clocks, and Cultures* (Boulder: University Press of Colorado, 2002), 91–92.

8. Aveni, *Empires of Time*, 89, 98.

9. Ibid., 99.

10. Ibid.

11. Ibid., 100.

12. Richards, *Mapping Time*, 215.

13. Aveni, *Empires of Time*, 100.

14. Ibid.

15. Richards, *Mapping Time*, 239–46.

16. This phrase is borrowed from the title of Charles Freeman's famous study of the European Middle Ages. *The Closing of the Western Mind: The Rise of Faith and the Fall of Reason* (New York: Alfred A. Knopf, 2003).

17. David Martel Johnson, *How History Made the Mind: The Cultural Origins of Objective Thinking* (Peru, IL: Open Court, 2003), 92.

18. Ibid.

19. David Allen Park, *The Grand Contraption: The World as Myth, Number, and Chance* (Princeton: Princeton University Press, 2005), 168.

20. Helge S. Kragh, *Conceptions of Cosmos: From Myths to the Accelerating Universe: A History of Cosmology* (New York: Oxford University Press, 2007), 34.

21. Adrian Keith Goldsworthy, *How Rome Fell: Death of a Superpower* (New Haven: Yale University Press, 2009), 2–3.

22. Kragh, *Conceptions of Cosmos*, 36–37.

23. Ibid., 37.

24. David C. Lindberg, *The Beginnings of Western Science* (Chicago: University of Chicago Press, 1992), 204.

25. Part of this grievance with Aristotelian cosmology resulted from the line in Genesis regarding the waters above the firmament that could not be accounted for if the outermost space was a vacuum. Kragh, *Conceptions of Cosmos*, 38.

26. Ibid, 42.

27. Ibid.

28. Ibid.

29. Ibid., 45.

30. Gilbert of Mons, *Chronicle of Hainaut*, trans. Laura Napran (Woodbridge: Boydell Press, 2005), 115.

31. Gerhard Dohrn-van Rossum, *History of the Hour: Clocks and Modern Temporal Orders*, trans. Thomas Dunlap (Chicago: University of Chicago Press, 1996).

32. Aveni, *Empires of Time*, 80.

33. Ibid., 80.

34. Ibid., 80–81.

35. Using differentiated time would have been characteristic of only the educated classes during this time period. Dohrn-van Rossum, *History of the Hour*, 18–19.

36. The Rostrum and the Grecostasis. Aveni, *Empires of Time*, 81.

37. Dohrn-van Rossum, *History of the Hour*, 57–59.

38. Lewis Mumford, "The Monastery and the Clock," in *Technics and Civilization* (New York: Harcourt Brace, 1934), 13–14.

39. Dohrn-van Rossum, *History of the Hour*, 29.

40. Ibid., 36.

41. Ibid.

42. Ibid., 37.

43. Ibid., 37–38.

44. Ibid., 208.

45. Ibid., 248.

46. Ibid., 200.

47. Jacques Le Goff, *Time, Work and Culture in the Middle Ages* (Chicago: University of Chicago Press, 1980), 295.

48. Dohrn-van Rossum, *History of the Hour*, 45.

49. Ibid.

50. Ibid., 129–33.

51. Ibid., 146

52. Ibid., 282

53. Mumford, "The Monastery and the Clock," 17.

54. Basil S. Yamey, "Double-Entry Bookkeeping, Luca Pacioli and Italian Renaissance Art," in *Art and Accounting* (New Haven: Yale University Press, 1989), 125.

55. Roland H. Bainton, *Here I Stand: A Life of Martin Luther* (Peabody, MA: Hendrickson, 1977), xvii.

56. Nancy Smiler Levinson, *Magellan and the First Voyage Around the World* (New York: Clarion, 2001), 52.

57. Kragh, *Conceptions of Cosmos*, 50.

58. John North, *A Norton History of Astronomy and Cosmology* (New York: W. W. Norton, 1995), 283.

59. Ibid., 319.

60. Henry C. King, *The History of the Telescope* (New York: Dover, 1955), 30.

61. Timothy Ferris, *Coming of Age in the Milky Way* (New York: HarperCollins, 1988), 101.

62. Nicolaus Copernicus, *On the Revolutions of the Heavenly Spheres*, trans. Charles Glenn Wallis (Philadelphia: Running Press, 2002), 14, 15.

63. Kragh, *Conceptions of Cosmos*, 57.

64. William Gilbert, *On the Loadstone and Magnetic Bodies and on the Great Magnet the Earth*, trans. Paul Fleury Mottelay (New York: John Wiley, 1893).

65. Kragh, *Conceptions of Cosmos*, 63.

66. Martin Gorst, *Measuring Eternity: The Search for the Beginning of Time* (New York: Broadway Books, 2001), 15.

67. Ibid., 40.

68. John North, *God's Clockmaker: Richard of Wallingford and the Invention of Time* (New York: Continuum, 2006), 201.

69. Arno Borst, *The Ordering of Time: From the Ancient Computus to the Modern Computer* (Chicago: University of Chicago Press, 1993), 97.

CHAPTER 4: COSMIC MACHINES, ILLUMINATED NIGHT AND THE FACTORY CLOCK

1. Alec Skempton, *A Biographical Dictionary of Civil Engineers in Great Britain and Ireland 1500–1830* (London: Thomas Telford, 2002), 160.

2. M. W. Flinn, *Men of Iron: The Crowleys in the Early Iron Industry* (Edinburgh: Edinburgh University Press, 1962), 200–2.

3. Crowley lived from 1658 to 1713, Newton from 1643 to 1727 by the Gregorian calendar.

4. Skempton, *Biographical Dictionary*, 160.

5. J. C. Davis, *Utopia and the Ideal Society: A Study of English Utopian Writing 1516–1700* (New York: Cambridge University Press, 1981), 351.

6. Ibid.

7. Flinn, *Men of Iron*, 53.

8. Skempton, *Biographical Dictionary*, 160.

9. Those who ran the ironworks after Crowley's death in 1713 continued to build on the *Law Book*.

10. Allen C. Bluedorn and Mary J. Waller, "The Stewardship of the Temporal Commons," in *Research in Organizational Behavior: An Annual Series of Analytical Essays and Critical Reviews*, vol. 27, ed. Barry Staw (San Diego: JAI Press, 2006), 377.

11. Ibid.

12. Davis, *Utopia and the Ideal Society*, 353.

13. Skempton, *Biographical Dictionary*, 160.

14. Edward Hughes, *North Country Life in the Eighteenth Century: The North-East, 1700–1750*. (New York: Oxford University Press, 1952), 341.

15. Paul Glennie and Nigel Thrift, *Shaping the Day: A History of Timekeeping in England and Wales, 1300–1800* (New York: Oxford University Press, 2009), 168–69.

16. Gerhard Dohrn-van Rossum, *History of the Hour: Clocks and Modern Temporal Orders*, trans. Thomas Dunlap (Chicago: University of Chicago Press, 1996), 287.

17. Newton was born on Christmas 1642 according to the Julian calendar, which was in use at the time. By his death, England had switched to the Gregorian calendar making the date of his birth January 4, 1643. David Berlinski, *Newton's Gift: How Sir Isaac Newton Unlocked the System of the World* (New York: Free Press, 2000).

18. Ibid., 152–54.

19. L. W. Johnson and M. L. Wolbarsht, "Mercury Poisoning: A Probable Cause of Isaac Newton's Physical and Mental Ills," *Notes and Records of the Royal Society of London* 34, no. 1 (July 1979).

20. Elizabeth Abbott, *A History of Celibacy* (New York: Da Capo, 2001), 345.

21. Stephen D. Snobelen, interviewed by Paul Newall, "Stephen D. Snobelen: Newton Reconsidered," 2005, http://www.galilean-library.org/site/index.php?/page/index.html/_/interviews/stephen-d-snobelen-newton-reconsidered-r39 (accessed September 21, 2010).

22. Ibid.

23. Ibid.

24. Edward Grand, *Much Ado About Nothing* (New York: Cambridge University Press, 1981), 10.

25. Ibid., 54.

26. Helge S. Kragh, *Conceptions of Cosmos: From Myths to the Accelerating Universe: A History of Cosmology* (New York: Oxford University Press, 2007), 68.

27. Ibid., 67–66.

28. Robert Rynasiewicz, "Newton's Views on Space, Time, and Motion," in *The Stanford Encyclopedia of Philosophy*, ed. Edward N. Zalta, Fall 2008 ed., http://plato.stanford.edu/archives/fall2008/entries/newton-stm/ (accessed October 20, 2010).

29. Samuel L. Macey, *Encyclopedia of Time* (New York: Garland, 1994), 426.

30. Karl Popper, *The World of Parmenides: Essays on the Presocratic Enlightenment* (New York: Routledge, 1998), 123; J. Hadamard, "Newton and the Infinitesimal Calculus," in *Newton Tercentenary Celebrations* (Cambridge: The Royal Society, 1947), 123.

31. Dava Sobel, *Longitude: The True Story of a Lone Genius Who Solved the Greatest Scientific Problem of His Time* (New York: Walker, 1995), 11.

32. For a guide to modern sextant use see Richard K. Hubbard, *Boater's Bowditch: The Small-Craft American Practical Navigator* (Camden, ME: International Marine/McGraw-Hill, 2000), 157.

33. Derek Howse, *Greenwich Time and Longitude* (London: Philip Wilson, 1997), 75.

34. Harrison's work on the chronometer was in quest of the so-called Longitude Prize. In 1714 the British government offered £20,000 as a prize for a solution to the longitude problem. Provide a longitude measurement system that was good to within half a degree (two minutes of time) and the prize could be claimed. Many inventors tried, and many inventors failed, including many of the age's scientific luminaries.

35. Howse, *Greenwich Time and Longitude*, 75.

36. Ibid.

37. Kragh, *Conceptions of Cosmos*, 76.

38. Ibid.

39. M. A. Hoskin, "Newton, Providence and the Universe of Stars," *Journal for the History of Astronomy* 8 (1977): 79.

40. Helge S. Kragh, "Cosmology and the Entropic Creation Argument," *Historical Studies in the Physical and Biological Sciences* 37, no. 2 (2007): 371.

41. Ibid.

42. Edward Harrison, *Cosmology: The Science of the Universe* (New York: Cambridge University Press, 2000), 74.

43. Kragh, *Conceptions of Cosmos*, 82.

44. Roger Hahn, "Laplace and the Mechanistic Universe," in *God and Nature: Historical Essays on the Encounter Between Christianity and Science*, ed. David C. Lindberg and Ronald L. Numbers (Berkeley: University of California Press, 1986), 256.

45. Rodney Carlisle, *Scientific American: Inventions and Discoveries* (Hoboken, NJ: John Wiley and Sons, 2004), 209.

46. Arthur Haberman, *The Making of the Modern Age* (Toronto: Gage, 1984).

47. Saskia Sassen, *Territory, Authority, Rights: From Medieval to Global Assemblages* (Princeton: Princeton University Press, 2006), 103.

48. A. Roger Ekirch, *At Day's Close: Night in Times Past* (New York: W. W. Norton, 2005), 3.

49. Ibid., xxxi.

50. Ibid., 15.

51. Ibid., 328.

52. Wax and spermaceti candles were particularly expensive. In 1765, Horace Walpole estimated that it cost the marquis de la Borde, a wealthy Parisian financier, more than 28,000 livres to heat and light his palatial home. Ibid., 103.

53. Ibid., 300.

54. Ibid., 302.

55. Ibid., 334.

56. Ibid., 93.

57. Wolfgang Schivelbusch, *Disenchanted Night: The Industrialization of Light in the Nineteenth Century*, trans. Angela Davies (Berkeley: University of California Press, 1988), 93.

58. Ibid., 110.

59. Ibid., 111–13.

60. Ibid., 115.

61. Ibid., 34.

62. Ekirch, *At Day's Close*, 333.

63. Russell Leigh Sharman and Cheryl Harris Sharman, *Nightshift NYC* (Berkeley: University of California Press, 2008), 113.

64. Schivelbusch, *Disenchanted Night*, 58.

65. Ibid., 65.

66. Ibid., 70.

67. Ibid., 71.

68. Sharman and Sharman, *Nightshift NYC*, 113.

69. The true nature of the connection between time and entropy flow is debated to this day, though there remains general agreement that they are somehow intertwined. (The phrase "arrow of time" was not coined until 1927 by Arthur Eddington.)

70. Kragh, *Conceptions of Cosmos*, 102.

71. Rudolf Clausius, "On the Second Fundamental Theorem of the Mechanical

Theory of Heat," *London, Edinburgh, and Dublin Philosophical Magazine and Journal of Science* 35, no. 239 (June 1868): 419.

72. In fact, Darwin and Lyell corresponded heatedly on the nature of evolution. Darwin was well versed in Lyell's writings and may have seen himself as the heir to Lyell's uniformitarianism. David N. Stamos, *Darwin and the Nature of Species* (Albany: State University of New York Press, 2007), 54.

73. Kragh, *Conceptions of Cosmos*, 102.

74. *Report of the Annual Meeting of the British Society for the Advancement of Science* (London: Taylor and Francis, 1872).

75. Kragh, *Conceptions of Cosmos*, 106.

CHAPTER 5: THE TELEGRAPH, THE ELECTRIC CLOCK AND THE BLOCK UNIVERSE

1. *Traveler's Official Railway Guide for the United States and Canada*, vol. 11 (New York: National Railway Publication Co., 1881), 108.

2. Donald McCloskey, "The Industrial Revolution 1780–1860: A Survey," in *The Economics of the Industrial Revolution*, ed. Joel Mokyr (Totowa, NJ: Rowman and Littlefield, 1985), 58.

3. In 1865, the United States was already crisscrossed with telephone lines. Western Union boasted 44,000 miles of telegraph lines, the American Telegraph Company had 23,000 miles and the United States Telegraph Company had 16,000 miles. Richard Allen Schwarzlose, *The Nation's Newsbrokers*, vol. 2: *The Rush to Institution, from 1865 to 1920* (Evanston, IL: Northwestern University Press, 1990), 10.

4. The "flying machine," operating during the 1770s, was advertised as a miracle of speed because it covered the 100 miles between New York City and Philadelphia in only two days. The website MapQuest says that a trip from New York City to Philadelphia today would take 2 hours. John Thomas Scharf, *A History of Philadelphia* (Philadelphia: J. H. Everts, 1884), 3:2159.

5. Express trains would make the trip in about three hours. With stops the trip would be closer to four hours. Emory Edwards, *Modern American Locomotive Engines, Their Design Construction and Management: A Practical Work for Practical Men* (Philadelphia: Henry Carey Baird, 1895), 130.

6. The Pennsylvania Railroad ran on Philadelphia time, which was five minutes slower than New York time and five minutes faster than Baltimore time. Carlton J. Corliss, *The Day of Two Noons*, 6th ed. (Washington, DC: Association of American Railroads, 1952), 2. Some cities, such as Albany

and Buffalo had local times more than twenty minutes apart on the same rail. Ian R. Bartky, *Selling the True Time: Nineteenth-Century Timekeeping in America* (Stanford: Stanford University Press, 2000), 22.

7. Peter Galison, *Einstein's Clocks, Poincaré's Maps* (New York: W. W. Norton, 2003), 125.

8. For example, by 1874, the Pennsylvania Railway operated on the times of Philadelphia and Columbus, Ohio. Bartky, *Selling the True Time*, 244.

9. Galison, *Einstein's Clocks*, 126.

10. Ibid., 99.

11. Ibid., 116.

12. Ibid., 122.

13. Ibid., 123.

14. Ibid., 127.

15. In 1881 the trip from Florida to New York could be made in about sixteen hours. *Traveler's Official Railway Guide for the United States and Canada*, 321.

16. Galison, *Einstein's Clocks*, 136.

17. Ibid., 141.

18. Ibid., 144.

19. Walter Isaacson, *Einstein: His Life and Universe* (New York: Simon & Schuster, 2007), 114.

20. The experiment used crossed beams of light that would be brought together and compared after they had moved against, with or perpendicular to the aether. By comparing shifts in the relative position of the light wave peaks, the experimenters could look for shifts in the speed of light due to motion with respect to the aether. This process is called interferometry.

21. Galison, *Einstein's Clocks*, 261–62.

22. Ibid., 258.

23. Ibid., 253.

24. Ibid., 262.

25. Ibid., 265.

26. The title in English translation is "The Foundation of the General Theory of Relativity," published in *Annalen der Physik* (1916). Alice Calaprice and Trevor Lipscombe, *Albert Einstein: A Biography* (Westport, CT: Greenwood Press, 2005), 70.

27. Ibid., 67.

28. Galison, *Einstein's Clocks*, 29.

29. Ibid., 34.

30. Ibid., 325.

31. Ibid., 312.

32. Ibid., 327–28.

CHAPTER 6: THE EXPANDING UNIVERSE, RADIO HOURS AND WASHING MACHINE TIME

1. Susan Strasser, *Never Done* (New York: Pantheon, 1982), 105.
2. Elaine Landau, *The History of Everyday Life* (Minneapolis, MN: Twenty-First Century Books, 2006), 26.
3. Ibid.
4. Ibid., 26–27.
5. Carolyn Dever, *Skeptical Feminism: Activist Theory, Activist Practice* (Minneapolis: University of Minnesota Press, 2004), 133.
6. Strasser, *Never Done*, 79.
7. David J. Cole, Eve Browning and Fred E. H. Schroeder, *Encyclopedia of Modern Everyday Inventions* (Westport, CT: Greenwood Press, 2003), 171.
8. Ibid., 11.
9. Peter Marber, *Seeing the Elephant: Understanding Globalization from Trunk to Tail* (Hoboken, NJ: John Wiley, 2009), 100.
10. David E. Kyvig, *Daily Life in the United States 1920–1939: Decades of Promise and Pain* (Westport, CT: Greenwood Press, 2002), 56.
11. Ibid.
12. Ibid., 55, 202.
13. Loretta Lorance, "Promise, Promises: The Allure of Household Appliances in the 1920s," *Part*, Spring 1998, http://web.gc.cuny.edu/dept/arthi/part/part2-3/house.html.
14. Some stars put out more energy per second than others. Thus a star that appears bright in the sky might be of inherently low energy output but just be closer to us than stars that are putting out more energy but are farther away. Bradley W. Carroll and Dale A. Ostlie, *An Introduction to Modern Astrophysics* (New York: Pearson, 2007).
15. Allen Sandage et al., *Centennial History of the Carnegie Institution of Washington*, vol. 1: *The Mt. Wilson Observatory* (New York: Cambridge University Press, 2004).
16. Helge S. Kragh, *Conceptions of Cosmos: From Myths to the Accelerating Universe: A History of Cosmology* (New York: Oxford University Press, 2007).
17. Ibid., 110.
18. Ibid.
19. Ibid., 80.
20. Ibid., 118.

21. Ibid., 117.
22. Ibid., 118.
23. Martin Gorst, *Measuring Eternity: The Search for the Beginning of Time* (New York: Broadway Books, 2001), 215.
24. A fellow Rhodes Scholar, Jakob Larsen, wrote of Hubble, "We laughed at his effort to acquire an extreme English pronunciation. . . . We always claimed that he could not be consistent, so that he might take a bäth in a bäth tub." Gale E. Christianson, *Edwin Hubble: Mariner of the Nebulae* (Chicago: University of Chicago Press, 1995), 65.
25. Kragh, *Conceptions of Cosmos*, 119.
26. Ibid., 115.
27. Ibid.
28. Ibid., 119.
29. Gorst, *Measuring Eternity*, 221.
30. Kragh, *Cosmology and Controversy: The Historical Development of Two Theories of the Universe* (Princeton, NJ: Princeton University Press, 1996), 7.
31. Astronomer Hugo von Seeliger had shown just two decades before that an infinite universe with constant density could not be brought into agreement with Newton's law. Kragh, *Conceptions of Cosmos*, 109.
32. D. Neuenschwander, "History of Big Bang Cosmology, Part 3: The De Sitter Universe and Redshifts," *Radiations Magazine* (Fall 2008): 26.
33. Gorst, *Measuring Eternity*.
34. Neuenschwander, "History of Big Bang Cosmology," 27.
35. John North, *God's Clockmaker: Richard of Wallingford and the Invention of Time* (New York: Continuum, 2006).
36. The exact quote is "We wonder why we should be shunned as though our system were a plague spot in the universe." Arthur Stanley Eddington, *The Expanding Universe* (New York: Macmillan, 1933).
37. Gorst, *Measuring Eternity*.
38. Kragh, *Conceptions of Cosmos*, 140.
39. Gorst, *Measuring Eternity*, 227.
40. Kragh, *Conceptions of Cosmos*, 141.
41. Gorst, *Measuring Eternity*, 233.
42. Joseph Cambray and David H. Rosen, *Synchronicity: Nature and Psyche in an Interconnected Universe* (New York: Cambridge University Press, 2009), 18.
43. Anthony J. Rudel, *Hello, Everybody! The Dawn of American Radio* (New York: Harcourt, 2008), 94–95.
44. Ibid.
45. For a detailed breakdown of NBC in its earliest phases see Christopher

H. Sterling, "Blue Network," in *Encyclopedia of Radio* (New York: Fitzroy Dearborn, 2004).

46. Manuel G. Doncel, "On Hertz's Conceptual Conversion: From Wire Waves to Air Waves," in *Heinrich Hertz: Classical Physicist, Modern Philosopher*, ed. Davis Baird, R. I. G. Hughes and Alfred Nordman (Hingham, MA: Kluwer Academic, 1998), 73.

47. Rudel, *Hello, Everybody*, 13.

48. Ibid.

49. Ibid.

50. Tom Lewis, *Empire of the Air: The Men Who Made Radio* (New York: Edward Burlingame Books, 1991), 89.

51. Rudel, *Hello, Everybody*, 28.

52. Ibid., 31.

53. Ibid., 33.

54. Ibid.

55. Ibid.

56. Ibid., 65.

57. Ibid., 66.

58. Ibid., 83.

59. R. Alton Lee, *The Bizarre Careers of John R. Brinkley* (Lexington: University Press of Kentucky, 2002), 97.

60. Rudel, *Hello, Everybody*, 209.

61. Christopher H. Sterling, "Amos 'n' Andy," in *Encyclopedia of Radio* (New York: Fitzroy Dearborn, 2004), 125–26.

62. Rudel, *Hello, Everybody*.

63. Ibid.

64. Luther F. Sies, *Encyclopedia of American Radio, 1920–1960*, 2nd ed. (New York: McFarland, 2008), 1:607.

65. Rudel, *Hello, Everybody*.

66. Ibid.

CHAPTER 7: THE BIG BANG, TELSTAR AND A NEW ARMAGEDDON

1. For a detailed account of the Castle Bravo shot see Richard Lee Miller, *Under the Cloud: The Decade of Nuclear Testing* (The Woodlands, TX: Two Sixty, 1991).

2. Ibid., 188–92.

3. Ibid. The fallout from the Castle Bravo test eventually covered a 7,000-square-mile area of the Pacific. See also U.S. Arms Control and

Disarmament Agency, *Worldwide Effects of Nuclear War: Some Perspectives* (Whitefish, MT: Kessinger, 2004), 7.

4. The blast was expected to reach a height of twenty-five miles, then "safely" scatter radioactive particulates over hundreds of miles. However, radiation levels in the immediate vicinity (the control bunker at Enyu, twenty miles away) rose to dangerous levels at an alarming rate, trapping researchers and enlisted men in the bunker until radiation levels were low enough for the men to be extracted. Miller, *Under the Cloud*, 191, 193.

5. Peter J. Coughtry, "Report of the Scientific Secretary," Paper presented at the Nato Advanced Research Workshop: Nuclear Tests: Long-term Consequences in the Semipalatinski/Altai Region, Barnaul, Russia, September 5–10, 1994, 1998.

6. Jonathan I. Katz, *The Biggest Bangs: The Mystery of Gamma-Ray Bursts, the Most Violent Explosions in the Universe* (New York: Oxford University Press, 2002), 35.

7. Edward L. Wolf, *Nanophysics and Nanotechnology: An Introduction to Modern Concepts in Nanoscience*, 2nd ed. (Moerlenbach: Wiley VCH, 2006), 3.

8. A detailed history of quantum mechanics along with a history of the atomic bomb can be found in Richard Rhodes, *The Making of the Atomic Bomb* (New York: Simon & Schuster, 1986).

9. Heisenberg was speaking to our inability to conceptualize the atom given our present language and conceptual vocabulary that deals with phenomena on an everyday, human scale. James D. Stein, *How Math Explains the World: A Guide to the Power of Numbers, from Car Repair to Modern Physics* (New York: HarperCollins/Smithsonian Books, 2008), 57.

10. Mark P. Silverman, *Quantum Superposition: Counterintuitive Consequences of Coherence, Entanglement, and Interference* (Berlin: Springer, 2008), vii.

11. Victor Mansfield, *Synchronicity, Science and Soul-Making: Understanding Jungian Synchronicity* (Peru, IL: Open Court, 2001), 32.

12. Carsten Reinhardt, *Chemical Sciences in the 20th Century: Bridging Boundaries* (Moerlenbach: Wiley VCH, 2001).

13. Eddington's reluctance to broach a scientific narrative of the beginning may have been due in part to his private beliefs. He was a Quaker by upbringing. While not a Christian apologist, Eddington warned against intermingling the scientific and spiritual realms: "I repudiate the idea of proving the distinctive beliefs of religion either from the data of physical science or by the methods of physical science." Larry Witham, *Measure of God: Our Century-Long Struggle to Reconcile Science and Religion* (San Francisco: HarperSanFrancisco, 2005), ix.

14. Helge S. Kragh, *Cosmology and Controversy: The Historical Development of*

Two Theories of the Universe, (Princeton, NJ: Princeton University Press, 1996), 46.

15. Howard Caygill, *A Kant Dictionary* (Padstow, Cornwall: Blackwell, 1995), 136.

16. Gerald James Holton and Stephen G. Brush, *Physics, the Human Adventure: From Copernicus to Einstein and Beyond* (Piscataway, NJ: Rutgers University Press, 2004), 497.

17. Georges Lemaître, "L'Univers en expansion," *Ann Soc Sci Bruxelles* A 21 (1933): 51–85.

18. J. S. Plastkett, "The Expansion of the Universe," *Journal of the Royal Astronomical Society of Canada* 27, no. 35 (1933): 252.

19. Helge S. Kragh, *Conceptions of Cosmos: From Myths to the Accelerating Universe: A History of Cosmology* (New York: Oxford University Press, 2007), 157.

20. Eddington, *The Expanding Universe*, 125.

21. Mark S. Madsen, *The Dynamic Cosmos: Exploring the Physical Evolution of the Universe* (Cornwall: CRC, 1995), 6.

22. Étienne Klein and Marc Lachièze-Rey liken opposition to Lemaître's big bang to the Catholic Church's opposition to cosmologies emerging in the sixteenth century in that the idea of the expanding universe "clashed with the centuries-old doctrine of a changeless universe." Étienne Klein and Marc Lachièze-Rey, *The Quest for Unity: The Adventure of Physics* (New York: Oxford University Press, 1999).

23. Paul Halpern, *Countdown to Apocalypse: A Scientific Exploration of the End of the World* (New York: Perseus, 1998), 11.

24. Citing an atomic weapon's power in equivalent TNT is misleading, as it does not embrace the new killing physics of radiation inherent to an atomic bomb. The neutrons, gamma rays and beta particles released during a nuclear explosion have the right mean free path to cause maximum damage to living tissue. For an in-depth study of the deleterious effects of nuclear explosions, see Leo Sartori, "Effects of Nuclear Weapons," in *Physics and Nuclear Arms Today: Readings from Physics Today*, ed. David Hafemeister (New York: American Institute of Physics, 1991), 2.

25. *America in the 20th Century: 1940–1949*, 2nd ed. (Tarrytown, NY: Marshall Cavendish, 2003), 640.

26. Richard Rhodes, *Dark Sun: The Making of the Hydrogen Bomb* (New York: Simon & Schuster, 1995), 497.

27. "Harold Agnew on: The 'Mike' Test," *American Experience*, PBS, http://www.pbs.org/wgbh/amex/bomb/filmmore/reference/interview/

agnewmiketest.html (accessed February 11, 2011).

28. Robert Frank Futrell, *Ideas, Concepts, Doctrine: Basic Thinking in the United States Air Force: 1907–1960* (Maxwell Air Force Base, AL: Air University Press, 2002), 1:513.

29. Ibid., 202.

30. Eric A. Croddy and James J. Wirtz, eds., *Weapons of Mass Destruction: An Encyclopedia of Worldwide Policy, Technology and History* (Santa Barbara, CA: ABC-CLIO, 2005), 26.

31. Marianne J. Dyson, *Space and Astronomy: Decade by Decade* (New York: Facts on File, 2007), 122–23.

32. Roger Handberg, *Seeking New World Vistas: The Militarization of Space* (Westport, CT: Praeger, 2000), 77.

33. Stephen Weiner, "Systems and Technology," in *Ballistic Missile Defence*, eds. Ashton B. Carter and David N. Schwartz (Washington, DC: Brookings Institution, 1984), 51–52.

34. For a complete breakdown of nuclear weapons by country and year, see Robert Norris, "Global Nuclear Stockpiles, 1945–2006," *Bulletin of the Atomic Scientists* 62, no. 4 (July–Aug, 2006): 66.

35. John H. Barton and Lawrence D. Weiler, eds., *International Arms Control: Issues and Agreements* (Stanford: Stanford University Press, 1976), 56.

36. Kragh, *Conceptions of Cosmos*, 178.

37. These are the most abundant isotopes of the elements listed.

38. Joseph A. Angelo Jr., "Gamow, George," in *Encyclopedia of Space and Astronomy* (New York: Facts on File, 2006), 257.

39. Ibid.

40. Kragh, *Conceptions of Cosmos*, 178.

41. Ibid.

42. Deborah Todd and Joseph A. Angelo Jr., *A to Z of Scientists in Space and Astronomy* (New York: Facts on File, 2005), 10.

43. Later versions would push this start time earlier. See R. Alpher and R. Herman, "Remarks on the Evolution of the Expanding Universe," *Physical Review*, 75 (1949): 1089–99, and Kragh, *Conceptions of Cosmos*, 183.

44. George Gamow, *Creation of the Universe* (New York: Viking Press, 1952), 62.

45. Mauro D'Onofrio and Carlo Burgiana, eds., *Questions of Modern Cosmology: Galileo's Legacy* (Berlin: Springer-Verlag, 2009), 48; Kragh, *Cosmology and Controversy*, 130.

46. Kragh, *Conceptions of Cosmos*, 183.

47. Todd and Angelo, *A to Z of Scientists in Space and Astronomy*, 11.

48. Rikky Rooksby, *Inside Classic Rock Tracks: Songwriting and Recording Secrets of 100 Great Songs from 1960 to the Present Day* (San Francisco: Backbeat Books, 2001), 21–22.

49. Richard Lewis, "Telstar: First with the Most," *Bulletin of the Atomic Scientists: The Magazine of Science and Public Affairs* 18, no. 10 (December 1962): 38.

50. The Thor-Delta rocket was of the same family of rockets as the Thor-Intercontinental ("Thoric"), which was a proposed ICBM. However the project was scrapped in favour of the Atlas and Titan ICBMs. David J. Darling, *The Complete Book of Space: From Apollo 1 to Zero Gravity* (Hoboken, NJ: John Wiley, 2003), 432–33.

51. Benjamin F. Shearer et al., eds., *Home Front Heroes: A Biographical Dictionary of Americans During Wartime* (Westport, CT: Greenwood Press, 2007), 1:231.

52. David K. van Keuren, "Moon in Their Eyes: Moon Communication Relay at the Naval Research Laboratory, 1951–1962," in *Beyond the Ionosphere: Fifty Years of Satellite Communication*, ed. Andrew J. Butrica (Washington, DC: NASA, Office of History, 1997), 9.

53. Mark Williamson, *Spacecraft Technology: The Early Years*, 2nd ed. (Bodmin, Cornwall: Institution of Engineering and Technology, 2008), 179.

54. America's first satellite, Explorer 1, was launched on January 31, 1958. Matthew A. Bille and Erika Lishock, *The First Space Race: Launching the World's First Satellites* (College Station: Texas A&M University Press, 2004), 130.

55. Donald H. Martin, *Communication Satellites*, 4th ed. (El Segundo, CA: Aerospace Corporation, 2000), 4, 7, 8.

56. John Bray, *Innovation and the Communications Revolution: From the Victorian Pioneers to Broadband Internet* (Bodmin, Cornwall: Institution of Engineering and Technology, 2009), 214.

57. Khi V. Thai et al., *Handbook of Globalization and the Environment* (Boca Raton, FL: CRC Press, 2007), 104.

58. Walter Cronkite, "The Day the World Got Smaller," NPR, July 23, 2002, http://www.npr.org/news/specials/cronkite (accessed February 23, 2001).

59. James Schwoch, *Global TV: New Media and the Cold War, 1946–69* (Urbana: University of Illinois Press, 2009), 1.

60. Kragh, *Cosmology and Controversy*, 129.

61. Corey Powell, *God in the Equation: How Einstein Transformed Religion* (New York: Free Press, 2003), 169.

62. Ibid.

63. Kragh, *Cosmology and Controversy*, 199.

64. Ibid., 85.

65. Simon Mitton, *Conflict in the Cosmos: Fred Hoyle's Life in Science* (Washington, DC: Joseph Henry, 2005), 134.

66. Martin Gorst, *Measuring Eternity: The Search for the Beginning of Time* (New York: Broadway Books, 2001), 254.

67. Ken Croswell, *The Universe at Midnight: Observation Illuminating the Cosmos* (New York: Free Press, 2001), 45.

68. Mitton, *Conflict in the Cosmos*, 204.

CHAPTER 8: INFLATION, MOBILE PHONES AND THE OUTLOOK UNIVERSE

1. Shunryu Suzuki, *Zen Mind, Beginner's Mind* (Boston: Shambhala, 2010), 12.

2. Steve Lohr, "Is Information Overload a 650 Billion Dollar Drag on the Economy?" *New York Times*, December 20, 2007, http://bits.blogs.nytimes.com/2007/12/20/is-information-overload-a-650-billion-drag-on-the-economy (accessed December 29, 2010).

3. Michael B. Schiffer, *Power Struggles: Scientific Authority and the Creation of Practical Electricity Before Edison* (Cambridge, MA: MIT Press, 2008), 144.

4. Robert V. Bruce, *Bell: Alexander Graham Bell and the Conquest of Solitude* (Ithaca, NY: Cornell University Press, 2000), 181.

5. Todd Campbell, "The First Email Message: Who Sent It and What It Said," *Pretext* (March 1998).

6. Ibid.

7. Claire Hewson et al., *Internet Research Methods: A Practical Guide for the Social and Behavioural Sciences* (Thousand Oaks, CA: Sage, 2003), 4.

8. Tom Van Vleck, "The History of Electronic Mail," http://www.multicians.org/thvv/mail-history.html (accessed February 24, 2011).

9. Tomlinson did not invent file transfer as a concept or the File Transfer Protocol (FTP) widely in use today. Rather, his experimental protocol CPYNET was instrumental to the file transfer protocol that became e-mail. Campbell, "The First Email Message."

10. Ibid.

11. Patrice Flichy, *The Internet Imaginaire* (Cambridge, MA: MIT Press, 2007), 46.

12. Gerard O'Regan, *A Brief History of Computing* (London: Springer, 1998), 188.

13. Thomas Streeter, *The Net Effect: Romanticism, Capitalism, and the Internet* (New York: New York University Press, 2011), 126.

14. See Mark Fischetti and Tim Berners-Lee, *Weaving the Web: The Original Design and Ultimate Destiny of the World Wide Web by Its Inventor* (San Francisco: HarperSanFrancisco, 1999), 68.

15. Lelia Green, *The Internet: An Introduction to New Media* (New York: Berg, 2010), 33.

16. Catherine C. Marshall, "How People Manage Information over a Lifetime," in *Personal Information Management*, ed. William P. Jones and Jaime Teevan (Seattle: University of Washington Press, 2008), 70.

17. Since the speed of light is an upper bound on motion, all accelerators bring their particles to velocities very close to c. Bigger accelerators add lots more energy to the particles, which, however, only increases their actual velocity by a small amount.

18. Richard Feynman, *Perfectly Reasonable Deviations from the Beaten Track: The Letters of Richard P. Feynman*, ed. Michelle Feynman (New York: Basic Books, 2006), 100.

19. Bogdan Pohv et al., *Particles and Nuclei: An Introduction to the Physical Concepts*, trans. Martin Lavelle, 6th ed. (Heidelberg: Springer-Verlag, 2008), 3.

20. Peter Coles, *The Routledge Companion to the New Cosmology* (New York: Routledge, 2001), 66.

21. If omega begins larger than 1, it remains larger than one 1 forever. If it begins less than one 1, then it remains so forever.

22. Helge S. Kragh, *Conceptions of Cosmos: From Myths to the Accelerating Universe: A History of Cosmology* (New York: Oxford University Press, 2007), 233.

23. See Alan Guth and Paul J. Steinhardt, "The Inflationary Universe," *Scientific American* 250 (1984): 90.

24. Steven Soter and Neil de Grasse Tyson, eds., *Cosmic Horizons: Astronomy at the Cutting Edge* (New York: New Press, 2000).

25. Kragh, *Conceptions of Cosmos*, 213.

26. Dava Sobel, *Longitude: The True Story of a Lone Genius Who Solved the Greatest Scientific Problem of His Time* (New York: Walker, 1995), 16.

27. Nel Samama, *Global Positioning: Technologies and Performance* (Hoboken, NJ: John Wiley, 2008), 21.

28. William Lowrie, *Fundamentals of Geophysics*, 3rd ed. (Cambridge: Cambridge University Press, 2002), 70.

29. Ibid.

30. Ibid. The system comprises twenty-four satellites such that five to eight are visible at any given time.

31. Ibid.
32. Ibid. A series of satellite-based measurements in 1989 and 1993 in Greece found that southwestern Greece moved to the southwest on the order of 20–40 mm per year.
33. Claude Audoin and Bernard Guinot, *The Measurement of Time: Time, Frequency, and the Atomic Clock* (Cambridge, UK: Cambridge University Press, 2001), 99; Tom Logsdon, *Understanding the Navstar: GPS, GIS, and IVHS*, 2nd ed. (New York: Chapman & Hall, 1992), 166.
34. Tony Jones, *Splitting the Second: The Story of Atomic Time* (Philadelphia: Institute of Physics, 2001), 136.
35. Ibid.
36. Richard Pogge, "Real-World Relativity: The GPS Navigation System," http://www.astronomy.ohio-state.edu/~pogge/Ast162/Unit5/gps.html (accessed February 18, 2011).
37. Mary Pat Kelly, *"Good to Go": The Rescue of Capt. Scott O'Grady, USAF, from Bosnia* (Annapolis: U.S. Naval Institute Press, 1996), 135.
38. Anne G. K. Solomon, "The Global Positioning System," in *Triumphs and Tragedies of the Modern Presidency: Seventy-Six Case Studies in Presidential Leadership*, ed. David Abshire (Westport, CT: Praeger, 2001), 108.
39. Mohinder S. Grewal et al., *Global Positioning Systems, Inertial Navigation, and Integration*, 2nd ed. (Hoboken, NJ: John Wiley, 2007), 144.
40. Woodrow Barfield and Thomas Caudell, *Fundamentals of Wearable Computers and Augmented Reality* (Mahwah, NJ: Lawrence Erlbaum, 2001), 374.
41. Richard Raysman, *Emerging Technologies and the Law: Forms and Analysis* (New York: Law Journal Press, 2003), 1–24.
42. Martin Gorst, *Measuring Eternity: The Search for the Beginning of Time* (New York: Broadway Books, 2001), 281.
43. Ibid., 282.
44. Marcelo Gleiser, *A Tear at the Edge of Creation: A Radical New Vision for Life in an Imperfect Universe* (New York: Free Press, 2010), 96.
45. S. Alan Stern, *Our Universe: The Thrill of Extragalactic Exploration as Told by Leading Experts* (New York: Cambridge University Press, 2001), 51.
46. Peter Coles and Francesco Lucchin, *Cosmology: The Origin and Evolution of Cosmic Structure*, 2nd ed. (Hoboken, NJ: John Wiley, 2003), 263.
47. Marcelo Gleiser, *The Prophet and the Astronomer: A Scientific Journey to the End of Time* (New York: W. W. Norton, 2003), 282.

CHAPTER 9: WHEELS WITHIN WHEELS

1. Cyclical cosmology is common in orthodox stoic thought. Universes end in complete, devastating conflagration. Chrysippus argues that conflagration did not involve substantial change, let alone all-consuming fire. Ricardo Salles describes Chrysippus' cycles: "The initial and final fire of any cosmic cycle is nothing but god himself in a completely undifferentiated state." Ricardo Salles, "Introduction: God and Cosmos in Stoicism," in *God and Cosmos in Stoicism*, ed. Ricardo Salles (New York: Oxford University Press, 2009), 3–4.

2. Paul J. Steinhardt and Neil Turok, *Endless Universe: Beyond the Big Bang* (New York: Doubleday, 2007), 175.

3. Helge S. Kragh, *Conceptions of Cosmos: From Myths to the Accelerating Universe: A History of Cosmology* (New York: Oxford University Press, 2007), 147.

4. R. H. Dicke and P. J. E. Peebles, "The Big Bang Cosmology—Enigmas and Nostrums," in *General Relativity: An Einstein Centenary Survey*, ed. Stephen Hawking and W. Israel (Cambridge: Cambridge University Press, 1979), 511.

5. Charles Misner et al., *Gravitation* (New York: Freeman, 1973), 805.

6. Kip S. Thorne, *Black Holes and Time Warps: Einstein's Outrageous Legacy* (New York: W. W. Norton, 1994).

7. Misner et al., *Gravitation*, 11.

8. See Lee Smolin, *Three Roads to Quantum Gravity* (New York: Basic Books, 2001).

9. Steinhardt and Turok, *Endless Universe*, 124.

10. Ibid., 125, 129.

11. A nontechnical description is in Brian Greene, *The Elegant Universe* (New York: W. W. Norton, 1999), 206. A hypertechnical description can be found in Joseph Polchinski, *String Theory: An Introduction to the Bosonic String* (New York: Cambridge University Press, 1998), 1:6.

12. In what follows I am taking the description from Steinhart and Turok, *Endless Universe*. For original papers on the subject, see Justin Khoury et al., "The Ekpyrotic Universe: Colliding Branes and the Origin of the Hot Big Bang," *Physical Review D* 64, no. 12 (November 28, 2001), and Paul J. Steinhardt and Neil Turok, "The Cyclic Model Simplified," *New Astronomy Reviews* 49, no. 2–6 (2005): 43–57.

13. Andreas Albrecht and Neil Turok, "Evolution of Cosmic Strings," *Physical Review Letters* 54 (1985): 1868; Alan Guth and Paul J. Steinhardt, "The Inflationary Universe," *Scientific American* 250 (1984): 90.

14. Steinhardt and Turok, *Endless Universe*, 148.

15. Ibid., 189.

16. I am using the term *ekpyrotic cyclic model* to distinguish it from other new versions of cyclic cosmology.

17. Steinhardt and Turok, *Endless Universe*, 197.

18. See Clifford M. Will, *Was Einstein Right?: Putting General Relativity to the Test* (New York: Perseus, 1984), 181, for a description of gravity waves and pulsars.

19. I have slightly modified this version of the story from these sources: Heinrich Robert Zimmer, *Myths and Symbols in Indian Art and Civilization*, ed. Joseph Campbell, 2nd ed. (New York: Bollingen, 1974), 3, and David Adams Leeming, *The World of Myth: An Anthology* (New York: Oxford University Press, 1990).

20. Peter Woit, *Not Even Wrong: The Failure of String Theory and the Search for Unity in Physical Law* (New York: Basic Books, 2006); Lee Smolin, *The Trouble with Physics: The Rise of String Theory, the Fall of a Science, and What Comes Next* (New York: Houghton Mifflin Harcourt, 2007).

21. Leonard Susskind, *The Cosmic Landscape: String Theory and the Illusion of Intelligent Design* (New York: Hachette, 2005).

22. Smolin, *Three Roads to Quantum Gravity*.

23. Martin Bojowald, *Once Before Time: A Whole Story of the Universe* (New York: Random House, 2010).

24. Paul H. Frampton, *Did Time Begin? Will Time End? Maybe the Big Bang Never Occurred* (Singapore: World Scientific, 2009).

25. Roger Penrose, *Cycles of Time: An Extraordinary New View of the Universe* (New York: Alfred A. Knopf, 2011).

CHAPTER 10: EVER-CHANGING ETERNITIES

1. Hoyle's lectures were broadcast on Saturdays during the "snowbound" months of January and February 1950. In the absence of an exact date for this particular broadcast I have chosen February 4, 1950, a Saturday. Jane Gregory, *Fred Hoyle's Universe* (New York: Oxford University Press, 2005), 47, 48.

2. For a more complete description, see Alan Guth, *The Inflationary Universe: The Quest for a New Theory of Cosmic Origins* (New York: Basic Books, 1998).

3. Readers who want more information should see Sean Carroll's wonderful book, *From Eternity to Here: The Quest for the Ultimate Theory of Time* (New York: Penguin, 2010).

4. Ibid.

5. See Guth, *The Inflationary Universe*; Alan Guth and Paul J. Steinhardt, "The

Inflationary Universe," *Scientific American* 250 (1984): 90.

6. Carroll, *From Eternity to Here*.

7. See Paul C. W. Davies, "John Archibald Wheeler and the Clash of Ideas," in *Science and Ultimate Reality: Quantum Theory, Cosmology, and Complexity*, ed. John D. Barrow, Paul C. W. Davies, and Charles L. Harper Jr. (New York: Cambridge University Press, 2004), 20.

8. Alexander Vilenkin, *Many Worlds in One: The Search for Other Universes* (New York: Hill and Wang, 2006), 170.

9. Quotes from Carroll come from an interview with the author.

10. Some parts of this article appeared in a story I wrote for *Discover*: "3 Theories That Might Blow Up the Big Bang," April 2008.

11. See Carroll, *From Eternity to Here*, 357.

12. Davies, "John Archibald Wheeler and the Clash of Ideas."

13. Gregory, *Fred Hoyle's Universe*.

14. Leonard Susskind, *The Cosmic Landscape: String Theory and the Illusion of Intelligent Design* (New York: Hachette, 2005), 79.

15. Davies, "John Archibald Wheeler and the Clash of Ideas," xv.

16. Susskind, *Cosmic Landscape*.

CHAPTER 11: GIVING UP THE GHOST

1. This account comes from an interview with Julian Barbour on September 17, 2010.

2. Julian Barbour, *The End of Time: The Next Revolution in Our Understanding of the Universe* (New York: Oxford University Press, 1999).

3. Parts of this chapter first appeared in my articles "3 Theories That Might Blow Up the Big Bang" and "Is the Search for Immutable Laws of Nature a Wild-Goose Chase?", *Discover* (April 2010).

4. Barbour, *The End of Time*, 47

5. Ibid., 229

6. Quotes with Albrecht come from interview with author.

7. Andreas Albrecht et al., "New Inflation in Supersymmetric Theories," *Nuclear Physics B* 229 (1983): 528.

8. See James Hartle, "Theories of Everything and Hawking's Wave Function of the Universe," in *The Future of Theoretical Physics and Cosmology: Celebrating Stephen Hawking*, ed. G. W. Gibbons, P. Shellard, and S. J. Rankin (New York: Cambridge University Press, 2003), 38.

9. Stephen Hawking, *A Brief History of Time* (New York: Bantam, 1988), 140.

10. Andreas Albrecht and Alberto Iglesias. "The Clock Ambiguity: Implications

and New Developments," in *The Origin of Time's Arrow* (New York: New York Academy of Sciences Press, 2008).

11. John Archibald Wheeler and Hubery Żurek Wojciech, eds., *Quantum Theory and Measurement* (New York: Princeton University Press, 1983).

12. Lee Smolin, *Three Roads to Quantum Gravity* (New York: Basic Books, 2001).

13. Roberto Mangabeira Unger, *Law in Modern Society: Toward a Criticism of Social Theory* (New York: Free Press, 1976); Roberto Mangabeira Unger, *The Self Awakened: Pragmatism Unbound* (New York: Harvard University Press, 2007).

14. Quotes from Smolin and Unger come from interviews with author.

15. Lee Smolin, *The Life of the Cosmos* (Oxford: Oxford University Press, 1998).

16. Lee Smolin, "On the Reality of Time and the Evolution of Laws," (Work in progress presented at the Perimeter Institute for Theoretical Physics, Ontario, Canada, October 2, 2008).

17. Lee Smolin, "The Unique Universe," *Physics World* (June 2009): 26.

CHAPTER 12: IN THE FIELDS OF LEANING GRASS

1. Jeremy Rifkin, *Time Wars: The Primary Conflict in Human History* (New York: Simon & Schuster, 1989), 123.

2. This quote comes from comes from English economist and philosopher Sir William Petty in 1682. Peter C. Dooley, *The Labour Theory of Value* (New York: Routledge, 2005), 30.

3. Rifkin, *Time Wars*, 127.

4. For an entertaining and somewhat curmudgeonly overview of oil supply and other resource depletion issues, see James Howard Kunstler, *The Long Emergency: Surviving the End of Oil, Climate Change, and Other Converging Catastrophes of the Twenty-First Century* (New York: Grove Press, 2006).

5. Maude Barlow, *Blue Covenant: The Global Water Crisis and the Coming Battle for the Right to Water* (New York: New Press, 2009).

6. M. Barret, "ATLAS Experiment Reports Its First Physics Results from the LHC," http://atlas.ch/news/2010/first-physics.html (accessed February 20, 2011); "The Sensitive Giant: At CERN, ATLAS Effort Emphasizing People Skills," U.S. Department of Energy Research News, http://www.eurekalert.org/features/doe/2004-03/dnal-tsg032604.php (accessed February 20, 2011).

7. Barret, "ATLAS Experiment Reports."

8. Ibid.

9. Jonathan Leake, "Big Bang at the Atomic Lab After the Scientists Got Their Maths Wrong," *Sunday Times* (London), April 8, 2007.

10. Alan Sokal, "A Physicist Experiments with Cultural Studies," *Lingua Franca* (May–June 1996).

11. Samir Okasha, *Philosophy of Science: A Very Short Introduction* (New York: Oxford University Press, 2002), 59.

12. Edward Harrison, *Masks of the Universe* (New York: Macmillan, 1985), 2, 1.

13. "Slow Food," http://www.slowfood.com/about_us/eng/mission.lasso (accessed February 20, 2011).

14. Alain Ducasse, "Carlo Petrini: The Slow Revolutionary," *Time Europe*, October 11, 2004.

15. The Sanskrit term is *pratītyasamutpāda*, sometimes translated as "dependent co-arising" or "dependent origination." For a scholarly examination of dependent arising and causation, see Jay L. Garfield, *Empty Words: Buddhist Philosophy and Cross-Cultural Interpretation* (New York: Oxford University Press, 2002), 24.

BIBLIOGRAPHY

Abbott, Elizabeth. *A History of Celibacy*. New York: Da Capo, 2001.

Albrecht, Andreas, and Neil Turok. "Evolution of Cosmic Strings." *Physical Review Letters* 54 (1985): 1868.

Albrecht, Andreas Dimopoulos, et al. "New Inflation in Supersymmetric Theories." *Nuclear Physics* B 229 (1983): 528.

Angelo, Joseph A., Jr. "Gamow, George." In *Encyclopedia of Space and Astronomy*. New York: Facts on File, 2006.

Armstrong, Karen. *A Short History of Myth*. New York: Canongate, 2005.

Audoin, Claude, and Bernard Guinot. *The Measurement of Time: Time, Frequency, and the Atomic Clock*. Cambridge: Cambridge University Press, 2001.

Aveni, Anthony. *Empires of Time: Calendars, Clocks, and Cultures*. Boulder: University Press of Colorado, 2002.

Bahn, Paul G., and Jean Vertut. *Journey Through the Ice Age*. Berkeley: University of California Press, 1997.

Baillargeon, Renée. "How Do Infants Learn About the Physical World?" *Current Directions in Psychological Science* 3, no. 5 (1994).

Bainton, Roland H. *Here I Stand: A Life of Martin Luther*. Peabody, MA: Hendrickson, 1977.

Barbour, Julian. *The End of Time: The Next Revolution in Our Understanding of the Universe*. New York: Oxford University Press, 1999.

Barfield, Woodrow, and Thomas Caudell. *Fundamentals of Wearable Computers and Augmented Reality*. Mahwah, NJ: Lawrence Erlbaum, 2001.

Barlow, Maude. *Blue Covenant: The Global Water Crisis and the Coming Battle for the Right to Water*. New York: New Press, 2009.

Barret, M. "ATLAS Experiment Reports its First Physics Results from the LHC." http://atlas.ch/news/2010/first-physics.html (accessed February 20, 2011).

Bartky, Ian R. *Selling the True Time: Nineteenth-Century Timekeeping in America*. Stanford: Stanford University Press, 2000.

Barton, Tamsyn. *Ancient Astrology*. New York: Routledge, 1994.

Benn, Charles D. *China's Golden Age: Everyday Life in the Tang Dynasty*. New York: Oxford University Press, 2004.

Berlinski, David. *Newton's Gift: How Sir Isaac Newton Unlocked the System of the World*. New York: Free Press, 2000.

Berners-Lee, Tim, and Mark Fischetti. *Weaving the Web: The Original Design and Ultimate Destiny of the World Wide Web by Its Inventor*. San Francisco: HarperSanFrancisco, 1999.

Berryman, Sylvia. "Leucippus." In *The Stanford Encyclopedia of Philosophy*, edited by Edward N. Zalta. Fall 2010 ed., *http://plato.stanford.edu/entries/leucippus/* (accessed November 8, 2010).

Bille, Matthew A., and Erika Lishock. *The First Space Race: Launching the World's First Satellites*. College Station: Texas A&M University Press, 2004.

Bodanis, David. *E=mc²: A Biography of the World's Most Famous Equation*. 2nd ed. New York: Walker, 2005.

Bojowald, Martin. *Once Before Time: A Whole Story of the Universe*. New York: Random House, 2010.

Borst, Arno. *The Ordering of Time: From the Ancient Computus to the Modern Computer*. Chicago: University of Chicago Press, 1993.

Brantingham, P. Jeffrey, Steven L. Kuhn, and Kristopher W. Kerry. *The Early Upper Paleolithic Beyond Western Europe*. Berkeley: University of California Press, 2004.

Bray, John. *Innovation and the Communications Revolution: From the Victorian Pioneers to Broadband Internet*. Bodmin, Cornwall: Institution of Engineering and Technology, 2009.

Bruce, Robert V. *Bell: Alexander Graham Bell and the Conquest of Solitude*. Ithaca, NY: Cornell University Press, 2000.

Burns, Edward McNall. *Western Civilizations: Their History and Their Culture*. New York: W. W. Norton, 1968.

Calaprice, Alice, and Trevor Lipscombe. *Albert Einstein: A Biography*. Westport, CT: Greenwood Press, 2005.

Cambray, Joseph, and David H. Rosen. *Synchronicity: Nature and Psyche in an Interconnected Universe*. New York: Cambridge University Press, 2009.

Campbell, Todd. "The First Email Message: Who Sent It and What It Said." *Pretext* (March 1998). http://web.archive.org/web/20030806031641/www .pretext.com/mar98/features/story2.htm (accessed December 20, 2010).

Carlisle, Rodney. *Scientific American: Inventions and Discoveries*. Hoboken, NJ: John Wiley, 2004.

Carroll, Sean. *From Eternity to Here: The Quest for the Ultimate Theory of Time*. New York: Penguin, 2010.

Carroll, Bradley W., and Dale A. Ostlie. *An Introduction to Modern Astrophysics.* New York: Pearson, 2007.

Cartledge, Paul, Paul Millett, and Sitta von Reden. *Kosmos: Essays in Order, Conflict and Community in Classical Athens.* New York: Cambridge University Press, 1998.

Castelden, Rodney. *The Stonehenge People: An Exploration of Life in Neolithic Britain, 4700–2000 BC.* New York: Routledge, 2002.

Caygill, Howard. *A Kant Dictionary.* Padstow, Cornwall: Blackwell, 1995.

Christianson, Gale E. *Edwin Hubble: Mariner of the Nebulae.* Chicago: University of Chicago Press, 1995.

Clausius, Rudolph. "Prof. R. Clausius on the Second Fundamental Theorem of the Mechanical Theory of Heat." *London, Edinburgh, and Dublin Philosophical Magazine and Journal of Science* 35 (January–June 1868).

Cole, David J., Eve Browning, and Fred E. H. Schroeder. *Encyclopedia of Modern Everyday Inventions.* Westport, CT: Greenwood Press, 2003.

Coles, Peter. *The Routledge Companion to the New Cosmology.* New York: Routledge, 2001.

Coles, Peter, and Francesco Lucchin. *Cosmology: The Origin and Evolution of Cosmic Structure.* 2nd ed. Hoboken, NJ: John Wiley, 2003.

Copernicus, Nicolaus, *On the Revolutions of the Heavenly Spheres,* Translated by Charles Glenn Wallis. Philadelphia: Running Press, 2002.

Corliss, Carlton J. *The Day of Two Noons.* 6th ed. Washington, DC: Association of American Railroads, 1952.

Coughtry, Peter J. "Report of the Scientific Secretary." Paper presented at the Nato Advanced Research Workshop: Nuclear Tests: Long-term Consequences in the Semipalatinski/Altai Region, Barnaul, Russia, September 5–10, 1994.

Croddy, Eric A., and James J. Wirtz, eds. *Weapons of Mass Destruction: An Encyclopedia of Worldwide Policy, Technology and History.* Santa Barbara, CA: ABC-CLIO, 2005.

Croswell, Ken. *The Universe at Midnight: Observation Illuminating the Cosmos.* New York: Free Press, 2001.

Dalley, Stephanie. *Myths from Mesopotamia: Creation, the Flood, Gilgamesh and Others.* Oxford: Oxford University Press, 1998.

Darling, David J. *The Complete Book of Space: From Apollo 1 to Zero Gravity.* Hoboken, NJ: John Wiley, 2003.

Davidson, Iain, and April Nowell. *Stone Tools and the Evolution of Human Cognition.* Boulder: University Press of Colorado, 2010.

Davies, Paul C. W. "John Archibald Wheeler and the Clash of Ideas." In *Science and Ultimate Reality: Quantum Theory, Cosmology, and Complexity,* edited by

John D. Barrow, Paul C. W. Davies, and Charles L. Harper Jr. New York: Cambridge University Press, 2004.

Davis, J. C. *Utopia and the Ideal Society: A Study of English Utopian Writing 1516–1700*. New York: Cambridge University Press, 1981.

Dever, Carolyn. *Skeptical Feminism: Activist Theory, Activist Practice*. Minneapolis: University of Minnesota Press, 2004.

Dicke, R. H., and P. J. E. Peebles. "The Big Bang Cosmology—Enigmas and Nostrums." In *General Relativity: An Einstein Centenary Survey*, edited by Stephen Hawking and W. Israel. Cambridge: Cambridge University Press, 1979.

Dohrn-van Rossum, Gerhard. *History of the Hour: Clocks and Modern Temporal Orders*. Translated by Thomas Dunlap. Chicago: University of Chicago Press, 1996.

Doncel, Manuel G. "On Hertz's Conceptual Conversion: From Wire Waves to Air Waves." In *Heinrich Hertz: Classical Physicist, Modern Philosopher*, edited by Davis Baird, R. I. G. Hughes, and Alfred Nordman. Hingham, MA: Kluwer Academic, 1998.

D'Onofrio, Mauro, and Carlo Burgiana, eds. *Questions of Modern Cosmology: Galileo's Legacy*. Berlin: Springer-Verlag, 2009.

Dooley, Peter C. *The Labour Theory of Value*. New York: Routledge, 2005.

Dreuille, Mayeul de. *The Rule of St. Benedict: A Commentary in Light of World Ascetic Traditions*. Leominster, Herefordshire: Newman Press, 2000.

Ducasse, Alain. "Carlo Petrini: The Slow Revolutionary." *Time Europe*, October 11, 2004.

Dusseldorp, G. L. *A View to a Kill: Investigating Middle Paleolithic Subsistence Using an Optimal Foraging Perspective*. Leiden: Sidestone Press, 2009.

Dyson, Marianne J. *Space and Astronomy: Decade by Decade*. Twentieth-Century Science. New York: Facts on File 2007.

Eddington, Arthur Stanley. *The Expanding Universe*. New York: Macmillan, 1933.

Edwards, Emory. *Modern American Locomotive Engines, Their Design, Construction and Management: A Practical Work for Practical Men*. Philadelphia: Henry Carey Baird, 1895.

Ekirch, A. Roger. *At Day's Close: Night in Times Past*. New York: W. W. Norton, 2005.

Eliade, Mircea. *Myth and Reality*. Prospect Heights, IL: Waveland, 1998.

———. *The Myth of the Eternal Return or Cosmos and History*. Translated by Willard R. Trask. Princeton, NJ: Princeton University Press, 1991.

Ferris, Timothy. *Coming of Age in the Milky Way*. New York: HarperCollins, 1988.

Feynman, Richard. *Perfectly Reasonable Deviations from the Beaten Track: The Letters of Richard P. Feynman*. Edited by Michelle Feynman. New York: Basic Books, 2006.

Fischer, Henry George. "The Origin of Egyptian Hieroglyphs." In *The Origins of Writing*, edited by Wayne M. Senner. Lincoln: University of Nebraska Press, 1991.

Flichy, Patrice. *The Internet Imaginaire*. Cambridge, MA: MIT Press, 2007.

Flinn, M. W. *Men of Iron: The Crowleys in the Early Iron Industry*. Edinburgh: Edinburgh University Press, 1962.

Frampton, Paul H. *Did Time Begin? Will Time End? Maybe the Big Bang Never Occurred*. Singapore: World Scientific, 2009.

Freeman, Charles. *The Closing of the Western Mind: The Rise of Faith and the Fall of Reason*. New York: Alfred A. Knopf, 2003.

Futrell, Robert Frank. *Ideas, Concepts, Doctrine: Basic Thinking in the United States Air Force: 1907–1960*. Vol. 1. Maxwell Air Force Base, AL: Air University Press, 2002.

Galison, Peter. *Einstein's Clocks, Poincaré's Maps*. New York: W. W. Norton, 2003.

Gamow, George. *Creation of the Universe*. New York: Viking Press, 1952.

Garfield, Jay L. *Empty Words: Buddhist Philosophy and Cross-Cultural Interpretation*. New York: Oxford University Press, 2002.

Gates, Charles. *Ancient Cities: The Archaeology of Urban Life in the Ancient Near East and Egypt, Greece, and Rome*. New York: Routledge, 2005.

Gilbert of Mons. *The Chronicle of Hainaut*. Translated by Laura Napran. Woodbridge, Suffolk: Boydell Press, 2005.

Gilbert, William. *On the Loadstone and Magnetic Bodies and on the Great Magnet the Earth*. Translated by Paul Fleury Mottelay. New York: John Wiley, 1893.

Giles, Bill. "Katabatic Winds." BBC. http://www.bbc.co.uk/weather/features/az/alphabet31.shtml (accessed August 24, 2010).

Gleiser, Marcelo. *The Dancing Universe: From Creation Myths to the Big Bang*. Lebanon, NH: Dartmouth College Press, 2005.

———. *The Prophet and the Astronomer: A Scientific Journey to the End of Time*. New York: W. W. Norton, 2003.

———. *A Tear at the Edge of Creation: A Radical New Vision for Life in an Imperfect Universe*. New York: Free Press, 2010.

Glennie, Paul, and Nigel Thrift. *Shaping the Day: A History of Timekeeping in England and Wales 1300–1800*. New York: Oxford University Press, 2009.

Goldsworthy, Adrian Keith. *How Rome Fell: Death of a Superpower*. New Haven: Yale University Press, 2009.

Gorst, Martin. *Measuring Eternity: The Search for the Beginning of Time.* New York: Broadway Books, 2001.

Graham, Daniel W. "Heraclitus." In *The Stanford Encyclopedia of Philosophy,* edited by Edward N. Zalta. Fall 2008 ed. http://plato.stanford.edu/archives/sum2011/entries/heraclitus/ (accessed November 10, 2010).

Grand, Edward. *Much Ado About Nothing: Theories of Space and Vacuum from the Middle Ages to the Scientific Revolution.* New York: Cambridge University Press, 1981.

Green, Lelia. *The Internet: An Introduction to New Media.* Berg New Media Series. New York: Berg, 2010.

Greene, Brian. *The Elegant Universe.* New York: W. W. Norton, 1999.

Gregory, Jane. *Fred Hoyle's Universe.* New York: Oxford University Press, 2005.

Grewal, Mohinder S., et al. *Global Positioning Systems, Inertial Navigation, and Integration.* 2nd ed. Hoboken, NJ: John Wiley, 2007.

Guth, Alan. *The Inflationary Universe: The Quest for a New Theory of Cosmic Origins.* New York: Basic Books, 1998.

Guth, Alan, and Paul J. Steinhardt. "The Inflationary Universe." *Scientific American* 250 (1984): 90.

Haberman, Arthur. *The Making of the Modern Age.* Toronto: Gage, 1984.

Hadamard, J., ed. "Newton and the Infinitesimal Calculus." In *Newton Tercentenary Celebrations.* Cambridge: The Royal Society, 1947.

Hahn, Roger. "Laplace and the Mechanistic Universe." In *God and Nature: Historical Essays on the Encounter between Christianity and Science,* edited by David C. Lindberg and Ronald L. Numbers. Berkeley: University of California Press, 1986.

Halpern, Paul. *Countdown to Apocalypse: A Scientific Exploration of the End of the World.* New York: Perseus, 1998.

Handberg, Roger. *Seeking New World Vistas: The Militarization of Space.* Westport, CT: Praeger, 2000.

Harrison, Edward. *Cosmology: The Science of the Universe.* New York: Cambridge University Press, 2000.

———. *Masks of the Universe.* New York: Macmillan, 1985.

"Harold Agnew on: The 'Mike' Test." *American Experience,* PBS.com. http://www.pbs.org/wgbh/amex/bomb/filmmore/reference/interview/agnewmiketest.html (accessed February 11, 2011).

Hartle, James. "Theories of Everything and Hawking's Wave Function of the Universe." In *The Future of Theoretical Physics and Cosmology: Celebrating Stephen Hawking,* edited by G. W. Gibbons, P. Shellard, and S. J. Rankin. New York: Cambridge University Press, 2003.

Hawking, Stephen. *A Brief History of Time*. New York: Bantam, 1988.

Hesiod. "Works and Days." In *Hesiod*. Translated by Richmond Lattimore. Ann Arbor: University of Michigan Press, 1972.

Hewson, Claire, et al. *Internet Research Methods: A Practical Guide for the Social and Behavioural Sciences*. Thousand Oaks, CA: Sage, 2003.

Holton, Gerald James, and Stephen G. Brush. *Physics, the Human Adventure: From Copernicus to Einstein and Beyond*. Piscataway, NJ: Rutgers University Press, 2004.

Hoskin, M. A., "Newton, Providence and the Universe of Stars." *Journal for the History of Astronomy* 8 (1977): 77–101.

Howse, Derek. *Greenwich Time and Longitude*. London: Philip Wilson, 1997.

Hubbard, Richard K. *Boater's Bowditch: The Small-Craft American Practical Navigator*. Camden, ME: International Marine/McGraw-Hill, 2000.

Huffman, Carl. "Pythagoras." In *The Stanford Encyclopedia of Philosophy*, edited by Edward N. Zatta. Winter 2009 ed. http://plato.stanford.edu/entries/pythagoras/#LifWor (accessed January 30, 2011).

Hughes, Edward. *North Country Life in the Eighteenth Century: The North-East, 1700–1750*. New York: Oxford University Press, 1952.

Humphrey, Paul, ed. *America in the 20th Century: 1940–1949*. 2nd ed. Tarrytown, NY: Marshall Cavendish, 2003.

Hunger, Herman, and David Pingree. *Astral Sciences in Mesopotamia*. Leiden: Brill, 1999.

Isaacson, Walter. *Einstein: His Life and Universe*. New York: Simon & Schuster, 2007.

Johnson, David Martel. *How History Made the Mind: The Cultural Origins of Objective Thinking*. Peru, IL: Open Court, 2003.

Johnson, L. W., and M. L. Wolbarsht. "Mercury Poisoning: A Probable Cause of Isaac Newton's Physical and Mental Ills." *Notes and Records of the Royal Society of London* 34, no. 1 (July 1979).

Johnston, Ian. "Device That Let the Greeks Decode the Solar System." *Scotsman*, November 20, 2006.

Jones, Tony. *Splitting the Second: The Story of Atomic Time*. Philadelphia: Institute of Physics, 2001.

Katz, Jonathan I. *The Biggest Bangs: The Mystery of Gamma-Ray Bursts, the Most Violent Explosions in the Universe*. New York: Oxford University Press, 2002.

Kelly, Mary Pat. *"Good to Go": The Rescue of Capt. Scott O'Grady, USAF, from Bosnia*. Annapolis: U.S. Naval Institute Press, 1996.

Kemp, Barry J. *Ancient Egypt: Anatomy of a Civilization*. New York: Routledge, 2006.

Khoury, Justin, et al. "The Ekpyrotic Universe: Colliding Branes and the Origin of the Hot Big Bang." *Physical Review* D 64, no. 12 (November 28, 2001).

King, Henry C. *The History of the Telescope*. New York: Dover, 1955.

Klein, Étienne, and Marc Lachièze-Rey. *The Quest for Unity: The Adventure of Physics*. New York: Oxford University Press, 1999.

Knowles, David. *The Monastic Order in England: A History of Its Development from the Times of St. Dunstan to the Fourth Lateran Council, 940–1216*. New York: Cambridge University Press, 1963.

Kragh, Helge S. *Conceptions of Cosmos: From Myths to the Accelerating Universe: A History of Cosmology*. New York: Oxford University Press, 2007.

———. *Cosmology and Controversy: The Historical Development of Two Theories of the Universe*. Princeton, NJ: Princeton University Press, 1996.

———. "Cosmology and the Entropic Creation Argument." *Historical Studies in the Physical and Biological Sciences* 37, no. 2 (2007): 369.

Kunstler, James Howard. *The Long Emergency: Surviving the End of Oil, Climate Change, and Other Converging Catastrophes of the Twenty-First Century*. New York: Grove, 2006.

Kyvig, David E. *Daily Life in the United States 1920–1939: Decades of Promise and Pain*. Westport, CT: Greenwood, 2002.

Landau, Elaine. *The History of Everyday Life*. Minneapolis, MN: Twenty-First Century Books, 2006.

Le Goff, Jacques. *Time, Work & Culture in the Middle Ages*. Chicago: University of Chicago Press, 1980.

Leake, Jonathan. "Big Bang at the Atomic Lab After the Scientists Got Their Maths Wrong." *Sunday Times* (London), April 8, 2007.

Lee, R. Alton. *The Bizarre Careers of John R. Brinkley*. Lexington: University Press of Kentucky, 2002.

Leeming, David Adams. *The World of Myth: An Anthology*. New York: Oxford University Press, 1990.

Lemaître, Georges. "L'Univers en expansion." *Ann. Soc. Sci. Bruxelles* A 21 (1933): 51–85.

Levinson, Nancy Smiler. *Magellan and the First Voyage Around the World*. New York: Clarion, 2001.

Lewis, Richard. "Telstar: First with the Most." *Bulletin of the Atomic Scientists: The Magazine of Science and Public Affairs* 18, no. 10 (December 1962).

Lewis, Tom. *Empire of the Air: The Men Who Made Radio*. New York: Edward Burlingame, 1991.

Lewis-Williams, David, and David Pearce. *Inside the Neolithic Mind: Consciousness, Cosmos, and the Realm of the Gods*. London: Thames and Hudson, 2005.

Lindberg, David C. *The Beginnings of Western Science*. Chicago: University of Chicago Press, 1992.

Logsdon, Tom. *Understanding the Navstar: GPS, GIS, and IVHS*. 2nd ed. New York: Chapman & Hall, 1992.

Lohr, Steve. "Is Information Overload a 650 Billion Dollar Drag on the Economy?" *New York Times*, December 20, 2007. http://bits.blogs.nytimes .com/2007/12/20/is-information-overload-a-650-billion-drag-on-the-economy (accessed December 29, 2010).

Lorance, Loretta. "Promise, Promises: The Allure of Household Appliances in the 1920s." *Part*, Spring 1998. http://web.gc.cuny.edu/dept/arthi/part/part2–3/house.html.

Lowrie, William. *Fundamentals of Geophysics*. 3rd ed. Cambridge: Cambridge University Press, 2002.

Macey, Samuel L. *Encyclopedia of Time*. New York: Garland, 1994.

Madsen, Mark S. *The Dynamic Cosmos: Exploring the Physical Evolution of the Universe*. Cornwall: CRC, 1995.

Mansfield, Victor. *Synchronicity, Science and Soul-Making: Understanding Jungian Synchronicity*. Peru, IL: Open Court, 2001.

Marber, Peter. *Seeing the Elephant: Understanding Globalization from Trunk to Tail*. Hoboken, NJ: John Wiley, 2009.

Marshack, Alexander. *The Roots of Civilization*. Mount Kisco, NY: Moyer Bell, 1991.

Marshall, Catherine C. "How People Manage Information over a Lifetime." In *Personal Information Management*, edited by William P. Jones and Jaime Teevan. Seattle: University of Washington Press, 2008.

Martin, Donald H. *Communication Satellites*. 4th ed. El Segundo, CA: Aerospace Corporation, 2000.

McCloskey, Donald. "The Industrial Revolution 1780–1860: A Survey." In *The Economics of the Industrial Revolution*, edited by Joel Mokyr. Totowa, NJ: Rowman & Littlefield, 1985.

Miller, Richard Lee. *Under the Cloud: The Decade of Nuclear Testing*. The Woodlands, TX: Two Sixty, 1991.

Milliken, E. K. *English Monasticism: Yesterday and Today*. London: George G. Harrap, 1967.

Misner, Charles, et al. *Gravitation*. New York: Freeman, 1973.

Mithen, Steven. *After the Ice: A Global Human History*. Cambridge, MA: Harvard University Press, 2003.

———. *The Prehistory of the Mind: The Cognitive Origins of Art, Religion and Science*. New York: Thames and Hudson, 1996.

Mitton, Simon. *Conflict in the Cosmos: Fred Hoyle's Life in Science.* Washington, DC: Joseph Henry, 2005.

Morris, I. *Why the West Rules—for Now.* New York: Farrar, Straus & Giroux, 2010.

Mumford, Lewis. "The Monastery and the Clock." In *Technic and Civilization.* New York: Harcourt Brace, 1934.

Neuenschwander, D. "History of Big Bang Cosmology, Part 3: The De Sitter Universe and Redshifts." *Radiations Magazine* (Fall 2008): 68.

Neugebauer, Otto. *The Exact Sciences in Antiquity.* New York: Dover, 1969.

Norris, Robert. "Global Nuclear Stockpiles, 1945–2006." *Bulletin of the Atomic Scientists* 62, no. 4 (July–August 2006): 64.

North, John. *God's Clockmaker: Richard of Wallingford and the Invention of Time.* London: Continuum, 2006.

———. *A Norton History of Astronomy and Cosmology.* New York: W. W. Norton, 1995.

Nowell, April, and Mark White. "Growing Up in Middle Pleistocene." In *Stone Tools and the Evolution of Human Cognition.* Boulder: University Press of Colorado, 2010.

Office of Science, U.S. Department of Energy. "The Sensitive Giant: At CERN, ATLAS Effort Emphasizing People Skills." U.S. Department of Energy Research News, http://www.eurekalert.org/features/doe/2004–03/dnal-tsg032604.php (accessed February 20, 2011).

Okasha, Samir. *Philosophy of Science: A Very Short Introduction.* New York: Oxford University Press, 2002.

O'Regan, Gerard. *A Brief History of Computing.* London: Springer, 1998.

Palmer, John. "Parmenides." In *The Stanford Encyclopedia of Philosophy*, edited by Edward N. Zalta. Fall 2008 ed., http://plato.stanford.edu/entries/parmenides/ (accessed October 21, 2010).

Park, David Allen. *The Grand Contraption: The World as Myth, Number, and Chance.* Princeton, NJ: Princeton University Press, 2005.

Penrose, Roger. *Cycles of Time: An Extraordinary New View of the Universe.* New York: Alfred A. Knopf, 2011.

Plastkett, J. S. "The Expansion of the Universe." *Journal of the Royal Astronomical Society of Canada* 27, no. 35 (1933).

Pogge, Richard. "Real-World Relativity: The GPS Navigation System." http://www.astronomy.ohio-state.edu/~pogge/Ast162/Unit5/gps.html (accessed February 18, 2011).

Pohv, Bogdan, et al. *Particles and Nuclei: An Introduction to the Physical Concepts.* Translated by Martin Lavelle. 6th ed. Heidelberg: Springer-Verlag, 2008.

Polchinski, Joseph. *String Theory: An Introduction to the Bosonic String, Volume 1.* New York: Cambridge University Press, 1998.

Popper, Karl. *The World of Parmenides: Essays on the Presocratic Enlightenment.* New York: Routledge, 1998.

Powell, Corey. *God in the Equation: How Einstein Transformed Religion.* New York: Free Press, 2003.

Raysman, Richard. *Emerging Technologies and the Law: Forms and Analysis.* New York: Law Journal Press, 2003.

Reinhardt, Carsten. *Chemical Sciences in the 20th Century: Bridging Boundaries.* Moerlenbach: Wiley VCH, 2001.

Renfrew, Colin. *Prehistory: The Making of the Human Mind.* New York: Random House, 2007.

Report of the Annual Meeting of the British Society for the Advancement of Science. London: Taylor and Francis, 1872.

Rhodes, Richard. *Dark Sun: The Making of the Hydrogen Bomb.* New York: Simon & Schuster, 1995.

———. *The Making of the Atomic Bomb.* New York: Simon & Schuster, 1986.

Richards, E. G. *Mapping Time: The Calendar and Its History.* New York: Oxford University Press, 1998.

Rifkin, Jeremy. *Time Wars: The Primary Conflict in Human History.* New York: Simon & Schuster, 1989.

Rist, John. *The Stoics.* Berkeley: University of California Press, 1998.

Rooksby, Rikky. *Inside Classic Rock Tracks: Songwriting and Recording Secrets of 100 Great Songs from 1960 to the Present Day.* San Francisco: Backbeat Books, 2001.

Rosenberg, Daniel. "Marking Time." *Cabinet* (Winter 2007).

Rudel, Anthony J. *Hello, Everybody! The Dawn of American Radio.* New York: Harcourt, 2008.

Rynasiewicz, Robert. "Newton's Views on Space, Time, and Motion." In *The Stanford Encyclopedia of Philosophy*, edited by Edward N. Zalta. Fall 2008 ed. http://plato.stanford.edu/entries/newton-stm/ (accessed October 20, 2010).

Sadava, David, et al. *Life: The Science of Biology.* 9th ed. Sunderland, MA: Sinauer Associates, 2009.

Salles, Ricardo. "Introduction: God and Cosmos in Stoicism." In *God and Cosmos in Stoicism*, edited by Ricardo Salles. New York: Oxford University Press, 2009.

Samama, Nel. *Global Positioning: Technologies and Performance.* Hoboken, NJ: John Wiley, 2008.

Sample, Ian. "Mysteries of Computer from 65 BC Are Solved." *Guardian*, November 30, 2006.

Sandage, Allen, et al. *Centennial History of the Carnegie Institution of Washington.* Volume 1: *The Mt. Wilson Observatory.* New York: Cambridge University Press, 2004.

Sartori, Leo. "Effects of Nuclear Weapons." In *Physics and Nuclear Arms Today: Readings from Physics Today*, edited by David Hafemeister. New York: American Institute of Physics, 1991.

Sassen, Saskia. *Territory, Authority, Rights: From Medieval to Global Assemblages.* Princeton, NJ: Princeton University Press, 2006.

Scharf, John Thomas. *A History of Philadelphia, Vol. III.* Philadelphia: J. H. Everts, 1884.

Schiffer, Michael B. *Power Struggles: Scientific Authority and the Creation of Practical Electricity Before Edison.* Cambridge, MA: MIT Press, 2008.

Schivelbusch, Wolfgang. *Disenchanted Night: The Industrialization of Light in the Nineteenth Century.* Translated by Angela Davies. Berkeley: University of California Press, 1988.

Schmandt-Besserat, Denise. *How Writing Came About.* Austin: University of Texas Press, 1996.

Schwarzlose, Richard Allen. *The Nation's Newsbrokers: The Rush to Institution, from 1865 to 1920.* Evanston, IL: Northwestern University Press, 1990.

Schwoch, James. *Global TV: New Media and the Cold War, 1946–69.* Urbana: University of Illinois Press, 2009.

Sharman, Russell Leigh, and Cheryl Harris Sharman. *Nightshift NYC.* Berkeley: University of California Press, 2008.

Shaw, Ian, and Robert Jameson. *A Dictionary of Archaeology.* Padstow, Cornwall: Blackwell, 2002.

Shearer, Benjamin F., et al. *Home Front Heroes: A Biographical Dictionary of Americans During Wartime*, edited by Benjamin F. Shearer. Westport, CT: Greenwood Press, 2007.

Shields, Christopher. "Aristotle." In *The Stanford Encyclopedia of Philosophy*, edited by Edward N. Zalta. Winter 2009 ed., http://plato.stanford.edu/archives/win2009/entries/aristotle/ (accessed September 27, 2010).

Sies, Luther F. *Encyclopedia of American Radio, 1920–1960.* Rev. ed. New York: McFarland & Co., 2008.

Silverman, Mark P. *Quantum Superposition: Counterintuitive Consequences of Coherence, Entanglement, and Interference.* Berlin: Springer, 2008.

Skempton, Alec. *A Biographical Dictionary of Civil Engineers in Great Britain and Ireland 1500–1830.* London: Thomas Telford, 2002.

Slow Food. "Our Mission." http://www.slowfood.com/about_us/eng/mission .lasso (accessed February 20, 2011).

Smith, Fred H. "Neandertal and Early Modern Human Interactions in Europe." *American Anthropologist* 110, no. 2 (June 2008).

Smolin, Lee. *The Life of the Cosmos*. Oxford: Oxford University Press, 1998.

———. "On the Reality of Time and the Evolution of Laws." Work in progress presented at the Perimeter Institute for Theoretical Physics, Ontario, Canada, October 2, 2008.

———. *Three Roads to Quantum Gravity*. New York: Basic Books, 2001.

———. *The Trouble with Physics: The Rise of String Theory, the Fall of a Science, and What Comes Next*. New York: Houghton Mifflin Harcourt, 2007.

———. "The Unique Universe." *Physics World* (June 2009): 21–26.

Snobelen, Stephen D., interviewed by Paul Newall. "Stephen D. Snobelen: Newton Reconsidered." http://www.galilean-library.org/site/index.php?/ page/index.html/_/interviews/stephen-d-snobelen-newton-reconsidered -r39 (accessed September 21, 2010).

Sobel, Dava. *Longitude: The True Story of a Lone Genius Who Solved the Greatest Scientific Problem of His Time*. New York: Walker, 1995.

Sokal, Alan. "A Physicist Experiments with Cultural Studies." *Lingua Franca* (May/June 1996).

Solomon, Anne G. K. "The Global Positioning System." In *Triumphs and Tragedies of the Modern Presidency: Seventy-Six Case Studies in Presidential Leadership*, edited by David Abshire. Westport, CT: Praeger, 2001.

Soter, Steven, and Neil de Grasse Tyson, eds. *Cosmic Horizons: Astronomy at the Cutting Edge*. New York: New Press, 2000.

Spelke, Elizabeth. "Nativism, Empiricism and the Origins of Knowledge." *Infant Behavior and Development* 21, no. 2 (1998): 181–200.

Stamos, David N. *Darwin and the Nature of Species*. Albany: State University of New York Press, 2007.

Stanford Arms Control Group. *International Arms Control: Issues and Agreements*, edited by John H. Barton and Lawrence D. Weiler. Stanford: Stanford University Press, 1976.

Staw, Barry M., ed. *Research in Organizational Behavior: An Annual Series of Analytical Essays and Critical Reviews*. Vol. 27. San Diego: JAI Press, 2006.

Stein, James D. *How Math Explains the World: A Guide to the Power of Numbers, from Car Repair to Modern Physics*. New York: HarperCollins/Smithsonian Books, 2008.

Steinhardt, Paul J., and Neil Turok. "The Cyclic Model Simplified." *New Astronomy Reviews* 49, no. 2–6 (2005): 43–57.

———. *Endless Universe: Beyond the Big Bang*. New York: Doubleday, 2007.

Sterling, Christopher H. "Amos 'n' Andy." In *Encyclopedia of Radio*. New York: Fitzroy Dearborn, 2004.

———. "Blue Network." In *Encyclopedia of Radio*. New York: Fitzroy Dearborn, 2004.

Stern, S. Alan. *Our Universe: The Thrill of Extragalactic Exploration as Told by Leading Experts*. New York: Cambridge University Press, 2001.

Strasser, Susan. *Never Done*. New York: Pantheon,1982.

Streeter, Thomas. *The Net Effect: Romanticism, Capitalism, and the Internet*. New York: New York University Press, 2011.

Susskind, Leonard. *The Cosmic Landscape: String Theory and the Illusion of Intelligent Design*. New York: Hachette, 2005.

Suzuki, Shunryu. *Zen Mind, Beginner's Mind*. Boston: Shambhala, 2010.

Swan, Laura, ed. *The Benedictine Tradition: Spirituality in History*. Collegeville, MN: Order of St. Benedict, 2007.

Tattersall, Ian. "Once We Were Not Alone." *Scientific American* (January 2000).

Tedlock, Barbara. *The Woman in the Shaman's Body: Reclaiming the Feminine in Religion and Medicine*. New York: Bantam, 2005.

Thai, Khi V., et al. *Handbook of Globalization and the Environment*. Boca Raton, FL: CRC, 2007.

Thorne, Kip S. *Black Holes and Time Warps: Einstein's Outrageous Legacy*. New York: W. W. Norton, 1994.

Thurston, Hugh. *Early Astronomy*. New York: Springer-Verlag, 1993.

Todd, Deborah, and Joseph A. Angelo Jr. *A to Z of Scientists in Space and Astronomy*. New York: Facts on File, 2005.

Traveler's Official Railway Guide for the United States and Canada. New York: National Railway Publication Co., 1881.

U.S. Arms Control and Disarmament Agency. *Worldwide Effects of Nuclear War: Some Perspectives*. Whitefish, MT: Kessinger, 2004.

Unger, Roberto Mangabeira. *Law in Modern Society: Toward a Criticism of Social Theory*. New York: Free Press, 1976.

———. *The Self Awakened: Pragmatism Unbound*. New York: Harvard University Press, 2007.

Van Keuren, David K. "Moon in their Eyes: Moon Communication Relay at the Naval Research Laboratory, 1951–1962." In *Beyond the Ionosphere: Fifty Years of Satellite Communication*, edited by Andrew J. Butrica. Washington, DC: NASA History Office, 1997.

Van Vleck, Tom. "The History of Electronic Mail." http://www.multicians.org/thvv/mail-history.html (accessed February 24, 2011).

Varela, Francisco J., Evan Thompson, and Eleanor Rosch. *The Embodied Mind: Cognitive Science and Human Experience*. Cambridge, MA: MIT Press, 1991.

Vilenkin, Alexander. *Many Worlds in One: The Search for Other Universes*. New York: Hill and Wang, 2006.

Weiner, Stephen. "Systems and Technology." In *Ballistic Missile Defence*, edited by Ashton B. Carter and David N. Schwartz. Washington, DC: Brookings Institution, 1984.

Wheeler, John Archibald, and Hubert Zurek Wojciech, eds. *Quantum Theory and Measurement*. Princeton, NJ: Princeton University Press, 1983.

Will, Clifford M. *Was Einstein Right? Putting General Relativity to the Test*. New York: Perseus, 1984.

Williamson, Mark. *Spacecraft Technology: The Early Years*. History of Technology Series. 2nd ed. Bodmin, Cornwall: Institution of Engineering and Technology, 2008.

Wilson, Peter. *The Domestication of the Human Species*. New Haven: Yale University Press, 1988.

Witham, Larry. *Measure of God: Our Century-Long Struggle to Reconcile Science and Religion*. San Francisco: HarperSanFrancisco, 2005.

Woit, Peter. *Not Even Wrong: The Failure of String Theory and the Search for Unity in Physical Law*. New York: Basic Books, 2006.

Wolf, Edward L. *Nanophysics and Nanotechnology: An Introduction to Modern Concepts in Nanoscience*. 2nd ed. Moerlenbach: Wiley VCH, 2006.

Yamey, Basil S. "Double-Entry Bookkeeping, Luca Pacioli and Italian Renaissance Art." In *Arts and Accounting*. New Haven: Yale University Press, 1989.

Yoffee, Norman. *Myths of the Archaic State: Evolution of the Earliest Cities, States and Civilizations*. New York: Cambridge University Press, 2005.

Zimmer, Heinrich Robert. *Myths and Symbols in Indian Art and Civilization*. Edited by Joseph Campbell. 2nd ed. New York: Bollingen, 1974.

ILLUSTRATION CREDITS

Figure 1.1: Courtesy of the Peabody Museum of Archaeology and Ethnology, Harvard University, 2005.16.318.38.

Figure 1.2: Source: Eric Lebrun.

Figure 1.3: Photo and diagram: Anthony Aveni.

Figure 2.1: Source: The British Museum.

Figure 2.2: Photo credit: Adam Frank.

Figure 2.3: Reproduced from J. Norman Lockyer, *The Dawn of Astronomy* (London: Cassell & Co., 1894), 35.

Figure 2.4: Source: Wikipedia.

Figure 2.5: Source: Vatican Museum drawing after Jane E. Harrison, *Epilegomena and Themis* (New Hyde Park, NY: University Books, 1962), 98. Reproduced in Anthony F. Aveni, *Empires of Time* (Boulder: University Press of Colorado, 2002), 45.

Figure 3.1: Source: Part of the Bibliothèque Nationale de France collection.

Figure 3.2: Source: Cosmas Indicopleustes, *Christian Topography* (c. 550), reproduced in Helge Kragh, *Conceptions of Cosmos* (New York: Oxford University Press, 2007).

Figure 3.3: Source: From Bartholomaeus Angelicus, *De Proprietatibus Rerum* (Lyons, 1485), the Huntington Library.

Figure 3.4: Source: Part of the Bayerische Staatsbibliothek collection.

Figure 3.5: Source: From Petrus Apianus, *Cosmographicum liber* (1533).

Figure 3.6: Reproduced from Helge Kragh, *Conceptions of Cosmos* (New York: Oxford University Press, 1996), 48.

Figure 4.1: Source: Part of the Deutsches Museum collection in Munich.

Figure 4.2: Source: René Descartes, *Le Monde* (1633).

Figure 4.3: Source: Charles J. E. Wolf, *Histoire de l'Observatoire de Paris de sa fondation à 1793* (Paris: Gauthier-Villars, 1902).

Figure 4.4: Source: *La lumière électrique* (Paris: Aux Bureaux Du Journal, 1885).

Figure 4.5: Courtesy of the Library of Congress.

Figure 5.1: Courtesy of the Library of Congress.

Figure 5.2: Source: Cartlon J. Corliss, *The Day of Two Noons* (Washington, DC: Association of American Railroads, 1952). Reproduced in Peter Galison, *Einstein's Clocks, Poincare's Maps* (New York: W.W. Norton, 2003), 124.

Figure 5.3: Courtesy of the Library of Congress.

Figure 5.4: Source: Max Flückiger, *Albert Einstein in Bern* (Bern: Verlag Paul Haupt, 1974), 61.

Figure 6.1: Photo credit: General Electric.

Figure 6.2: Source: The Huntington Library.

Figure 6.6: Courtesy of the Library of Congress.

Figure 7.1: Photo courtesy of National Nuclear Security Administration/Nevada Site Office.

Figure 7.2: Source: National Museum of the U.S. Air Force.

Figure 7.3: Source: *Buffalo News* Archive.

Figure 7.4: Used with permission of the Estate of Ralph A. Alpher, Ph.D., Victor S. Alpher, Ph.D., Executor. Use of this photograph does not imply endorsement of views or opinions of the author of this work by Dr. Ralph A. Alpher nor Dr. Victor S. Alpher.

Figure 7.6: Copyright Decca Records and Universal Music.

Figure 7.7: Source: The Emilio Segrè Visual Archives of the Niels Bohr Library and Archives.

Figure 8.1: Source: www.slipstick.com/images/maccalendarlg.jpg.

Figure 8.2: Source: Fermilab.

Figure 8.4: Source: Caltech Archives.

Figure 8.5: Source: M. Blanton and the SDSS Collaboration.

Figure 8.6: Source: RAND MR614-A.2. Reprinted with permission.

Figure 8.7: Source: NASA/WMAP Science Team.

Figure 10.1: Reproduced with permission of the Master and Fellows of St. John's College, Cambridge.

Figure 11.4: Cover of Lee Smolin, *The Trouble with Physics* (New York: Houghton Mifflin, 2006).

All other illustrations by Sameer Zavery.

INDEX

Page numbers in *italics* refer to illustrations.

ABOUT THE AUTHOR

Adam Frank is a theoretical astrophysicist at the University of Rochester in New York, where he runs a research group focusing on supercomputer studies of star formation and stellar death. Along with his research, Professor Frank has held a lifelong interest in the relationship between science and culture. A seasoned contributor to *Discover* magazine, his work has also appeared in *Astronomy, Scientific American* and *Tricycle*. Frank is cofounder of the National Public Radio blog 13.7 Cosmos and Culture, where he contributes weekly. His first book, *The Constant Fire: Beyond the Science vs. Religion Debate* was picked as one of *Seed* magazine's "Books to Read Now". His writing has earned him an American Astronomical Society writing award and has been included in *Best American Science and Nature Writing 2009*.